PRODUCTIVITY EQUIPMENT SERIES

Cutting Fluids & Lubricants

CUTTING FLUIDS
LUBRICANTS
CUTTING FLUID SYSTEMS
METALFORMING COMPOUNDS
TREATMENT AND
APPLICATION EQUIPMENT

Published by
Society of Manufacturing Engineers
Publications Development Department
One SME Drive
P.O. Box 930
Dearborn, Michigan 48121

Cutting Fluids & Lubricants

Productivity Equipment Series

First Edition

First Printing

Library of Congress Catalog Card Number: 85-50994
International Standard Book Number: 0-87263-193-1
Manufactured in the United States of America

CUTTING FLUIDS & LUBRICANTS
Productivity Equipment Series

This volume of the Productivity Equipment Series is designed to aid managers, engineers, and technicians in the purchasing of industrial cutting fluids and lubricants, systems, and equipment. This book will prove to be an important time-saver, placing current information in the hands of the reader.

More than 600 suppliers of cutting fluids and lubricants were contacted in the development of this volume. Each contribution was carefully analyzed as to its industrial usefulness and timeliness before being accepted into this volume.

In the first section, "Cutting Fluids," 37 companies are represented.

Section Two, "Lubricants," brings the reader information on 15 companies.

The third section, "Cutting Fluid Systems," displays seven companies.

Section Four, "Metalforming Compounds," includes product sheets from seven companies; and the final section, "Treatment and Application Equipment," presents information from six companies.

Within each section, product sheets are grouped alphabetically by manufacturer. As a special addition to aid in the ordering process, supplier names, addresses, and telephone numbers are featured on every page.

Product data in this volume does not constitute a product endorsement by SME or its affiliated associations. The number of products listed for each manufacturer does not imply or indicate the manufacturer's share of the market, product quality, nor any other differentiation among manufacturers.

Special thanks are extended to the many companies who participated in the development of this unique publication.

TABLE OF CONTENTS

SECTION 1

CUTTING FLUIDS

SECTION 2

LUBRICANTS

SECTION 3

CUTTING FLUID SYSTEMS

SECTION 4

METALFORMING COMPOUNDS

SECTION 5

TREATMENT AND APPLICATION EQUIPMENT

SECTION 1

CUTTING FLUIDS

COOLANT - Model 82115
and SAFETY PARTS CLEANER - Model 81115

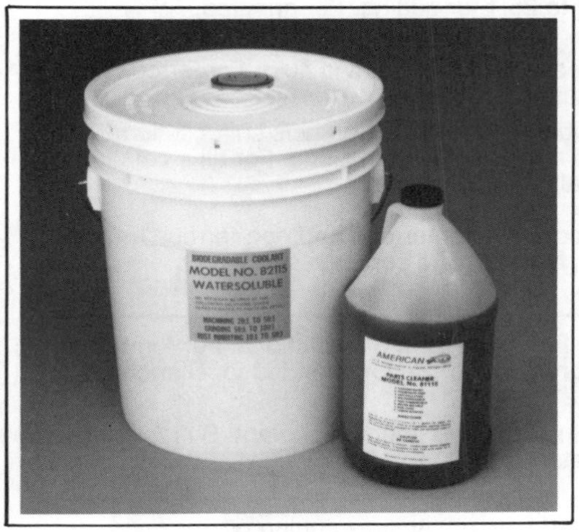

Part No. 81115 - SAFETY PARTS CLEANER, A heavy duty, fast acting grease and soil emulsifier. A water soluble concentrated degreaser, works on an entirely different principle of soil removal, which is that of transforming soilings into water soluble solutions thus making their removal easy. This is done by its unique and very rapid emulsifying process. Features include: Phosphate Free, Anti-Polluting, Biodegradable, Non-Combustible and Non-Toxic. Mixing requirements: 16 oz. of concentrate to 1 gallon of water for degreasing of parts. Available in 1 or 5 gallons.

Part No. 82115 - BIODEGRADABLE COOLANT, is a highly efficient universal, fool-proof biodegradable metal working fluid. It can be used for both machining and grinding on Ferrous and Non-Ferrous materials. It is an excellent rust inhibitor even on cast iron, at dilutions up to 50:1. #82115 is an excellent additive to other coolants to improve finish and rust protection. This product settles grit and sludge, particles are not recirculated through the machine. #82115 does a good job of keeping the machines clean and does not turn rancid which means that the machines do not have to be cleaned as frequently. The concentrated coolant is clear which enables the operator to see his work. It also is not slippery which makes it easier to handle wet parts. #82115 can be used at the following dilutions (given in parts water to #82115).
Available in 1 or 5 gallons

AMERICAN

WATERSOLUBLE
BIODEGRADABLE COOLANT
MODEL NO. 82115

NO. 82115 CAN BE USED AT THE FOLLOWING DILUTIONS (GIVEN IN PARTS WATER TO PARTS NO. 82115.)

MACHINING 20:1

GRINDING 30:1

RUST INHIBITING 30:1

AMERICAN GATOR
111 E. MICHIGAN AVE. • AUGUSTA, MICH. 49012
(616) 731-4177

American Oil and Supply Company
238 Wilson Avenue
Newark, NJ 07105
(201) 589-0250

TYPE	PRODUCT	DESCRIPTION
Water Soluble Cutting & Grinding Fluids	Lafayette® #60	Combination cutting fluid and grinding coolant. Use on ferrous and non-ferrous metals. Mix 10 to 25:1 for cutting. For grinding, mix 30 to 50:1.
	Lafayette® #70	Sulfur and chlorine free. Heavy duty type cutting fluid. For ferrous and non-ferrous metals. Non-staining. Mix 15 to 40:1.
	Lafayette® #71	Heavy duty type water soluble cutting fluid. Contains sulfur, chlorine and fatty oils to form stable non-staining emulsions. For all steel alloys as well as non-ferrous metals. Mix 15 to 40:1.
	Lafayette® #80	General purpose grinding coolant. For ferrous and non-ferrous metals. Use for cylindrical, surface, internal, and centerless grinding. Mix 30 to 60:1.
Synthetic Coolants	Lafayette® #91	Synthetic grinding coolant. Forms transparent solutions. For ferrous and non-ferrous metals. Non-foaming. Recommended for Blanchard type surface grinders. Mix 20 to 40:1.
	Lafayette® #93	Combination synthetic cutting fluid and grinding coolant. Suitable for ferrous and non-ferrous metals. Mix 20 to 50:1.
Straight Cutting & Grinding Oils	Lafayette® Base A	Dark, active sulfur-chlorinated type oil. Acts as cutting oil additive — can be used as supplied for difficult operations, i.e., threading, broaching-cold forming. Recommended for use in automatic screw machine.
	Lafayette® A-13	Low viscosity sulfurized cutting oil for use on stainless steel, steel, other alloyed steels. Not for use on non-ferrous metals. Use in automatic screw machines, turret lathes, drill presses and millers.
	Lafayette® 599	General purpose — transparent, non-staining. Use on ferrous and non-ferrous metals, screw machines, multiple spindle automatics, turret lathes, etc.
	Lafayette® B-115	Low viscosity — Transparent oil for aluminum and brass.
	Lafayette® Base C	Highly sulfur-chlorinated mineral oil combined with fatty oils for severe metal cutting operations. Not for use on copper and brass alloys. Recommended for broaching, gear cutting, gun drilling.
	Lafayette® C-13	Sulfur-chlorinated fatty oil — for stainless steel, high alloyed steel. Not for use on brass or copper. Use in screw machines, multiple spindle automatics, turret lathes, etc.
	Lafayette® CM-64	Low viscosity sulfurized fatty oil type for wide range of steel alloys. For both cutting and grinding. Recommended for turning, facing, drilling, trepanning, etc.
	Lafayette® C-75	Sulfurized cutting oil medium viscosity. For use on steel, stainless steel, and high alloyed steels. Not for copper and brass. Use in screw machines, multiple spindle automatics, turret lathes, etc. Recommended for deep hole drilling (gun drilling), broaching, gear hobbing.
	Lafayette® KM-16	Low viscosity, honing oil for steel and non-ferrous metals. Keeps stones clean and free cutting. Rapid heat removal controls size.
Drawing Compound, Straight and Water Soluble	Lafayette® SCF-30	Low viscosity, pale straw colored oil. Mineral oil and fatty oil blend. Contains no sulfur or chlorine. Will not stain copper or brass. Recommended for light drawing and stamping operations.
	Lafayette® P-12	Highly polar lubricant for difficult drawing operations. Especially recommended for use on stainless steel. Ideal for use in high-speed transfer presses.
	Lafayette® #71	Heavy duty water soluble drawing compound. For ferrous and non-ferrous metals. Recommended for use in presses with progressive dies and also, high-speed transfer presses, by circulating on the slides. Mix 4 to 10:1.
	Lafayette® 745-90	Medium viscosity drawing lubricant. Non-staining. Use for drawing, blanking and stamping.

	FERROUS METALS				NON-FERROUS METALS		
OPERATIONS	CARBON STEELS	ALLOY STEELS	STAINLESS STEELS	CAST IRON	ALUMINUM	BRASS & COPPER	MAGNESIUM
Automatic Screw Machines	A-13 599 C-13	A-13 599 C-13	C-13 C-75		B-115	599 B-115	599
Turret Lathes	599 C-13	C-13 C-75	C-13 C-75		B-115	599 B-115	599
Boring Milling Drilling	599 A-13	C-13 C-75	C-13 C-75		B-115	B-115 599	599
Broaching	C-75 C-13 Base C	C-75 C-13 Base C	C-13 C-75 Base C		599 B-115	599 B-115	599
Gun Drilling	C-75	C-75	C-75	C-75	C-75	599 SCF-10	599
Gear Cutting	C-13 C-75 599	Base C C-75 599	C-13 C-75 599	C-13 599	599	599 SCF-20	599
Threading and Tapping	A-13 C-75 C-13	C-13 C-75 Base C	C-13 C-75 Base C	C-13 C-75	599	599 SCF-20	599
Grinding					B-115		B-115
Thread Grinding	B-16 C-75	C-75 CM-64	C-75 CM-64	CM-64 599	599 B-115	599 SCF-10	B-115
Honing	KM-16	KM-16	KM-16	KM-16	KM-16	KM-16	KM-16

Bijur Lubricating Corporation
50 Kocher Drive
Bennington, VT 05201-1994
(802) 447-2174

Bijur Lubricating Corporation

50 Kocher Drive, Bennington, Vermont 05201-1994 802 447-2174 Telex: 6817445

APPLICATIONS:

A semi-synthetic, heavy duty, non foaming coolant and cutting fluid for use in Spraymist Mist Coolant Systems for general shopwide chip-making and grinding operations. Fluid is multi-metal safe with special passivates to allow use on both ferrous and non-ferrous metals. It contains chlorine-free and sulphur-free EP (extreme pressure) additives to facilitate the machining of difficult alloys. It also has the lubricity needed to cut aluminum without galling on tooling.

This highly concentrated formulation contains inhibitors for effective rust control, even at the recommended dilution of 50-80 parts water to 1 part concentrate.

TYPICAL INSPECTIONS:

Specific Gravity 60/60ºF 1.025
Lbs/Gal 8.55
pH, 2% Dilution 9.3
Color . . . . Yellow Translucent

RECOMMENDED DILUTIONS WITH WATER:

Cutting 50:1
Grinding 70-80:1

NOTE: Spraymist uses diluted coolant at the rate of 2-3 ounces per hour, per jet, or approximately 1 gallon per 40 hour week.

STORAGE:

Spraymist coolant is stable through freezing/thaw cycles, but it is recommended that it be kept from freezing.

SAFETY:

This non-flammable fluid is considered non-hazardous in normal use and *does not* contain nitrites, phenols, PCB's, mercurials, sulphur or chlorine. It *does* contain a unique synthetic ingredient to protect against dermititis. OSHA 20 sheet is available on request.

PACKAGING:

55 gallon drums, 5 gallon pails and 1 gallon containers (6 per case).

The information contained in this bulletin is believed to be accurate, but all recommendations or suggestions are made without guarantee and there is no implied warranty of merchantability or fitness for purpose of the product described herein. In submitting this information no liability is assumed, or license or other rights expressed or implied given with respect to any existing or pending patent, patent applications or trademarks.

Printed In U.S.A.

Sterling Gun Drills
Subsidiary of Bijur Lubricating Corporation
50 Kocher Drive Bennington, Vt. 05201-1994 802 447-2174 Telex: 6817445

PRODUCT INFORMATION

SPEEDBIT CUTTING FLUID

APPLICATIONS:

Heavy duty soluble oil for use in milling, drilling, turning and most other chip forming operations. It contains large amounts of EP additive to provide good tool life. Speedbit Cutting Fluid has rust inhibitors, defoamer and germicide to give good corrosion protection, low foam and excellent microbial resistance in either Flood or Mist Coolant applications.

TYPICAL INSPECTIONS:

Appearance Neat	Green Oil
Appearance Diluted	Green Emulsion
Lb/Gal 60ºF	8.53
Foaming Characteristics	Low Foaming
10% Emulsion Stability	Perfect After 1 Hr.
Odor	Mild
4% pH	8.7 - 9.0

MIXING INSTRUCTIONS:

Concentrate should always be added to water. Warm water is preferred (70-80ºF). Mixing Ratios 10:1 to 75:1 dependent on type and severity of application.

STORAGE:

Store drums at temperatures between 40º to 120ºF.

SAFETY:

Speedbit Cutting Fluid is considered non-hazardous in normal use. It does not contain sulfur, PCB's, mercurials or phenols.

This product is not classified as a "Hazardous Material" in normal use as defined in U.S. Dept. of Labor Regulation 29 CFR Parts 1501, 1502, and 1503.

OSHA 20 Sheet is available upon request.

Blaser Swisslube Incorporated
1 Holland Avenue
White Plains, NY 10603
(914) 997-6931

Blaser
Swisslube Inc.

Blaser Swisslube Inc., White Plains, N.Y. 10603 USA
11, Virginia Road, Tel. (914) 997-6931 / 997-6932
Tx. 131482

High performance cutting oils also found under the Blasocut® trade name

For cutting and forming – for all materials.

Blasocut® Automatic 840, art. no. 840

Especially recommended for threading, rolling and tapping from .2 inch to 6 inch diameter, broaching, drilling, pressing and less demanding forming operations.

Blasocut® Automatic 846, art. no. 846

For automatic lathes and wherever non water-soluble lubricating coolants are applied.

Blasocut® Compound 850, art. no. 850

For drilling, pressing, drawing and less demanding operations, as well as threading of difficult materials. Viscosity allows use of gear pumps.

Blasocut® Compound 851, art. no. 851

Usually applied by brush. Used for threading, pressing, deep drawing, forming, bending etc., or as an additive to other products improving considerably their film resistance and lubricity.

Safety

Research, production and worldwide distribution are all under one roof. Painstaking raw material and manufacturing controls guarantee a constant, high level of quality of all Blasocut metal working coolants. We attach great importance to a safe and hygienic work place and are proud to say that our Blasocut products have achieved the highest marks with regard to environment and health requirements, as many attests will show. The LD 50 value of all Blasocut concentrates is > 15 g/kg.

For further assistance, please contact our local representative:

Blaser Swisslube Incorporated
1 Holland Avenue
White Plains, NY 10603
(914) 997-6931

Blaser

Mineral Oil Base Water-Soluble Blasocut® Coolants

General Machining

Blasocut® 2000 Universal, art. no. 870

Universal cutting fluid, mineral oil based, for all chip-removing operations and for all materials (except magnesium).
Color: milky green.

Blasocut® 3000 Ferro, art. no. 873

For machining of ferro metals, mineral oil based. Especially for rust susceptible materials, such as spheroidal cast iron, grey cast iron, sulphur-alloyed steels etc.
Color: milky green.

Blasocut® 4000 Strong, art. no. 872

High performance mineral oil based coolant for demanding operations, such as hobbing, deep hole drilling, sawing, broaching and less demanding forming operations. Often replaces a cutting oil.
Color: milky green.

...and for Grinding:

Blasocut® Grindex 7804, art. no. 882

Universal grinding coolant, mineral oil based. Successfully replaces synthetic and semi-synthetic grinding fluids. For all materials and grinding operations. Semi-transparent. Color: yellow.

Blasocut® Grindex 883, art. no. 883

Grinding coolant, mineral oil based for high performance grinding. Also suitable for general machining, if a lubricating coolant void of halogen is needed. Used as press water additive. Semi-transparent. Color: reddish.

Blasocut® Grindex 884, art. no. 884

Similar qualities as Grindex 883. Contains high pressure additives for more demanding grinding operations, centerless grinding. Also used as a universal emulsion for general machining. Semi-transparent.
Color: reddish

Useful Tools For Blasocut® Coolants Mixing and Checking

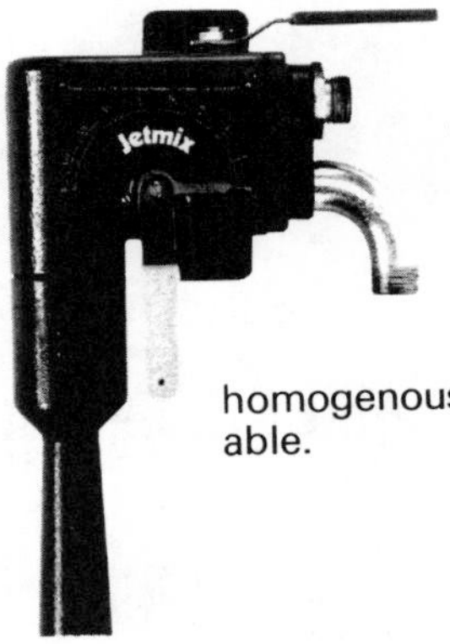

JETMIX

A unique mixing device designed and produced in our factories. This device allows quick, clean and accurate preparation of emulsions in a homogenous form. Detailed literature available.

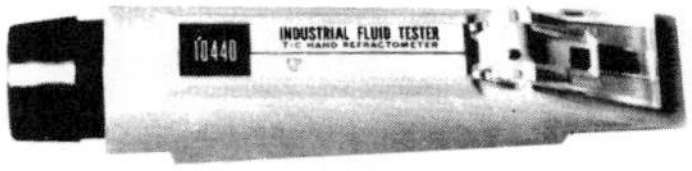

Refractometer

A practical instrument well known for quick and accurate checking of concentration for water-soluble coolants.

Blaser Swisslube Incorporated
1 Holland Avenue
White Plains, NY 10603
(914) 997-6931

Excellent compatibility with human skin

Chemical-technical data

Concentrate

Description	Unit	BLASOCUT 2000 Universal	BLASOCUT 3000 Ferro	BLASOCUT 4000 Strong	BLASOCUT Grindex 7804	BLASOCUT Grindex 883	BLASOCUT Grindex 884
Appearance		transparent clear	transparent clear	transparent clear	transparent clear	transparent clear	transparent clear
Color		green	green	green	green/yellow	red	red
Density/20°C/68°F	g/ml	0,966	0,967	0,986	0,962	0,976	0,977
Viscos./20°C/68°F	mm/s	168	156	285	133	200	202
Viscosity 100°F	sus	269	259	366	251	315	334
Viscosity 100°F	mm/s	58	56	79	54	68	72
Chlor cont. (inact.)	%	3	3	6	0	0	2
Active sulphur	%	0	0	0	0	0	0
Flash point	°F	334	302	367	275	367	363
Pour point	°F	–22	–22	–22	–22	–22	–22

Emulsion prepared with tap water with a hardness of 267 mg/l $CaCO_3$

Concentration	5%			10%			3%		
Description	BLASOCUT 2000	BLASOCUT 3000	BLASOCUT 4000	BLASOCUT 2000	BLASOCUT 3000	BLASOCUT 4000	BLASOCUT 7804	BLASOCUT 883	BLASOCUT 884
Appearance Color	milky green	milky green	milky green	milky green	milky green	milky green	semi-tr. yellow	semi-tr. pink	semi-tr. pink
pH-value at preparation	8,8	9,3	8,8	8,9	9,4	8,9	9,1	8,9	8,9
pH-value after 24 hours	8,6	9,0	8,7	8,7	9,1	8,7	8,9	8,7	8,7
Stability (DIN 51 367)	100%	100%	100%	100%	100%	100%	100%	100%	100%

All data in this catalog are based on known and tested applications and conditions and represent mean values. They are not legally binding and are subject to change.
Because conditions of use and application are beyond our control, no warranty or representation is made or intended in connection with the use of these products.

Heavy-duty cutting oils

For our *non water-soluble* metal working products, BLASOCUT-Automatic and BLASOCUT-Compound, please request detailed literature.

For further assistance,
please contact our local representative:

Blaser Swisslube Incorporated
1 Holland Avenue
White Plains, NY 10603
(914) 997-6931

Blaser Swisslube Inc.

Swisslube®
Blasocut
Recommended concentrations

Machinability of ferrous and non ferrous materials

Material Category 1

Tensile Strength
7,110 lbs. / sq. in.

Carbon Steel
Free Cutting Steel
Molybdenum (cast)
Grey Cast Iron
Malleable Iron

Material Category 2

Tensile Strength
72,522-142,200 lbs. / sq. in.

Carbon Steel
Molybdenum
Manganese
Nickel
Chrome-Nickel Alloy
Chrome
Stainless Steels
Spheroidal Cast Iron

Material Category 3

Tensile Strength
over 142,200 lbs. / sq. in.

Carbon Steel
Molybdenum
Manganese Steel
Nickel
Chrome-Nickel Alloy
Chrome Vanadium Alloy
Manganese-Silicon
Stainless Steels
Titanium and Titanium alloys,
Inconel, Monel, Hastelloy, etc.

Material Category 4

Easier Machining

Tin
Copper
Phosphorous-Bronze

Material Category 5

Medium Machining

Nickel-Bronze
Brass-Copper
Silicon-Bronze
Nickel-Silver
Aluminium and their alloys

Material Category 6

Difficult Machining

Manganese-Bronze
Manganese-Aluminium-Bronze
Aluminium-Bronze
Beryllium-Bronze
Phosphorous-Bronze
Titanium and Titanium alloys,
Inconel, Monel, Hastelloy, etc.

Blaser Swisslube Incorporated
1 Holland Avenue
White Plains, NY 10603
(914) 997-6931

Metalworking applications of BLASOCUT water-soluble coolants

Type of Machining	Material Category 1	Material Category 2	Material Category 3	Material Category 4	Material Category 5	Material Category 6
Surface and Cylindrical Grinding	Grindex-7804 3 %	Grindex-7804 3-4 %	Grindex-884 4-6 % (Grindex-7804) 4-6 %	Grindex-7804 4 %	Grindex-7804 5 %	Grindex-7804 5 % Grindex-884 3-5 %
Form Grinding	Grindex-7804 3 %	Grindex-7804 3-6 % Grindex-884 3-6 %	Grindex-884 -10 %	Grindex-7804 3-5 %	Grindex-7804 5 % Grindex-884 4 %	Grindex-7804 5 % Grindex-884 3-5 % 2000 Universal 5-7 %
Turning	2000 Universal 6 % 3000 Ferro 6 %	2000 Universal 8 % 3000 Ferro 8 % 4000 Strong 8 %	4000 Strong 10 %	2000 Universal 6 %	2000 Universal 8 %	2000 Universal 10 % 4000 Strong 10 %
Drilling conv. deep hole 1"+	2000 Universal 6 % 3000 Ferro 6 % 4000 Strong 10 %	2000 Universal 8 % 3000 Ferro 8 % 4000 Strong 15 %	4000 Strong 10 % 4000 Strong 20 %	2000 Universal 6 % 4000 Strong 10 %	2000 Universal 8 % 4000 Strong -15 %	4000 Strong 10 % 4000 Strong 15 %
Thread Cutting	2000 Universal 8 % 3000 Ferro 8 %	2000 Universal 8-12 % 3000 Ferro 8-12 % 4000 Strong 10 %	4000 Strong 10-15 %	2000 Universal -10 %	2000 Universal 8 % 4000 Strong 8 %	4000 Strong 10 %
Reaming	2000 Universal 6-8 % 3000 Ferro 6-8 %	2000 Universal 8 % 3000 Ferro 8 % 4000 Strong 8 %	4000 Strong 10 %	2000 Universal 8 %	2000 Universal 8-10 %	2000 Universal 10 % 4000 Strong 10 %
Milling	2000 Universal 6-8 % 3000 Ferro 6-8 %	2000 Universal 10 % 3000 Ferro 10 % 4000 Strong 8 %	4000 Strong 12 %	2000 Universal 6-10 %	2000 Universal 10 %	4000 Strong 8-10 %
Gear Hobbing Generating	4000 Strong 10 % 3000 Ferro 10 %	4000 Strong 12 % 3000 Ferro 10 %	4000 Strong 12-15 %	2000 Universal -15 % 4000 Strong -15 %	4000 Strong 12 %	4000 Strong 12-15 %
Sawing	2000 Universal 10 % 3000 Ferro 10 %	2000 Universal 10 % 3000 Ferro 10 % 4000 Strong 10 %	4000 Strong 12 %	2000 Universal 10 %	2000 Universal 10 %	4000 Strong 10 %
Autom. Tapping	2000 Universal 8 % 3000 Ferro 8 %	2000 Universal 10 % 3000 Ferro 10 % 4000 Strong 8 %	4000 Strong 12 %	2000 Universal 10-15 %	2000 Universal 12 % 4000 Strong 10 %	4000 Strong 10-12 %
Broaching	4000 Strong 10 %	4000 Strong 12-15 %	4000 Strong 15-20 %	2000 Universal 12 %	4000 Strong 12 %	4000 Strong 15 %
Stamping to .06" thick	4000 Strong 15 %	4000 Strong 15-25 %	4000 Strong 20 % (ideal: Automatic 840 / 842)	4000 Strong 10 %	4000 Strong -15 %	4000 Strong 10-20 % Automatic-848 Compound-851
Forming to 50% workpiece	4000 Strong 10-15 %	4000 Strong -25 %	Automatic-840 Compound-850 / -851	4000 Strong -15 % Automatic-848	4000 Strong 15 % Automatic-848	

12.75 (9.84)

For corrosion susceptible materials such as cast and spheroidal cast: 3000 Ferro

All data in this catalog are based on known and tested applications and conditions. They represent mean values and are not legally binding but are subject to change. Because conditions of use and applications are beyond our control, no warranty or representation is made or intended in connection with the use of this product.

Bruce Products Corporation
500 West Street/P.O. Box 99
Howell, MI 48843
(517) 546-0110

PRODUCT DATA

BRUCE PRODUCTS CORPORATION • P.O. BOX 99 • HOWELL, MICHIGAN 48843 • (517) 546-0110

II (d) Premium
Soluble Oil

BRUKO Brusol "C"
Hi E. P. Machining & Grinding Coolant

BRUKO Brusol "C" is a water soluble concentrate, resistant to odor causing bacteria, which is suitable for the entire range of general purpose machining and grinding of ferrous metals as well as certain alloys of aluminum.

Upon proper dilution with water, Brusol "C" can also be used for reaming, drilling, tapping, threading, serrating and sawing, as well as broaching of all ferrous metals.

It is outstanding at 1:45 dilutions for use in multiple head Kingsbury, Cross and similar equipment for high speed automatic drilling, tapping, threading, chamfering and reaming operations of ferrous and non-ferrous applications where chlorine and/or sulfur are metallurgically acceptable.

At 1:45 dilutions, Brusol "C" is useful for rough turning, finish turning, nose boring, base facing, internal thread hobbing and for turning copper rotating band grooves on 105 mm shell cases from SAE 1050; black jogged chips, rough surface finishes, wavy patterns in hobbled threads are eliminated and tool life is extended approximately 284%. (Considerable attention devoted to changing coolant flow-patterns produced these results).

Brusol "C" is particularly valuable, too, at 1:45 dilutions in automatics, bar machines, turret lathes, gear shapers and equipment that has been converted or is convertible from a cutting oil to a water dispersible product. It is excellent in roll forming mills, pipe making mills, and Grotnes passenger car, truck and agricultural rim rolling equipment at 1:45 dilutions. At 1:6 ratios, water extensions of Brusol "C" perform excellently in the spinning of passenger car wheel spiders.

Typical Specifications:

Viscosity @ 100°F	1500 - 1600
Specific Gravity	.98
Flash Point	735°F
Wt./Gal. =	8.20#

ds51883ew

CUTTING FLUIDS

Bruce Products Corporation
500 West Street/P.O. Box 99
Howell, MI 48843
(517) 546-0110

PRODUCT DATA

BRUCE PRODUCTS CORPORATION • P.O. BOX 99 • HOWELL, MICHIGAN 48843 • (517) 546-0110

BRUKO BRUSOL D

Metalworking Coolant

I Soluble Oil

A general - purpose emulsifying oil that covers a large part of the metalworking spectrum. It is a very versatile, economical product possessing great durability. It is used for cutting and grinding steels, and cast and nodular irons. Although Brusol D is used to machine aluminum and copper alloys, the alkaline nature of the emulsions tend to cause staining unless emulsions are removed soon after machining. Additional uses of Brusol D are:

2 - 4% emulsions at 160-180°F in industrial washers to clean and give short-term rust protection of ferrous parts.

15-20% emulsions at 140-180°F for temporary rust protection of steel plate.

Brusol D contains a mixture of anionic surfactants having a broad range of affinities for both oil and water which impart excellent shelf life to the concentrate and a broad emulsification temperature range (35-190°F). When properly prepared, the emulsions have an average particle diameter of one micron (1/25,000") imparting great stability. Brusol D emulsions are alkaline buffer systems that tend to resist acidification which, in turn, makes the emulsion more resistant to bacterial spoilage and rusting.

Typical Data:

Gravity, API/60F	28.6
Neat Oil Appearance	Dark
Emulsion Appearance	Off-white
Wt./Gal.	7.36 lbs.

mfr82083cjew

Bruce Products Corporation
500 West Street/P.O. Box 99
Howell, MI 48843
(517) 546-0110

PRODUCT DATA

BRUCE PRODUCTS CORPORATION • P.O. BOX 99 • HOWELL, MICHIGAN 48843 • (517) 546-0110

BRUKO Brusol G

Premium Soluble Oil

Brusol G is a heavy duty, high EP soluble oil which will perform difficult machining operations such as reaming, broaching and tapping.

Description: Brusol G is a brown liquid which forms a tan emulsion in water. It contains chlorine and sulfur EP agents, bacteriacides, corrosion preventatives and emulsifiers.

Application Concentration:

For broaching, tapping, threading	1-10 to 1-20
For heavy machining	1-20 to 1-50
For general machining	1-25 to 1-60

Characteristics: Brusol G possesses the high detergency and corrosion protection needed for good chip settling and removal. Excellent finishes are obtained from all machining operations, due to the high lubricity and extreme pressure properties of the product. Brusol G contains bacteriacides which are effective against a broad spectrum of microorganisms, thereby prolonging the life of the emulsion.

Storage: Brusol G requires no special provisions for storage but some increase in viscosity will be encountered in drums which are stored under low temperature conditions.

dls31183cj

Bruce Products Corporation
500 West Street/P.O. Box 99
Howell, MI 48843
(517) 546-0110

PRODUCT DATA

BRUCE PRODUCTS CORPORATION • P.O. BOX 99 • HOWELL, MICHIGAN 48843 • (517) 546-0110

BRUKO Brusol K

Hi E. P. Machining & Grinding Coolant

Bruko Brusol K is a water soluble concentrate, resistant to odor causing bacteria, which is suitable for the entire range of general purpose machining and grinding of ferrous and non-ferrous metals.

Upon proper dilution with water, Brusol K can be used for reaming, drilling, tapping, threading, serrating and sawing.

Its high polarity causes rapid plate-out of the lubricant on metal surfaces, normally difficult to wet out, and insures good film integrity and rapid carry-off of chips and swarf. Brusol K keeps drills and taps from loading up.

It is outstanding at 1:25 dilutions for use in multiple head Kingsbury, Cross and similar equipment for high speed automatic drilling, tapping, threading, chamfering and reaming operations of ferrous and non-ferrous applications where chlorine is metallurgically acceptable.

At 1:45 dilutions, Brusol K is useful for rough turning, finish turning, nose boring, base facing, internal thread hobbing and for turning copper rotating band groves on 105 mm shell cases from SAE 1050; black jagged chips, rough surface finishes, wavy patterns in hobbled threads are eliminated and tool life is extended.

It is excellent in roll forming mills, pipe making mills, and Grotnes passenger car, truck and agricultural rim rolling equipment at 1:45 dilutions. At 1:6 ratios, water extensions of Brusol K perform excellently in the spinning of passenger car wheel spiders.

<u>Typical Specifications</u>

Wt./Gallon	7.8#
SUS @ 100°F	425

mfr111183ew

Bruce Products Corporation
500 West Street/P.O. Box 99
Howell, MI 48843
(517) 546-0110

PRODUCT DATA

BRUCE PRODUCTS CORPORATION • P.O. BOX 99 • HOWELL, MICHIGAN 48843 • (517) 546-0110

BRUKO Brukool #85

IV Synthetic

Description: Brukool #85 is a synthetic coolant for grinding steel and cast iron with most types of grinders. It prevents rust in most types of grinders. It prevents rust in dilutions as high as one hundred to one and does not foam in high pressure, high capacity systems. Brukool #85 is a blue liquid soluble in all proportions in water. It does not contain nitrites.

Recommended Concentrations:

Light Grinding	1 - 100
Normal Grinding	1 - 75
Severe Rusting	1 - 50

Before charging a system with Brukool #85 it is recommended that the system be thoroughly cleaned or purged to remove residues from previously used coolants.

The system should be filled with hot water to which 2 ounces per gallon of trisodium phosphate has been added. This solution should be circulated for 30 to 60 minutes and then dumped. The system should then be rinsed with plain hot water.

Characteristics:

Color	Blue
Odor	nil
Specific Gravity	1.055
Weight per gallon	8.8#
pH	8.5 min.

Brukool #85 provides excellent cooling and prevents heat distortion and burning of work pieces. Excellent detergency keeps wheels and stones clean, prevents loading and drops chips and fines fast. Solution is clean and transparent and will not turn rancid when used in clean equipment.

Storage:

Freezes but thaws without separation.

Non-flammable.

ds72983cjew

Bruce Products Corporation
500 West Street/P.O. Box 99
Howell, MI 48843
(517) 546-0110

PRODUCT DATA

BRUCE PRODUCTS CORPORATION • P.O. BOX 99 • HOWELL, MICHIGAN 48843 • (517) 546-0110

BRUKO Brukool 100F

Soluble Tapping, Machining and Grinding Coolant

Brukool 100F is a concentrated water emulsifiable lubricant recommended for tapping, reaming, broaching, etc., of cold rolled and/or high nickel alloy steels.

When replacing ordinary soluble oils in machining operations, Brukool 100F will show a multi-fold improvement in tool life, finishes, and dimensional accuracy. It contains a high percentage of extreme pressure additives blended with excellent lubricants.

The obviously superior performance of Brukool 100F is readily noticeable, even during the first hour of production. It will not leave a residue that will gum or inhibit the operation of any machine designed to use water soluble coolants. The built-in bactericide incorporated in Brukool 100F has been proven to be non-toxic and extremely effective against rancidity and dermatitis. As with all emulsion products of this type, it is recommended that you add the product to the water in order to avoid an invert emulsion.

When used as a grinding fluid, the extreme pressure ingredients in Brukool100F react to keep the wheel open and free cutting, reducing smoke and excessive heat. You will realize longer wheel life, finer finishes, and greater dimensional accuracy. Brukool 100F keeps the wheel open, resulting in fewer dressings, longer wheel life and substantial savings.

Brukool 100F is recommended as a water soluble drawing and stamping compound that offers excellent results, reducing die wear, scrap and scoring, due to its superior combination of anti-weld lubricants. It recommended for high speed progressive die operations, transfer press operations, heavy blanking, piercing, and in some areas, even extruding operations. Brukool 100F may be used neat for the more severe blanking, piercing, and drawing operations, or may be blended with water for the less severe operations depending upon the application and the severity of the work.

Recommended Dilutions

Broaching, Tapping:	1 part Brukool 100F to 5-10 parts water.
Heavy Duty Machining:	1 part Brukool 100F to 10-40 parts water.
General Machining:	1 part Brukool 100F to 30-50 parts water.
Grinding:	1 part Brukool 100F to 40-60 parts water.
Drawing and Forming:	Neat, or diluted with water as required.

ds51883ew

CUTTING FLUIDS

Bruce Products Corporation
500 West Street/P.O. Box 99
Howell, MI 48843
(517) 546-0110

PRODUCT DATA

BRUCE PRODUCTS CORPORATION • HOWELL, MICHIGAN 48843 • (517) 546-0110

BRUKO Brukool 106

Machining & Grinding Coolant,
& Corrosion Inhibitor

Brukool 106 is a completely organic synthetic machining and grinding coolant and rust inhibitor designed for use on all ferrous metals as well as brass, bronze and copper alloys, and on galvanically active copper/magnesium/aluminum alloys where other types of coolants would stain the workpiece and turn it black or would itself turn rancid.

Brukool 106 contains no chlorine, no sulfur, no phosphate esters and no inorganic salts. It is a true chemical solution that is transparent amber in color and odorless both in the concentrate and at all water dilutions. It is considerably less toxic than sodium nitrite based coolants and displays greater rust inhibiting properties, particularly in the longer water dilutions. It contains sequestering agents for use in hard water localaties and is particularly effective in sequestering iron, calcium and magnesium ions, thereby eliminating hard water soaps, scum and scale formation, water spotting and streaking.

When appropriate ratios are employed and maintained, wet residual films will protect metal surfaces from rusting when stacked one upon the other (nested) and dry, transparent, non-tacky residual films provide short term internal and intransit protection in highly humid atmospheres.

Brukool 106 is recommended for . . .

1) machining and grinding cast iron, grey iron and Meehanite

2) honing of gears, brake drums, engine blocks, automatic transmission governor valving, water pump housings, etc. at 1:25

3) vapor blasting at 1:100

4) barrel tumbling and belt polishing of ferrous metals at 1:50

5) grinding of steel rule/epoxy die and trim die inserts for the die casting, paper and paper box industries

6) belt polishing of assembled pocket knives (steel, brass, plastic, wood and bone) at 1:50

7) use in "leak-testers" tank filled with water to detect leaks in pressure equipment.

Continued . . . .

Bruce Products Corporation
500 West Street/P.O. Box 99
Howell, MI 48843
(517) 546-0110

Page 2
BRUKO Brukool 106

8) at 0.2 to 1.0% dilutions in water, residual films protect metal substrates against corrosion for long periods of indoor protected storage.

Brukool 106 has no flash point nor fire point; is non-toxic and is emollient to skin tissues, of particular importance when replacing kerosene or Ucon Fluid in honing operations or where misting is generated. Because Brukool 106 is 100% biodegradeable it is almost always permissible to dump spent solutions into municipal and in-plant sewer systems.

Residual films are not suitable for introduction to silver brazing or soldering operations. Residual films need not be removed, however, prior to buffing or annealing and heat treating operations in controlled furnaces.

Weight per gallon = 9.15 lbs.

dls

111279

Bruce Products Corporation
500 West Street/P.O. Box 99
Howell, MI 48843
(517) 546-0110

PRODUCT DATA

BRUCE PRODUCTS CORPORATION • P.O. BOX 99 • HOWELL, MICHIGAN 48843 • (517) 546-0110

BRUKO Brukool 121 III(b) Semi-Synthetic

MACHINING AND GRINDING COOLANT

BRUKO Brukool 121 is a water soluble grinding coolant containing special corrosion and rust inhibitors, lubricants and extreme pressure lubricants, natural and synthetic anionic emulsifiers and wetting agents plus a bactericide.

BRUKO Brukool 121 is recommended for machining and grinding of cast iron, carbon and most alloy steels. It is generally used at concentrations of 3 - 5 percent for grinding and 5 - 10 percent for cutting and machining.

BRUKO Brukool 121 is compounded to impart good rust inhibition and bacteria control in the coolant.

TYPICAL PROPERTIES

Color	Pink
Appearance	Clear Liquid
Specific Gravity	1.02 at 60°F.
Pounds per Gallon	8.4 at 60°F.
Flash, °F.	none
Cold Stability	Good
5% Emulsion	Clear, low foaming
pH of 5% Emulsion	9.5 - 10.0
Falex Value	50 lbs. maximum torque

Bruce Products Corporation
500 West Street/P.O. Box 99
Howell, MI 48843
(517) 546-0110

PRODUCT DATA

BRUCE PRODUCTS CORPORATION • P.O. BOX 99 • HOWELL, MICHIGAN 48843 • (517) 546-0110

III(a) Semi-Synthetic

BRUKO Brukool #127
Water Soluble All-Purpose Coolant For
Grinding and Machining Operations

BRUKO Brukool 127 is semi-synthetic product developed to give the utmost in tool life, especially on the more difficult machining operations.

The action of this product is based on metal surface chemistry. Special polar-type extreme pressure agents in BRUKO Brukool 127 are attracted to metal surfaces. The resulting surface oiliness gives a high degree of lubricity in a relatively small area of extreme pressure contact between the tool and the work piece. The extreme pressure additives then act to reduce the severe friction between the tool cutting edge and the metal. The end result is increased tool life and finer finishes.

BRUKO Brukool 127 is especially recommended for aluminum in addition to ferrous metals.

BRUKO Brukool 127 also has the following features:

Highly resistant to bacteria that causes odors

Prevents rust

Non-foaming

Keeps machines clean

BRUKO Brukool 127 is suitable for the following machining operations at dilutions of 20:1 to 50:1.

boring	broaching	cutoff sawing
drilling	gear cutting	hobbing
milling	reaming	turning...tapping
threading		

BRUKO Brukool 127 is also suitable for grinding operations (including centerless) at dilutions of 50:1 and 100:1.

52876
mew

Information on this data sheet is presented to assist you in evaluating our product. It is intended for use by persons having technical skill at their own discretion and risk. We do not guarantee favorable results, and we assume no liability in connection with its use. This information is not intended as a license to operate under, or a recommendation to infringe, any patent covering material or use.

CUTTING FLUIDS

PRODUCT DATA

BRUCE PRODUCTS CORPORATION • P.O. BOX 99 • HOWELL, MICHIGAN 48843 • (517) 546-0110

BRUKO Brukool #130

Synthetic Coolant

BRUKO Brukool #130 is a synthetic machining and grinding coolant concentrate for use on all types of machining and grinding applications of ferrous and non-ferrous metals and excellent for EXOTIC metals such as titanium.

ADVANTAGES

(1) Mixes readily with hard or soft water

(2) Keeps grinding wheels clean and free cutting

(3) Reduces wheel wear, extending wheel life

(4) Will not gum, rust machine ways or work pieces

(5) Produces fine finish on both machining or grinding applications.

(6) Imparts a film between the tool and work piece removing heat instantaneously and developing the needed EP factor (lubrication) necessary in the metal working applications especially EXOTIC metals.

(7) When machining ferrous, non-ferrous or exotic metals, an invisible film is imparted on the work piece and chips preventing galvanic rust in the sump or on the work piece. This condition prevails in job shop procedures when the above metals are machined in one machine without changing the coolant in the sump after each metal is machined.

(8) By eliminating bacteria-producing raw materials, BRUKO Brukool #130 will not turn rancid. It is non-toxic and not injurious to plant personnel.

PROPERTIES

Wt./Gallon	8.66 lbs.	
pH of 10% Solution	8	
Falex 3000# gage	Jaw Load 3000#	Torque 38 in. lb.

continued...

Bruce Products Corporation
500 West Street/P.O. Box 99
Howell, MI 48843
(517) 546-0110

Page 2

SUGGESTED MIXTURE FOR BRUKOOL 130 SYNTHETIC COOLANT

General Machining	50 parts water to 1 part concentrate (replenish 60/1)
(Grey iron or Cast)	40 parts water to 1 part concentrate (replenish 50/1)
General Grinding	60 parts water to 1 part concentrate (replenish 70/1)
Drilling & Tapping	25 parts water to 1 part concentrate (replenish 30/1)
Milling & Reaming	40 parts water to 1 part concentrate (replenish 45/1)
Screw Machines	10 parts water to 1 part concentrate where seals will tolerate a water extendable coolant.

Because of its unique chemical characteristics, titanium cannot be worked the same way as aluminum and steel.

Titanium has a selective affinity for certain other materials. This characteristic can destroy titanium by stress corrosion cracking when it combines with such elements as iodine, chlorine, bromine and fluorine.

When titanium is cut, high pressure and temperatures up to 1000 degrees Fahrenheit develop. Under these conditions, the titanium chips weld together on the cutting tool.

Coolant fluids containing anti-weld additives for use with steel and aluminum are an aid in the cutting process, but increase the danger of stress-corrosion cracking. Use of these additives also requires carefully controlled cleaning of the titanium parts after machining. With the new formula fluids, no cleaning is required, and the risk of stress-corrosion cracking is eliminated.

mfr51184clj

PRODUCT DATA

BRUCE PRODUCTS CORPORATION • P.O. BOX 99 • HOWELL, MICHIGAN 48843 • (517) 546-0110

BRUKO Brukool 133-F

IV Synthetic

Machining and Grinding Coolant

DESCRIPTION: Brukool 133-F is a highly concentrated premium synthetic machining and grinding coolant for all ferrous metals and many non-ferrous alloys including brass, bronze and aluminum. It combines the cooling power of chemical coolants with the extreme pressure lubricity and anti-weld of heavy duty soluble oils.

APPLICATIONS: A synergistic balance of EP and anti-weld additives combines to give exceptional surface finish during high stock removal turning, threading, tapping and broaching of steels as well as internal grinding and centerless grinding of low machinability ferrous and non-ferrous metals.

Brukool 133-F gives outstanding performance in pipe threading to prevent chip welding, rough threads, and keeps collapsible die heads operating freely. It is also excellent for use in sawing tough alloy billets.

Brukool 133-F contains no petroleum oils, phenols, mercurials, nitrosamines, silicones or PTBBA. It forms clear solutions in water in all proportions and is not affected by hard water. It will reject most tramp oils while emulsifying only a minimum quantity thereby making it easy to skim sumps and reservoirs.

RANCIDITY CONTROL: Brukool 133-F possesses excellent rancidity control that protects against "Monday Morning Stink" and insures long fluid life and minimizes the use of additives. This in turn assures excellent rust protection and detergency which keeps machines and tools cleaner. It washes chips and grit away from the areas where the fluid flows across the machine tool.

DILUTION:

Operation	Carbon Steels Cast Steels	High Alloy and Stainless Steel	Copper, Brass Aluminum
Milling, Drilling, Turning	1:25	1:20-1:25	1:25
Heavy Duty Machining	1:10-1:20	1:10-1:20	1:10-1:25
Reaming, Tapping, Broaching Form Milling	1:10	1:10	1:10-1:15
Grinding	1:30-1:50	1:25-1:40	1:30-1:50
Cut Off & Sawing	1:10-1:15	1:10	1:15

Note: Concentrations may be controlled by refractometry or acid-base titrations. Concentration charts and methods are available upon request.

Continued . .

Bruce Products Corporation
500 West Street/P.O. Box 99
Howell, MI 48843
(517) 546-0110

Page 2
Brukool 133-F

PHYSICAL PROPERTIES:

Appearance	Dark Green Liquid
Odor	Bland, Pleasant Chemical
Specific Gravity	1.03
Wt./Gallon	8.6#
Flash Point	None
Solubility in Water	Infinite
Freezing Point	32°F
pH (1:10 or 9%)	9.0

STORAGE: Keep from freezing. If frozen, thaw completely at room temperature and agitate before withdrawing any quantity from container.

AVAILABILITY: Available in 5 gallon pails, 55 gallon drums, Liqua Bins and tank truck quantities. Smaller samples for testing available upon request.

dls32983cjew

Bruce Products Corporation
500 West Street/P.O. Box 99
Howell, MI 48843
(517) 546-0110

PRODUCT DATA

BRUCE PRODUCTS CORPORATION • P.O. BOX 99 • HOWELL, MICHIGAN 48843 • (517) 546-0110

TYPE IV
TRUE CHEMICAL SOLUTION

Brukool 140 Synthetic Coolant

Brukool 140 is a concentrated synthetic coolant for use in machining operations. It is not recommended for grinding.

DESCRIPTION:

Brukool 140 is a tan transparent liquid which dissolves readily in water. It contains no sulfur or chlorine additives but performs extremely well under extreme pressure conditions. All ingredients are biodegradable except the bacteriacides.

DILUTION:

Light machining 1:20 - 1:25, heavy machining 1:10 - 1:20, very heavy machining 1:10 - 1:15. Brukool 140 is soluble in hard and soft waters and will reject most tramp oils making it easy to skim sumps and reservoirs.

CHARACTERISTICS:

Brukool 140 has excellent lubricity and provides rust protection to both finished parts and machine tools. Good wetting and chip settling properties keep machines cool and clean.

Wt./ Gallon = 8.6#

Bruce Products Corporation
500 West Street/P.O. Box 99
Howell, MI 48843
(517) 546-0110

BRUKO CUTTING OILS AND COOLANTS

Product	Wt./Gal.	SUS @ 100°F	Flash °F	Anti-Weld	E.P.	Fat *
Neat Oils						
Brukut 4	7.4	160-180	340	Lo	No	Lo
Brukut 4LV	7.3	65-80	290	Lo	No	Lo
Brukut 7	7.6	180-210	340	Lo	Med.	Lo
Brukut 7A	7.6	180-210	340	Lo		Lo
Brukut 9	7.5	260-290	340	Med.	Lo	Med.
Brukut 11	7.56	200-220	430	No	No	Hi
Brukut 20	7.3	40-50	325	No	No	Hi
Brukut 22	7.3	50-60	290	Lo	Lo	Lo
Brukut 23	7.4	65-75	300	Lo	Lo	Med.
Brukut 102	7.5	180-190	340	Lo	Lo	Lo
Brukut 177X	7.6	175-200	340	Lo	Lo	Lo
Brukut 421	8.2	400-600	340	Med.	Hi	Lo
Brukut 421RS	8.2	400-600	340	Med.	Hi	Lo
Soluble Oils						
Brusol A	7.85	350	340	No	Med.	17
Brusol AL	7.6	225	-	No	Med.	12.5
Brusol C	8.2	1600	340	Med.	Hi	7
Brusol D	7.36	200	340	No	No	5.3
Brusol G	7.8	220	340	Lo	Lo	10.2
Brukool 66	7.8	1300	340	Lo	Med.	8.9
Brukool 100F	7.84	880	340	Hi	Lo	17
Brukool 127	7.8	900	340	Hi	Lo	11
Bruko D-633	8.58	6000	340	Lo	Hi	7
Bruko D-666	8.3	1700	340	Med.	Hi	13.5
Bruko D-667	8.4	3700	340	Hi	Hi	17
Bruko D-676	8.0	900	340	Hi	No	13.8
Bruko D-695	7.8	430	340	No	Hi	6
Bruko D-1059	8.15	1400	340	Hi	Hi	17
Synthetics						
Brukool 85	8.8			Lo	Lo	Lo
Brukool 106	9.15			Lo	Lo	Med.
Brukool 121	8.4			Lo	Med.	Med.
Brukool 127	7.86			Med.	Med.	Med.
Brukool 130	9.20			Med.	Med.	Lo
Brukool 133	8.8			Lo	Lo	Hi
Brukool 133AL	8.6			Lo	Lo	Hi
Brukool 133F	8.6			Lo	Med.	Hi
Brukool 140	8.60			Lo	Med.	Med.
Brukool 999	9.0			Lo	Lo	Lo

* Note: "Fat" may include fats, fatty acids, soaps, or other fatty materials and rosin.

Buckeye Lubricants
20801 Salisbury Road
Bedford, OH 44146
(216) 581-3600

BUCKEYE LUBRICANTS

20801 SALISBURY RD. **BEDFORD, OHIO 44146**

Phone (216) 581-3600 **Telex 98-0614**

PRODUCT DATA BULLETIN

BUCKEYE #320 SOLUBLE COOLANT

#320 is a semi-synthetic product developed to give the utmost in tool life, particularly on the more difficult machining operations. This product is highly effective for operations where dissimilar metals in contact are being machined. In addition, it is excellent for all operations on Aluminum, giving superior finish and tool life.

The action of #320 is based on metal surface chemistry. Special polar-type extreme pressure agents in this product are attracted to metal surfaces. The resulting surface oiliness gives a high degree of lubricity in a relatively small area of extreme pressure contact between the tool and the work piece. The extreme pressure additives then act to reduce the severe friction between the tool cutting edge and the metal. The end result is increased tool life and finer finishes.

#320 also has the following additional features:

Transparent - operators can see the work
Prevents rust
Non-foaming
Keeps machines clean

This product is suitable for the following operations at dilutions of 20:1 to 50:1.

Boring — Broaching — Cut Off Sawing
Drilling — Gear Cutting — Hobbing
Milling — Reaming — Turning
Tapping — Threading

#320 can also be used for grinding operations (including centerless) at dilutions of 50:1 to 100:1.

Samples are available on request.

Buckeye Lubricants
20801 Salisbury Road
Bedford, OH 44146
(216) 581-3600

BUCKEYE LUBRICANTS

20801 SALISBURY RD. **BEDFORD, OHIO 44146**
Phone (216) 581-3600

PRODUCT DATA BULLETIN

BUCKEYE # 333 HEAVY DUTY SOLUBLE COOLANT

Buckeye # 333 is a white emulsion, all purpose product that can be used on all metals. This product is particularly effective on aluminum and stainless steel.

By using # 333, staining of metals and welding or pick up on cutting tools, can be eliminated.

Recommended dilutions range from 5:1 for the most severe operations up to 50:1 for grinding.

SPECIFICATIONS

Viscosity	210 SSU @ 100°F
Flash Point	330°F
Fire Point	370°F
Chlorine	Nil
Sulfur	No active sulfur
Color	4½
Proprietary Lubricity Agents	18%
Gravity	24.3

Samples are available on request.

Buckeye Lubricants
20801 Salisbury Road
Bedford, OH 44146
(216) 581-3600

BUCKEYE LUBRICANTS
20801 SALISBURY RD. BEDFORD, OHIO 44146
Phone (216) 581-3600

PRODUCT DATA BULLETIN

BUCKEYE #991 GRINDING COOLANT

#991 is a biodegradable, synthetic, grinding coolant that can be used with both ferrous and non-ferrous metals. This product will not stain copper and is an excellent rust inhibitor, even on cast iron at dilutions up to 100:1.

#991 keeps wheels open. This reduces the frequency that wheels must be dressed and consequently increases wheel life.

This product will not turn rancid. Machines need only be cleaned periodically as a matter of good housekeeping. #991 also will not re-circulate grinding grit and sludge which eliminates scratching of the work piece.

#991 is a clear solution (water-white) which enables the operator to see his work. It is also non-foaming and non-slippery, which add to operator's acceptance.

This product can be used in all types of water, hard and soft. It also keeps the machines clean and will not allow sludge to harden in the sump.

#991 can be used at extended dilutions of 50:1 to 100:1 with water.

#991 can also be disposed of into any sewer system that has secondary treatment facilities, after tramp oil has been skimmed off.

Samples are available on request.

Buckeye Lubricants
20801 Salisbury Road
Bedford, OH 44146
(216) 581-3600

BUCKEYE LUBRICANTS

20801 SALISBURY RD. **BEDFORD, OHIO 44146**

Phone (216) 581-3600 Telex 98-0614

PRODUCT DATA BULLETIN

#17NNA-9 No-Nitrite Biodegradeable Coolant

This New Development has proven to be excellent when used in systems with machines working on more than one metal. It is particularly effective in eliminating welding and/or pick up on tools when working with aluminum.

#17NNA-9 is a highly effective, all-round metal working fluid that can be used on both machining and grinding operations. This product contains no Nitrite and consequently is suited for applications where Nitrites are not desired.

#17NNA-9 can be used with both ferrous and non-ferrous metals on a wide variety of applications. Typical usage dilutions are as follows. (Given in parts water to part #17' NNA-9)

Machining	10:1 to 40:1
Grinding	40:1 to 60:1
Rust Inhibiting	10:1 to 40:1

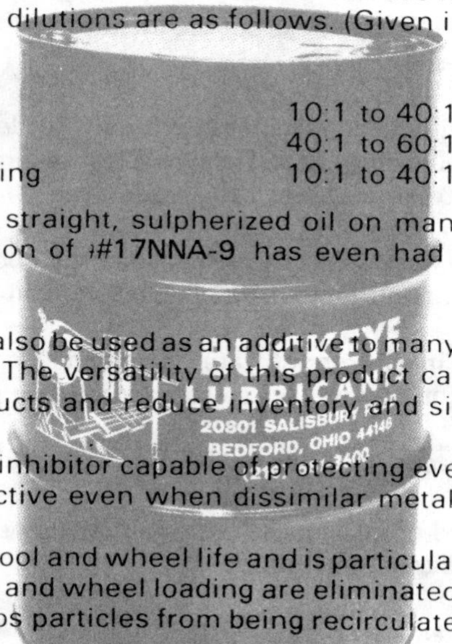

This product has replaced straight, sulpherized oil on many severe operations with excellent results. A variation of #17NNA-9 has even had superior performance on honing operations.

In addition, #17NNA-9 can also be used as an additive to many other products to improve finish and rust protection. The versatility of this product can enable the customer to replace several other products and reduce inventory and simplify operations.

This product is a good rust inhibitor capable of protecting even cast iron at dilutions up to 60:1. #17NNA-9 is effective even when dissimilar metals are in contact.

#17NNA-9 gives excellent tool and wheel life and is particularly effective on aluminum. In addition, pick up on tools and wheel loading are eliminated. This product also settles grit and sludge, which keeps particles from being recirculated through the machine or system.

Clean machines and systems are typical when #17NNA-9 is used. In addition, rancidity is virtually eliminated with this product, extending the time period before the machines must be cleaned. #17NNA-9 is also clear, enabling the operator to see his work. and non-slippery, preventing pieces from slipping out of the operators hands or flying off magnetic chucks.

#17NNA-9 contains no undesirable materials including halogens, silicones, phenols and mercury.

Samples of #17NNA-9 are available on request.

Buckeye Lubricants
20801 Salisbury Road
Bedford, OH 44146
(216) 581-3600

BUCKEYE LUBRICANTS

20801 SALISBURY RD. **BEDFORD, OHIO 44146**

Phone (216) 581-3600 **Telex 98-0614**

NEWS RELEASE

1973SP Blue Biodegradable Coolant

Buckeye #1973 Blue is a new biodegradable, synthetic coolant developed to replace heavy duty, soluble oil products on a wide variety of applications. This product contains proprietary EP additives, has excellent lubricity characteristics and is non-gumming. #1973SP Blue is a superior rust inhibitor, keeps the machine clean, has excellent operator's acceptance, and contains no oils, phenols, halogens or mercury.

This product, while economical in price, is suitable for most machining operations, and most metals, including stainless steel. It also can be used on many forming operations.

#1973SP Blue has many significant advantages over soluble oil type products. One primary benefit is longevity of life in the machine or system. Even when the machine is idle for a large percentage of the time there is no problem with rancidity. Life expectancy is measured in terms of months and even years, rather than days or weeks. Even when the machine must be cleaned #1973SP Blue offers further advantages. After allowing tramp oil to be skimmed and contaminants to settle, it can be eliminated into any water treatment system having secondary treatment facilities. Additional benefits include extended dilutions, superior rust protection, cleanliness, operators' acceptance, excellent tool life and low micro finish.

Samples of #1973SP Blue are available on request.

Buckeye Lubricants
20801 Salisbury Road
Bedford, OH 44146
(216) 581-3600

BUCKEYE LUBRICANTS

20801 SALISBURY RD. BEDFORD, OHIO 44146
Phone (216) 581-3600 Telex 98-0614

PRODUCT DATA BULLETIN

#SPLB-TLS-G "Tool Life Saver" Coolant
Heavy-Duty, No Nitrite Metalworking Fluid

This product has been developed for the most severe operations. On tough applications, where tool life or finish are problems, where production must be increased, or where it is desirable to replace a straight oil product, #SPLB-TLS-G can be the solution. It can be used on machining, grinding, and many forming operations.

Examples of the performance benefits of "Tool Life Saver" Coolant include the following:

A manufacturer of machine gun parts was able to increase carbide insert life by 400% on forged steel parts by switching to "TLS".

Wheel life in a Blanchard Grinder, operating on 1120 Steel, was doubled by a major fork lift manufacturer who changed to "Tool Life Saver." In addition a minimum of a 60% increase in carbide insert life (Carboloy PRO-MAX 570) was achieved on both low and high speeds with no performance problems. This eliminated both difficulties that had been experienced with the high speed tooling as well as corrosion that had been encountered on cast iron.

A pump manufacturer went to "TLS" and achieved multiple performance benefits. On 2011-T3 aluminum a significant increase in tap life was realized, along with better finishes.

Besides offering exceptional performance characteristics on the most severe applications, #SPLB-TLS-G also eliminates almost all need for disposal of spent coolant because of the extended life it offers in the machine or system, and its ability to be recycled.

"TLS" also exhibits other properties of a quality metalworking fluid to include:

Suitable for both Ferrous and Non-Ferrous metals
Excellent results have been obtained on aluminum
Clean
Excellent Finish
No Pickup on tools and wheels
Good operator acceptance
Mild to the skin and not too slippery
No gumming or sticking
Sludge remains soft

#SPLB-TLS-G contains no Nitrite or other known or suspected carcinogenic material. It also contains no Halogens, Oils, Silicones, Phenols or Mercury and is considered biodegradable.

SAMPLES OF #SPLB-TLS-G ARE AVAILABLE ON REQUEST

Butterfield/Litton
134 Chestnut Hill Avenue
Athol, MA 01331
(617) 249-3221

Introducing the tapping fluid that can increase work flow.

Now there's another choice in tapping fluids. And it's from Butterfield, the name industry knows for the finest taps and accessories.

Put this new tapping fluid up against brands of conventional formulations, and you'll discover that it's very *un*conventional.

Safety, Performance, Productivity all in one.

As soon as you open the can you'll notice the difference. But this amazing new tapping fluid has much more to offer besides pleasant fragrance.

- It contains no methyl chloroform or trichloroethane.
- It relies on superior lubrication to prevent high heat generation. Requires less tapping torque.
- It lubricates and prevents corrosion.
- It forms metallic chlorides for cutting purposes.
- It is effective in steel and a wide range of aerospace materials. (Greatly extended tap life is its main attribute.)

Butterfield tapping fluid is a clear, heavy oil which is chlorinated but does not contain chlorinated free solvents. It is a very pure, low acidity fluid. The following are some typical specifications:

Viscosity, Gardner – Holdt	U-V
Free Fatty Acids	1% Max.
Saponification Value	176-184
Moisture & Volatiles	.02% Max.
Hazardous Ingredients	N/A
Reactivity	Stable
Special Protection or Precautions	None

The proof is in the testing.

After extensive testing in the field and laboratory in a variety of applications, Butterfield tapping fluid was found to offer these user benefits:

- Reduced tapping torque
- Fewer scrapped parts
- Improved surface finish on threads
- Accuracy and size control
- Fewer tool changes (longer tool life)

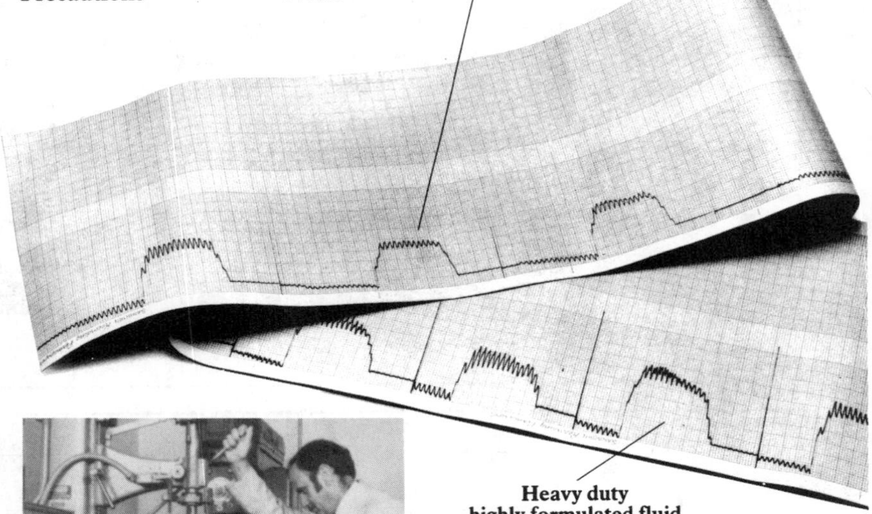

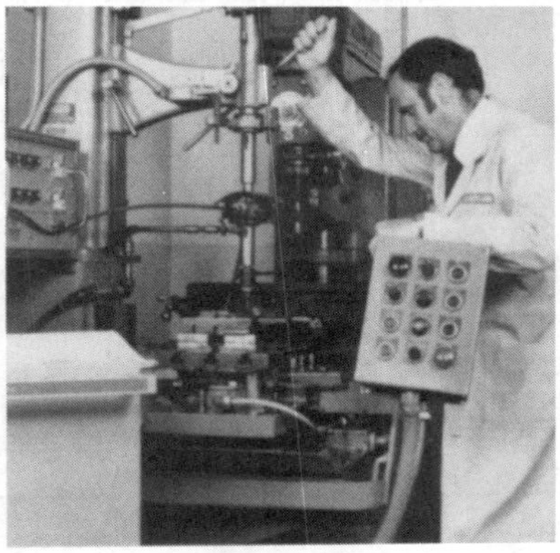

A 33% reduction in tapping torque. Butterfield fluid outperformed a heavy duty tapping fluid normally used by field service engineers. It produced a 33% reduction in tapping torque and noticeable heat reduction while tapping #3 drill holes in 6AL4V Titanium at 150 RPM with a 1/4-28NF-GH4 spiral point tap.

- Less regrinding and resharpening time
- Excellent boundary lubrication
- Corrosion retardation
- Application to application versatility
- Increased production

Add it all up. Then consider the irritating problems associated with conventional fluids, and the choice is clear. The next time you buy tapping fluid, ask for the best by name. Butterfield. Available through your local distributor now.

BUTTERFIELD
Litton Athol, Massachusetts 01331

INDUSTRIAL DISTRIBUTORS

Chem-Trend Incorporated
3205 East Grand River
Howell, MI 48843
(517) 546-4520

Safety-Cool® 407H

Soluble Oil Metalworking Fluid

Description

SAFETY-COOL® 407H is especially well suited for heavy-duty machining and grinding of both ferrous and nonferrous metals. This unique combination of proprietary ingredients provides exceptional tool life and long-term, trouble-free performance. SAFETY-COOL® 407H is recommended for use in individual sumps and central systems where exceptional rancidity control is required.

Benefits

- Exceptionally good rancidity control
- Excellent rust protection
- Extends tool life
- No nitrites or phenols
- Superior tramp oil rejection
- Forms stable emulsion in hard or soft water
- Operator compatible
- Dilutions easily monitored by refractive index
- Resists depletion
- Excellent foam control

Application/Dilutions

OPERATION	Carbon Steels Malleable Iron Cast Iron	High Alloy Steels Stainless Steel	Aircraft and Tool Steels	Nonferrous Aluminum, Brass and Copper	Machining with Carbide Tooling
Milling, Drilling, Turning	1:35 (3%)	1:25 (4%)	1:25 (4%)	1:40 (2.5%)	1:35 (3%)
Reaming, Light Duty Broaching Form Milling, Tapping, Sawing, Screw Machines	1:20 (5%)	1:20 (5%)	1:20 (5%)	1:30 (3.5%)	1:30 (3.5%)
Grinding: I.D., O.D., Centerless	1:40 (2.5%)	1:30 (3.5%)	1:30 (3.5%)	1:40 (2.5%)	----

NOTE: COOLANT SHOULD ALWAYS BE ADDED TO WATER. DO NOT ADD WATER TO CONCENTRATE.

Packaging

Available in 5-, 55-, 400-gallon, and bulk quantities.

HELPING INDUSTRY PRODUCE — Chem-Trend INCORPORATED

3205 E. Grand River • Howell, Michigan 48843 • Telephone (517) 546-4520 • Toll Free Number: 800-248-4056

Look for newer and better things from Chem-Trend, where constant research brings you tomorrow's products...today.

Chem-Trend Incorporated
3205 East Grand River
Howell, MI 48843
(517) 546-4520

Safety-Cool® 407H

Soluble Oil Metalworking Fluid

Typical Properties

Appearance: Concentrate	Dark, amber liquid
Dilution:	Opaque, yellowish liquid
pH, 1:40 (2.5%) dilution	9.0 ± 0.5
Density, lbs/gal; kg/l	8.2; .98
Residue	Oily
Odor	Neutral
Flash point (COC)	287°F/141°C
Freeze point, Concentrate	None
Nitrites, phenols	None

Disposal

Methods such as chemical treatment (acid-alum) and incineration may be used to meet local, state, and federal requirements.

Product Safety

Emergency and First Aid Procedures

Eyes: Flush with water for 15 minutes. Call physician.

Skin: Wash off contacted skin with soap and water.

Ingestion: Call physician immediately.

Inhalation: If affected by vapors, move to fresh air. Call physician.

This product is not hazardous under current EPA-RCRA regulations. Toxicity tests were conducted in accordance with techniques specified in the Regulations for the Enforcement of the Federal Hazardous Substances Act (16 CFR 1500). The material is not classified as toxic by oral or dermal administration. Information is available on request.

Foremost consideration is given to environmental and worker safety in the formulation of all CHEM-TREND products.

Data may vary slightly due to minor formulation changes.

Chem-Trend Incorporated
3205 East Grand River
Howell, MI 48843
(517) 546-4520

Safety-Cool 414

Soluble Oil Metalworking Fluid

Description

SAFETY-COOL 414 is a soluble oil metalworking fluid fortified with inactive sulfur and chlorine. It is specially formulated for severe machining and grinding operations. SAFETY-COOL 414 is ideal for broaching, tapping, turning, and grinding where a good microfinish is required on ferrous or nonferrous metals. This unique combination of proprietary ingredients provides exceptional tool life and long-term, trouble-free performance. SAFETY-COOL 414 is recommended for use in individual sumps, central systems, and screw machines.

Benefits

- High lubricity characteristics
- Exceptionally good rancidity control
- Excellent rust protection
- Extends tool life
- Reduces wheel wear
- No nitrites or phenols
- Superior tramp oil rejection
- Forms stable emulsion in hard or soft water
- Operator compatible
- Dilutions easily monitored by refractive index
- Resists depletion
- Excellent foam control

Application/ Dilutions

OPERATION	Carbon Steels Malleable Iron Cast Iron	High Alloy Steels Stainless Steel	Aircraft and Tool Steels	Nonferrous Aluminum, Brass and Copper	Machining with Carbide Tooling
Milling, Drilling, Turning	1:35 (3%)	1:25 (4%)	1:25 (4%)	1:40 (2.5%)	1:35 (3%)
Reaming, Broaching, Form Milling, Tapping, Sawing, Screw Machines	1:20 (5%)	1:20 (5%)	1:20 (5%)	1:30 (3.5%)	1:30 (3.5%)
Grinding: I.D., O.D., Centerless	1:40 (2.5%)	1:30 (3.5%)	1:30 (3.5%)	1:40 (2.5%)	----

NOTE: COOLANT SHOULD ALWAYS BE ADDED TO WATER. DO NOT ADD WATER TO CONCENTRATE.

Packaging

Available in 5-gallon pails, 55-gallon drums, tote tanks, and bulk quantities.

Chem-Trend Incorporated
3205 East Grand River
Howell, MI 48843
(517) 546-4520

Safety-Cool 414

Soluble Oil Metalworking Fluid

Typical Properties

Appearance: Concentrate	Dark-brown fluid
Dilution	Tan emulsion
pH, 1:40 (2.5%) dilution	8.5 ± 0.3
Density, lbs/gal; kg/l	8.0; 0.96
Residue	Oily
Odor	Neutral
Flash point (COC)	>385°F/196°C
Freeze point	None
Nitrites, phenols	None

Handling/Storage

No special precautions are necessary. SAFETY-COOL 414 can be stored indoors or outdoors. Product becomes viscous at cold temperatures. If stored outdoors, allow to warm to room temperature before mixing.

Disposal

Methods such as chemical treatment (acid-alum) and incineration may be used to meet local, state, and federal requirements.

Chem-Trend Incorporated — Helping Industry Produce

MATERIAL SAFETY DATA SHEET

CHEM-TREND INCORPORATED 3205 E. GRAND RIVER HOWELL, MICHIGAN 48843 PHONE 1 (517) 546-4520

SECTION I PRODUCT

TRADE NAME AND SYNONYMS: SAFETY-COOL 414
EMERGENCY TELEPHONE NO.: 517-546-4520
CHEMICAL FAMILY: Emulsifiable oil

SECTION II INGREDIENTS

MATERIAL	%	TOXICITY DATA
Proprietary blend of petroleum oil, emulsifiers, lubricity additives, polyol, couplers, and biocide	100	Not established

SECTION III PHYSICAL DATA

BOILING POINT (°F) (initial)	169	SPECIFIC GRAVITY (H_2O = 1)	0.96
VAPOR PRESSURE (mm Hg.)	Unknown	PERCENT VOLATILE BY WEIGHT (%)	≅ 8.0
VAPOR DENSITY (AIR = 1)	Unknown	EVAPORATION RATE (ether = 1)	<1
SOLUBILITY IN WATER	Dispersible		

APPEARANCE AND ODOR: Dark-brown liquid; bland odor

SECTION IV FIRE AND EXPLOSION HAZARD DATA

FLASH POINT (METHOD USED): > 385°F (COC)
FLAMMABLE LIMITS: Lel — Uel —
EXTINGUISHING MEDIA: Use foam, carbon dioxide or dry chemical
SPECIAL FIRE FIGHTING PROCEDURES: As for petroleum products
UNUSUAL FIRE AND EXPLOSION HAZARDS: None known

SP-106
0882

SAFETY-COOL 414

SECTION V HEALTH HAZARD DATA

THRESHOLD LIMIT VALUE: Not established for the product.
EMERGENCY AND FIRST AID PROCEDURES: Wash off contacted skin with soap and water. Launder contaminated clothing before reuse. In case of eye contact, flush eyes with water for at least 15 minutes. Call physician. If swallowed, call physician immediately. If affected by vapors, move to fresh air. Call physician.

SECTION VI REACTIVITY DATA

STABILITY: UNSTABLE — ; STABLE X. CONDITIONS TO AVOID: None known
INCOMPATIBILITY (MATERIALS TO AVOID): Store away from strong oxidizers.
HAZARDOUS DECOMPOSITION PRODUCTS: Hydrocarbon decomposition products at elevated temperatures.
HAZARDOUS POLYMERIZATION: MAY OCCUR — ; WILL NOT OCCUR X. CONDITIONS TO AVOID: None known

SECTION VII SPILL OR LEAK PROCEDURES

STEPS TO BE TAKEN IN CASE MATERIAL IS RELEASED OR SPILLED: Soak up with absorbent material and scrape up dried residue.
WASTE DISPOSAL METHOD: Dispose of in accordance with local, state, and federal regulations.
RCRA HAZARDOUS WASTE DESIGNATION: This product does not fall under current EPA RCRA definitions of hazardous waste.

SECTION VIII SPECIAL PROTECTION INFORMATION

RESPIRATORY PROTECTION (SPECIFY TYPE): Not when used under recommended conditions.
VENTILATION: LOCAL EXHAUST: As required to prevent exposure to mist. MECHANICAL (GENERAL): Recommended
PROTECTIVE GLOVES: Resistant gloves
EYE PROTECTION: Chemical goggles
OTHER PROTECTIVE EQUIPMENT: Appropriate clothing to avoid skin contact.

SECTION IX SPECIAL PRECAUTIONS

PRECAUTIONS TO BE TAKEN IN HANDLING AND STORING: Do not inhale mist. Do not take internally. Avoid skin contact. Do not transfer to unlabeled containers.

APPROVAL: KCK 08302

Chem-Trend Incorporated — Helping Industry Produce
CHEM-TREND INCORPORATED 3205 E. GRAND RIVER HOWELL, MICHIGAN 48843 PHONE 1 (517) 546-4520

Chem-Trend Incorporated
3205 East Grand River
Howell, MI 48843
(517) 546-4520

Safety-Cool® 801

Synthetic Metalworking Fluid

Description

SAFETY-COOL® 801 is a clear, water-extendable, synthetic surface grinding fluid. This nonfoaming fluid is especially recommended for Blanchard-type grinding of ferrous metals. SAFETY-COOL® 801 provides superior rust protection and excellent rancidity control.

Benefits

- Excellent foam control
- Excellent rust protection
- Exceptionally good rancidity control
- Prevents residue buildup
- Permits rapid settling of fines
- No nitrites or phenols
- Retains clarity throughout life of coolant
- Excellent for central systems
- Operator compatible
- Dilutions easily monitored by refractive index

Application/ Dilutions

OPERATION	Carbon Steels Malleable Iron Cast Iron	High Alloy Steels Stainless Steel	Aircraft and Tool Steels
Grinding: Surface, Blanchard, & Double-Disc	1:40 (2.5%)	1:40 (2.5%)	1:40 (2.5%)

NOTE: COOLANT SHOULD ALWAYS BE ADDED TO WATER. DO NOT ADD WATER TO CONCENTRATE.

Packaging

Available in 5-, 55-, 400-gallon, and bulk quantities.

3205 E. Grand River • Howell, Michigan 48843 • Telephone (517) 546-4520 • Toll Free Number: 800-248-4056

Chem-Trend Incorporated
3205 East Grand River
Howell, MI 48843
(517) 546-4520

Safety-Cool® 801

Synthetic Metalworking Fluid

Typical Properties

Appearance: Concentrate	Transparent fluid
Dilution	Transparent fluid
pH, 1:40 (2.5%) dilution	9.0 ± 0.5
Density, lbs/gal; kg/l	9.3; 1.1
Residue	Washable
Odor	Neutral
Flash point	None
Freeze point, Concentrate	-6°F/-21°C
Nitrites, phenols	None

Product Safety

Emergency and First Aid Procedures

Eyes: Flush with water for 15 minutes. Call physician.

Skin: Wash off contacted skin with soap and water.

Ingestion: Call physician immediately.

Inhalation: If affected by vapors, move to fresh air. Call physician.

This product is not hazardous under current EPA-RCRA regulations. Toxicity tests were conducted in accordance with techniques specified in the Regulations for the Enforcement of the Federal Hazardous Substances Act (16 CFR 1500). The material is not classified as toxic by oral or dermal administration. Information is available on request.

Foremost consideration is given to environmental and worker safety in the formulation of all CHEM-TREND products.

Data may vary slightly due to minor formulation changes.

Chem-Trend Incorporated
3205 East Grand River
Howell, MI 48843
(517) 546-4520

Safety-Cool® 802

Synthetic Metalworking Fluid

Description

SAFETY-COOL® 802 is a low-foaming version of SAFETY-COOL® 808. This new approach to synthetic coolant formulation has produced a unique and versatile metalworking fluid that leaves no residue. SAFETY-COOL® 802 is ideal for central system duty, grinding operations, and light-to-medium machining operations.

Benefits

- Prevents residue build-up
- Exceptionally good rancidity control
- Excellent rust protection
- Reduces wheel wear
- No nitrites or phenols
- Superior tramp oil rejection
- Excellent for I.D., O.D., and centerless grinding
- Forms stable dilution in hard or soft water
- Operator compatible
- Dilutions easily monitored by refractive index
- Resists depletion
- Excellent foam control

Application/ Dilutions

OPERATION	Carbon Steels Malleable Iron Cast Iron	High Alloy Steels Stainless Steel	Aircraft and Tool Steels	Nonferrous Aluminum, Brass and Copper	Machining with Carbide Tooling
Milling, Drilling, Turning	1:35 (3%)	1:25 (4%)	1:25 (4%)	1:40 (2.5%)	1:35 (3%)
Reaming, Light Duty Broaching Form Milling, Tapping, Sawing	1:20 (5%)	1:20 (5%)	1:20 (5%)	1:30 (3.5%)	1:30 (3.5%)
Grinding: I.D., O.D., Centerless	1:40 (2.5%)	1:30 (3.5%)	1:30 (3.5%)	1:40 (2.5%)	----

NOTE: COOLANT SHOULD ALWAYS BE ADDED TO WATER. DO NOT ADD WATER TO CONCENTRATE.

Packaging

Available in 5-gallon pails, 55-gallon drums, tote tanks, and bulk quantities.

HELPING INDUSTRY PRODUCE

Chem-Trend Incorporated
3205 East Grand River
Howell, MI 48843
(517) 546-4520

Safety-Cool® 802

Synthetic Metalworking Fluid

Typical Properties

Appearance:	Concentrate	Transparent fluid
	Dilution	Clear fluid
pH, Concentrate		9.6 ± 0.5
1:40 (2.5%) dilution		9.0 ± 0.2
Density, lbs/gal; kg/l		8.7; 1.04
Residue		Fluid
Odor		Neutral
Flash point		None
Freeze point, Concentrate		15°F/-9°C
Nitrites, phenols		None

Handling/ Storage

No special precautions are necessary. Indoor storage preferred. Product will freeze at cold temperatures (15°F/-9°C). If SAFETY-COOL® 802 should freeze, allow to warm to room temperature and mix thoroughly before use.

Disposal

Methods such as chemical treatment (acid-alum) and incineration may be used to meet local, state, and federal requirements.

Factory code 1

Chem-Trend INCORPORATED — HELPING INDUSTRY PRODUCE

MATERIAL SAFETY DATA SHEET

CHEM-TREND INCORPORATED 3205 E. GRAND RIVER HOWELL, MICHIGAN 48843 PHONE 1 (517) 546-4520

SECTION I PRODUCT

TRADE NAME AND SYNONYMS: SAFETY-COOL® 802	EMERGENCY TELEPHONE NO. 517 546-4520
CHEMICAL FAMILY: Synthetic	

SECTION II INGREDIENTS

MATERIAL	%	TOXICITY DATA
Proprietary blend of amines, corrosion inhibitors, lubricity additives, biocide, defoamer, and water.	100	Not established

SECTION III PHYSICAL DATA

BOILING POINT (°F)	≅ water	SPECIFIC GRAVITY (H₂O = 1)	1.04
VAPOR PRESSURE (mm Hg.)	≅ water	PERCENT VOLATILE BY WEIGHT (%)	67
VAPOR DENSITY (AIR = 1)	≅ water	EVAPORATION RATE (water = 1)	≅ 1
SOLUBILITY IN WATER	Miscible	pH (Concentrate)	9.6 ± 0.5
APPEARANCE AND ODOR	Transparent fluid; neutral odor.		

SECTION IV FIRE AND EXPLOSION HAZARD DATA

FLASH POINT (METHOD USED): None	FLAMMABLE LIMITS	Lel	Uel

EXTINGUISHING MEDIA: Use foam, carbon dioxide, or dry chemical.

SPECIAL FIRE FIGHTING PROCEDURES: As for petroleum products

UNUSUAL FIRE AND EXPLOSION HAZARDS: None known

SP-106
0882

SAFETY-COOL® 802 — **SECTION V HEALTH HAZARD DATA** — Factory code 1

THRESHOLD LIMIT VALUE: Not established for the product.

EMERGENCY AND FIRST AID PROCEDURES: Wash off contacted skin with soap and water. Launder contaminated clothing before reuse. In case of eye contact, flush eyes with water for 15 minutes. Call physician. If swallowed, call physician immediately. If affected by vapors, move to fresh air. Call physician.

SECTION VI REACTIVITY DATA

STABILITY	UNSTABLE		CONDITIONS TO AVOID: None known
	STABLE	X	
HAZARDOUS POLYMERIZATION	MAY OCCUR		CONDITIONS TO AVOID: None known
	WILL NOT OCCUR	X	

INCOMPATIBILITY (MATERIALS TO AVOID): Avoid materials incompatible with water.

HAZARDOUS DECOMPOSITION PRODUCTS: Hydrocarbon decomposition products at elevated temperatures.

SECTION VII SPILL OR LEAK PROCEDURES

STEPS TO BE TAKEN IN CASE MATERIAL IS RELEASED OR SPILLED: Soak up with absorbent material and scrape up dried residue.

WASTE DISPOSAL METHOD: Dispose of in accordance with local, state, and federal regulations.

RCRA HAZARDOUS WASTE DESIGNATION: This product does not fall under current EPA RCRA definitions of hazardous waste.

SECTION VIII SPECIAL PROTECTION INFORMATION

RESPIRATORY PROTECTION (SPECIFY TYPE): Not when used under recommended conditions.

VENTILATION — LOCAL EXHAUST: As required to prevent exposure to vapors and mist.

VENTILATION — MECHANICAL (GENERAL): Recommended

PROTECTIVE GLOVES: Resistant gloves — EYE PROTECTION: Chemical goggles

OTHER PROTECTIVE EQUIPMENT: Appropriate clothing to avoid skin contact.

SECTION IX SPECIAL PRECAUTIONS

PRECAUTIONS TO BE TAKEN IN HANDLING AND STORING: Do not inhale vapors or mist. Do not take internally. Avoid skin contact. Do not transfer to unlabeled containers.

APPROVAL: KCS 3283

Chem-Trend INCORPORATED — HELPING INDUSTRY PRODUCE — CHEM-TREND INCORPORATED 3205 E. GRAND RIVER HOWELL, MICHIGAN 48843 PHONE 1 (517) 546-4520

Chem-Trend Incorporated
3205 East Grand River
Howell, MI 48843
(517) 546-4520

Safety-Cool® 815

Synthetic Metalworking Fluid

Description

SAFETY-COOL® 815 is a general-purpose machining and grinding synthetic metalworking fluid. This unique formulation is a versatile metalworking fluid with superior rust protection and excellent rancidity control. SAFETY-COOL® 815 can be used in individual sumps and central systems.

Benefits

- Excellent rust protection
- Exceptionally good rancidity control
- Excellent for cast iron machining
- Superior tramp oil rejection
- Reduces wheel wear
- Resists buildup
- No nitrites or phenols
- Improves tool life
- Excellent for central systems
- Excellent for I.D., O.D., and centerless grinding
- Retains clarity throughout life of coolant
- Excellent foam control
- Operator compatible
- Dilutions easily monitored by refractive index

Application/ Dilutions

OPERATION	Carbon Steels Malleable Iron Cast Iron	High Alloy Steels Stainless Steel	Aircraft and Tool Steels	Nonferrous Aluminum, Brass and Copper	Machining with Carbide Tooling
Milling, Drilling, Turning	1:35 (3%)	1:25 (4%)	1:25 (4%)	1:40 (2.5%)	1:35 (3%)
Reaming, Light Duty Broaching, Form Milling, Tapping, Sawing,	1:20 (5%)	1:20 (5%)	1:20 (5%)	1:30 (3.5%)	1:30 (3.5%)
Grinding: I.D., O.D., Centerless, Surface	1:40 (2.5%)	1:30 (3.5%)	1:30 (3.5%)	1:40 (2.5%)	----

NOTE: COOLANT SHOULD ALWAYS BE ADDED TO WATER. DO NOT ADD WATER TO CONCENTRATE.

Packaging

Available in 5-, 55-, 400-gallon, and bulk quantities.

CUTTING FLUIDS

Chem-Trend Incorporated
3205 East Grand River
Howell, MI 48843
(517) 546-4520

Safety-Cool® 815

Synthetic Metalworking Fluid

Typical Properties

		SAFETY-COOL® 815	SAFETY-COOL® 815 BLUE
Appearance:	Concentrate	Clear, yellow	Dark blue
	Dilution	Clear liquid	Clear, blue
pH, 1:40 (2.5%) dilution		9.0 ± 0.5	9.0 ± 0.5
Density, lbs/gal; kg/l		8.1; 1.07	8.1; 1.07
Residue		Fluid	Fluid
Odor		Bland	Bland
Flash point		None	None
Freeze point, Concentrate		15°F/-9°C	15°F/-9°C
Nitrites, phenols		None	None

Product Safety

Emergency and First Aid Procedures

Eyes: Flush with water for 15 minutes. Call physician.

Skin: Wash off contacted skin with soap and water.

Ingestion: Call physician immediately.

Inhalation: If affected by vapors, move to fresh air. Call physician.

This product is not hazardous under current EPA-RCRA regulations. Toxicity tests were conducted in accordance with techniques specified in the Regulations for the Enforcement of the Federal Hazardous Substances Act (16 CFR 1500). The material is not classified as toxic by oral or dermal administration. Information is available on request.

Foremost consideration is given to environmental and worker safety in the formulation of all CHEM-TREND products.

Data may vary slightly due to minor formulation changes.

Chem-Trend Incorporated
3205 East Grand River
Howell, MI 48843
(517) 546-4520

Safety-Cool® 820

Synthetic Metalworking Fluid

Description

SAFETY-COOL® 820 is a synthetic metalworking fluid formulated to be used at lean dilutions and in applications where total rejection of tramp oil is desirable. This unique formulation demonstrates outstanding performance on grinding and machining operations of light-to-medium difficulty. SAFETY-COOL® 820 will not leave residue on machines or fixtures. This product is ideal for individual sumps and central coolant systems.

Benefits

- Superior tramp oil rejection
- Extends tool life
- Reduces wheel wear
- Excellent rust protection
- Exceptionally good rancidity control
- Prevents residue buildup
- No nitrites or phenols
- Dilutions easily monitored by refractive index
- Operator compatible
- Resists depletion

Application/ Dilutions

OPERATION	Carbon Steels Malleable Iron Cast Iron	High Alloy Steels Stainless Steel	Aircraft and Tool Steels	Nonferrous Aluminum, Brass and Copper	Machining with Carbide Tooling
Milling, Drilling, Turning	1:35 (3%)	1:25 (4%)	1:25 (4%)	1:40 (2.5%)	1:35 (3%)
Reaming, Light Duty Broaching, Form Milling, Tapping, Sawing,	1:20 (5%)	1:20 (5%)	1:20 (5%)	1:30 (3.5%)	1:30 (3.5%)
Grinding: I.D., O.D., Centerless	1:40 (2.5%)	1:30 (3.5%)	1:30 (3.5%)	1:40 (2.5%)	----

NOTE: COOLANT SHOULD ALWAYS BE ADDED TO WATER. DO NOT ADD WATER TO CONCENTRATE.

Packaging

Available in 5-, 55-, 400-gallon, and bulk quantities.

3205 E. Grand River • Howell, Michigan 48843 • Telephone (517) 546-4520 • Toll Free Number: 800-248-4056

Chem-Trend Incorporated
3205 East Grand River
Howell, MI 48843
(517) 546-4520

Safety-Cool® 820

Synthetic Metalworking Fluid

Typical Properties

Appearance: Concentrate	Light-brown fluid
Dilution	Clear fluid
pH, 1:40 (2.5%) dilution	9.1 ± 0.5
Density, lbs/gal; kg/l	8.7; 1.05
Residue	Fluid
Odor	Neutral
Flash point	None
Freeze point, Concentrate	15°F/-9°C
Nitrites, phenols	None

Product Safety

Emergency and First Aid Procedures

Eyes: Flush with water for 15 minutes. Call physician.

Skin: Wash off contacted skin with soap and water.

Ingestion: Call physician immediately.

Inhalation: If affected by vapors, move to fresh air. Call phusician.

This product is not hazardous under current EPA-RCRA regulations. Toxicity tests were conducted in accordance with techniques specified in the Regulations for the Enforcement of the Federal Hazardous Substances Act (16 CFR 1500). The material is not classified as toxic by oral or dermal administration. Information is available on request.

Foremost consideration is given to environmental and worker safety in the formulation of all CHEM-TREND products.

Data may vary slightly due to minor formulation changes.

Chem-Trend Incorporated
3205 East Grand River
Howell, MI 48843
(517) 546-4520

Mono-Lube® 3000H

Semisynthetic Metalworking Fluid

Description

MONO-LUBE® 3000H represents a new generation of semisynthetic metalworking fluids based on microemulsion technology. It combines the ease of maintenance of synthetic fluids and the economy of oil emulsions. MONO-LUBE® 3000H can be used for machining and grinding operations of ferrous and nonferrous materials. It is recommended for operations in which excellent rancidity control is required.

Benefits

- Exceptionally good rancidity control
- Extends tool life
- Superior tramp oil rejection
- Excellent foam control
- No nitrites or phenols
- Forms stable emulsion in hard or soft water
- Reduces wheel wear
- Excellent rust protection
- Operator compatible
- Resists depletion

Application/ Dilutions

OPERATION	Carbon Steels	High Alloy Steels Stainless Steel	Aircraft and Tool Steels	Nonferrous Aluminum, Brass and Copper	Machining with Carbide Tooling
Milling, Drilling, Turning	1:30 (3.5%)	1:20 (5%)	1:20 (5%)	1:25 (4%)	1:25 (4%)
Reaming, Tapping, Boring, Sawing, Screw Machines	1:25 (4%)	1:20 (5%)	1:20 (5%)	1:20 (5%)	1:25 (4%)
Grinding: I.D., O.D., Centerless	1:40 (2.5%)	1:30 (3.5%)	1:30 (3.5%)	1:40 (2.5%)	----

NOTE: COOLANT SHOULD ALWAYS BE ADDED TO WATER. DO NOT ADD WATER TO CONCENTRATE.

Packaging

Available in 5-, 55-, 400-gallon, and bulk quantities.

HELPING INDUSTRY PRODUCE Chem-Trend INCORPORATED

3205 E. Grand River • Howell, Michigan 48843 • Telephone (517) 546-4520 • Toll Free Number: 800-248-4056

Look for newer and better things from Chem-Trend, where constant research brings you tomorrow's products... today.

CUTTING FLUIDS

Chem-Trend Incorporated
3205 East Grand River
Howell, MI 48843
(517) 546-4520

Mono-Lube® 3000H

Semisynthetic Metalworking Fluid

Typical Properties

Appearance:	Concentrate	Clear, amber liquid
	Dilution	Opaque, yellowish liquid
pH, 1:40 (2.5%) dilution		9.1 ± 0.5
Density, lbs/gal; kg/l		8.3; 0.99
Residue		Oily
Odor		Mild
Flash point		None
Freeze point, Concentrate		32°F/0°C
Nitrites, phenols		None

Product Safety

Emergency and First Aid Procedures

Eyes: Flush with water for 15 minutes. Call physician.

Skin: Wash off contacted skin with soap and water.

Ingestion: Call physician immediately.

Inhalation: If affected by vapors, move to fresh air. Call physician.

This product is not hazardous under current EPA-RCRA regulations. Toxicity tests were conducted in accordance with techniques specified in the Regulations for the Enforcement of the Federal Hazardous Substances Act (16 CFR 1500). The material is not classified as toxic by oral or dermal administration. Information is available on request.

Foremost consideration is given to environmental and worker safety in the formulation of all CHEM-TREND products.

Data may vary slightly due to minor formulation changes.

Chem-Trend Incorporated
3205 East Grand River
Howell, MI 48843
(517) 546-4520

Mono-Lube® 3500

Semisynthetic Metalworking Fluid

Description

MONO-LUBE® 3500 represents a new generation of semisynthetic metalworking fluids. This product has extremely high lubricity characteristics. MONO-LUBE® 3500 can be used for severe machining of ferrous and nonferrous materials. The product is ideal for use in individual sumps and central coolant systems.

Benefits

- Excellent foam control
- Rejects tramp oil
- Does not corrode aluminum, brass, or zinc
- Operator compatible
- Excellent rust protection
- Fast fine-settling characteristics
- Dilutions easily monitored by refractive index
- No nitrites or phenols

Application/ Dilutions

OPERATION	Carbon Steels Malleable Iron Cast Iron	High Alloy Steels Stainless Steel	Aircraft and Tool Steels	Nonferrous Aluminum, Brass and Copper	Machining with Carbide Tooling
Milling, Drilling, Turning	1:35 (3%)	1:25 (4%)	1:25 (4%)	1:40 (2.5%)	1:35 (3%)
Reaming, Light Duty Broaching, Form Milling, Tapping, Sawing,	1:20 (5%)	1:20 (5%)	1:20 (5%)	1:30 (3.5%)	1:30 (3.5%)
Grinding: I.D., O.D., Centerless	1:40 (2.5%)	1:30 (3.5%)	1:30 (3.5%)	1:40 (2.5%)	----

NOTE: COOLANT SHOULD ALWAYS BE ADDED TO WATER. DO NOT ADD WATER TO CONCENTRATE.

Packaging

Available in 5-, 55-, 400-gallon, and bulk quantities.

CUTTING FLUIDS

Chem-Trend Incorporated
3205 East Grand River
Howell, MI 48843
(517) 546-4520

Mono-Lube® 3500

Semisynthetic Metalworking Fluid

Typical Properties

Appearance:	Concentrate	Clear, amber fluid
	Dilution	Clear, yellowish fluid
pH, 1:40 (2.5%) dilution		9.1 ± 0.5
Densith, lbs/gal; kg/l		8.36; 1.00
Residue		Fluid
Odor		Mild
Flash point		None
Freeze point, Concentrate		32°F/0°C
Nitrites, phenols		None

Product Safety

Emergency and First Aid procedures

Eyes: Flush with water for 15 minutes. Call physician.

Skin: Wash off contacted skin with soap and water.

Ingestion: Call physician immediately.

Inhalation: If affected by vapors, move to fresh air. Call physician.

This product is not hazardous under current EPA-RCRA regulations. Toxicity tests were conducted in accordance with techniques specified in the Regulations for the Enforcement of the Federal Hazardous Substances Act (16 CFR 1500). The material is not classified as toxic by oral or dermal administration. Information is available on request.

Foremost consideration is given to environmental and worker safety in the formulation of all CHEM-TREND products.

Data may vary slightly due to minor formulation changes.

Cincinnati Milacron Marketing Company
P.O. Box 9013
Cincinnati, OH 45209
(513) 841-8100

SYNTHETIC METALWORKING FLUID FOR FERROUS APPLICATIONS

CIMPLUS® 15

(Tested as CX-15X)

DESCRIPTION

CIMPLUS 15 metalworking fluid is a water-based, synthetic product.

CIMPLUS 15 metalworking fluid is a *nonnitrite, nonphenolic, nonsilicone,* product. It is also free of diethanolnitrosamine, mineral oil, mercurials, phosphates, PCB's, and PTBBA.

CIMPLUS 15 metalworking fluid has excellent ferrous corrosion control and rancidity control.

CIMPLUS 15 metalworking fluid can be recycled for reuse by using the CIMCOOL Full Cycle™ Module.

APPLICATION

Recommended for grinding and machining of cast iron and light-duty machining and grinding of carbon steel.

Especially well-suited for use in central systems on cast iron operations.

It is not recommended for use on aluminum or copper alloys.

CINCINNATI
MILACRON

Cincinnati Milacron Marketing Company
P.O. Box 9013
Cincinnati, OH 45209
(513) 841-8100

CIMPLUS 15

RECOMMENDED STARTING DILUTIONS

For machining: 1:25 (4.0%) to 1:35 (2.9%).
For grinding: 1:30 (3.3%) to 1:35 (2.9%).

CIMPLUS 15 metalworking fluid is to be mixed with water for use (add concentrate to water). Add no other substances to the concentrate or mix unless approved by CIMCOOL Technical Services **(1-800-On 2 time, or in Ohio 1-800-582-7522).**

RECOMMENDED CIMCOOL MIX MASTER PROPORTIONER TIP SIZES

Dilution	1:25 (4.0%)	1:30 (3.3%)	1:35 (2.9%)	1:40 (2.5%)	1:50 (2.0%)
Tip Size	56	57	60	62	64

After installation of the tip, titrate the mix to be sure the concentration is correct. Leaner dilutions may be needed for makeup.

For concentration analysis, use the CIMCOOL "Mini" Kit III, the Total Alkalinity Titration Procedure, or a refractometer.

TYPICAL PHYSICAL AND CHEMICAL PROPERTIES

Physical state .... liquid
Appearance and odor .... clear; chemical
Colors available .... green, undyed
Solubility in water 100%
Weight, lb/gal, 60°F (15.6°C) 9.06
Specific gravity (H_2O = 1) 1.087
Boiling point, °F (°C) 212 (100)
Flash point, COC, °F (°C) .... none; self-extinguishing
Fire point, COC, °F (°C) .... none; self-extinguishing
  Extinguishing media .... no fire hazard
  Unusual fire & explosion hazards .... none
Freezing point (or pour point), °F (°C) .... 6 (-14)
  If frozen, product separates. Thaw completely at room temperature and stir thoroughly.
pH, concentrate 10.4
pH, 1:30 (3.3%) mix, typical operating conditions 9.4
Total chlorine/chloride, wt %, calculated none/0.0026
Total sulfur, wt %, calculated .... none
Diethanolnitrosamine, mercurials, phenols, PCB's, nitrites, silicones,
  phosphates, PTBBA, mineral oil .... none

PACKAGING

Available in 5-gallon pails, 55-gallon drums, Liqua Bins, and in tank truck quantities.

Do not reuse container before cleaning. After cleaning, remove label and relabel appropriately.

DOT LABELING REQUIREMENTS

Hazardous Materials Description and Proper Shipping Name (49 CFR 172.101):	Hazard Class (49 CFR 172.101):
Not a hazardous material	Not applicable

Cincinnati Milacron Marketing Company
P.O. Box 9013
Cincinnati, OH 45209
(513) 841-8100

SYNTHETIC METALWORKING FLUID FOR FERROUS GRINDING APPLICATIONS

CIMPLUS® 80
(Tested as CX-80A)

DESCRIPTION
CIMPLUS 80 metalworking fluid is a transparent, water-based synthetic product.

CIMPLUS 80 metalworking fluid provides good rust and rancidity control, and excellent foam control.

CIMPLUS 80 metalworking fluid is a *nonnitrite, nonphenolic, nonsilicone* product. It is also free of diethanolnitrosamine, mineral oil, mercurials, phosphates, PCB's and PTBBA.

CIMPLUS 80 metalworking fluid can be recycled for reuse by using the CIMCOOL Full Cycle™ Module.

APPLICATION
Recommended for surface, double-disc, center-type, and light-duty centerless work on ferrous metals. It is not recommended for grinding of aluminum or copper alloys.

CINCINNATI
MILACRON

Cincinnati Milacron Marketing Company
P.O. Box 9013
Cincinnati, OH 45209
(513) 841-8100

CIMPLUS 80

RECOMMENDED STARTING DILUTIONS

Use at 1:30 (3.3%) for machining and grinding.

CIMPLUS 80 metalworking fluid is to be mixed with water for use (add concentrate to water). Add no other substances to the concentrate or mix unless approved by CIMCOOL Technical Services **(1-800-On 2 time, or in Ohio 1-800-582-7522).**

RECOMMENDED CIMCOOL MIX MASTER PROPORTIONER TIP SIZES

Dilution	1:10 (10.0%)	1:15 (6.7%)	1:20 (5.0%)	1:30 (3.3%)	1:40 (2.5%)	1:50 (2.0%)
Tip Size	48	52	55	58	64	68

After installation of the tip, titrate the mix to be sure the concentration is correct. Leaner dilutions may be needed for makeup.

For concentration analysis, use the CIMCOOL "Mini" Kit III, the Total Alkalinity Titration Procedure, or a refractometer.

TYPICAL PHYSICAL AND CHEMICAL PROPERTIES

Physical state .... liquid
Appearance and odor .... clear; chemical
Colors available .... undyed
Solubility in water 100%
Weight, lb/gal, 60°F (15.6°C) 9.02
Specific gravity (H_2O = 1) 1.082
Boiling point, °F (°C) 212 (100)
Flash point, COC, °F (°C) .... none; self-extinguishing
Fire point, COC, °F (°C) .... none; self-extinguishing
Extinguishing media .... no fire hazard
Unusual fire & explosion hazards .... none
Freezing point (or pour point), °F (°C) .... 10 (−12.2)
If frozen, product separates. Thaw completely at room temperature and stir thoroughly.
pH, concentrate 10.5
pH, 1:30 (3.3%) mix, typical operating conditions 9.5
Total chlorine/chloride, wt %, calculated none/0.0029
Total sulfur, wt %, calculated .... none
Diethanolnitrosamine, mercurials, phenols, PCB's, nitrites, silicones, phosphates, PTBBA, mineral oil .... none

PACKAGING

Available in 5-gallon pails, 55-gallon drums, Liqua Bins, and in tank truck quantities.

Do not reuse container before cleaning. After cleaning, remove label and relabel appropriately.

DOT LABELING REQUIREMENTS

Hazardous Materials Description and Proper Shipping Name (49 CFR 172.101):	Hazard Class (49 CFR 172.101):
Not a hazardous material	Not applicable

Cincinnati Milacron Marketing Company
P.O. Box 9013
Cincinnati, OH 45209
(513) 841-8100

SYNTHETIC METALWORKING FLUID WITH SYNTHETIC LUBRICANT FOR MODERATE-DUTY MACHINING AND GRINDING APPLICATIONS

CIMCOOL® 250 WITH MSL®

(Tested as CX-250)

DESCRIPTION

CIMCOOL 250 metalworking fluid contains a truly unique synthetic lubricant, MSL, developed and patented by Cincinnati Milacron.

CIMCOOL 250 metalworking fluid is a "true" synthetic developed to meet the needs of the metalworking industry today. This reliable and quality fluid gives the user a tool to increase productivity without sacrificing part quality.

CIMCOOL 250 metalworking fluid is a *nonnitrite, nonphenolic, nonsilicone,* synthetic product. CIMCOOL 250 metalworking fluid is also free of mineral oil, diethanolnitrosamine, mercurials, phosphates, PCB's, and PTTBA.

Easy to maintain and control. CIMCOOL 250 metalworking fluid can be recycled for reuse by using the CIMCOOL Full Cycle™ Module.

APPLICATION

Recommended for moderate-duty machining and grinding of ferrous metals. CIMCOOL 250 metalworking fluid is especially well suited for central system and transfer line applications. Can also be used in individual sump applications.

CINCINNATI MILACRON

Cincinnati Milacron Marketing Company
P.O. Box 9013
Cincinnati, OH 45209
(513) 841-8100

CIMCOOL® 250 WITH MSL®

RECOMMENDED STARTING DILUTIONS

For moderate-duty machining and grinding operations: 1:20 (5.0%) to 1:30 (3.3%).

CIMCOOL 250 metalworking fluid is to be mixed with water for use (add concentrate to water). Add no other substances to the concentrate or mix unless approved by CIMCOOL Technical Services **(1-800-On 2 time, or in Ohio 1-800-582-7522).**

RECOMMENDED CIMCOOL MIX MASTER PROPORTIONER TIP SIZES

Dilution	1:30 (3.3%)	1:40 (2.5%)	1:45 (2.2%)	1:50 (2.0%)	1:60 (1.7%)
Tip Size	56	62	64	65	66

After installation of the tip, titrate the mix to be sure the concentration is correct. Leaner dilutions may be needed for makeup.

For concentration analysis, use the CIMCOOL "Mini" Kit III, Total Alkalinity Titration Procedure, or a refractometer.

TYPICAL PHYSICAL AND CHEMICAL PROPERTIES

Physical state .... liquid
Appearance and odor .... clear; chemical
Colors available .... blue, undyed
Solubility in water 100%
Weight, lb/gal, 60°F (15.6°C) 8.71
Specific gravity (H_2O = 1) 1.045
Boiling point, °F (°C) 212 (100)
Flash point, COC, °F (°C) .... none; self-extinguishing
Fire point, COC, °F (°C) .... none; self-extinguishing
  Extinguishing media .... no fire hazard
  Unusual fire & explosion hazards .... none
Freezing point (or pour point), °F (°C) .... 22 (−5.6)
  If frozen, product separates. Thaw completely at room temperature and stir thoroughly.
pH, concentrate 9.60
pH, 1:30 (3.3%) mix, typical operating conditions 9.10
Total chlorine/chloride, wt %, calculated none/0.0034
Total sulfur, wt %, calculated .... none
Diethanolnitrosamine, mercurials, phenols, PCB's, nitrites, silicones,
  phosphates, PTBBA, mineral oil .... none

PACKAGING

Available in 5-gallon pails, 55-gallon drums, Liqua Bins, and in tank truck quantities.

Do not reuse container before cleaning. After cleaning, remove label and relabel appropriately.

DOT LABELING REQUIREMENTS

Hazardous Materials Description and Proper Shipping Name (49 CFR 172.101):	Hazard Class (49 CFR 172.101):
Not a hazardous material	Not applicable

Cincinnati Milacron Marketing Company
P.O. Box 9013
Cincinnati, OH 45209
(513) 841-8100

SYNTHETIC METALWORKING FLUID WITH SYNTHETIC LUBRICANT FOR HEAVY-DUTY APPLICATIONS

CIMCOOL® 400 WITH MSL®

DESCRIPTION

CIMCOOL 400 metalworking fluid contains a unique synthetic lubricant, MSL, developed and patented by Cincinnati Milacron. CIMCOOL 400 is a clear, water-based synthetic product which is free of mineral oil, phenols, mercurials, phosphates, PCB's, PTBBA, diethanolnitrosamine, chlorine, sulfur, and boron compounds.

APPLICATION

Recommended for heavy-duty machining and grinding of steel and cast iron to replace cutting oils and soluble oils. Especially effective for machining and grinding alloy steels and exotic metals. Excellent choice for central system applications.

MAJOR ADVANTAGES

- Very economical — low makeup rates
- Provides exceptional tool life — patented synthetic lubricant outperforms the cut zone lubricity obtained with oil-based extreme pressure cutting fluids
- Results in very high productivity — infrequent tool changes; higher speeds and feeds than with heavy-duty soluble oils
- Reduces heat (the enemy of tool life) during chip formation
- Eliminates oil mists associated with petroleum products; excellent cleanliness properties
- Provides extraordinary mildness properties
- Rejects high percentage of tramp oil
- Leaves a soft or slightly liquid residual film that is rinsable
- The *true* synthetic — contains a new proprietary synthetic lubricant developed by CIMCOOL chemists

CINCINNATI
MILACRON

Cincinnati Milacron Marketing Company
P.O. Box 9013
Cincinnati, OH 45209
(513) 841-8100

CIMCOOL® 400 WITH MSL®

RECOMMENDED STARTING DILUTIONS

Operation	Cast Iron	Carbon Steels, Cast Steels	High Alloy Steels, Stainless Steels	Tool Steels
Milling, Drilling, Turning	1:25 (4.0%)	1:25 (4.0%)	1:20 (5.0%)	1:20 (5.0%)
Machining - heavy duty	1:25 (4.0%)	1:20 (5.0%)	1:20 (5.0%)	1:15 (6.7%)
Reaming, Tapping, Broaching, Form Milling	1:20 (5.0%)	1:15 (6.7%)	1:10 (10.0%)	1:10 (10.0%)
Grinding	1:25 (4.0%)	1:30 (3.3%)	1:25 (4.0%)	1:20 (5.0%)
Cutoff & Sawing	1:15 (6.7%)	1:15 (6.7%)	1:10 (10.0%)	1:10 (10.0%)

CIMCOOL 400 is to be mixed with water for use (add concentrate to water). Add no other substances to the concentrate or mix unless approved by Cimcool Technical Services (1-800-On 2 Time, or in Ohio 1-800-582-7522).

RECOMMENDED CIMCOOL MIX MASTER PROPORTIONER TIP SIZES

Dilution	1:10 (10.0%)	1:15 (6.7%)	1:20 (5.0%)	1:30 (3.3%)	1:40 (2.5%)	1:50 (2.0%)	1:60 (1.7%)	1:70 (1.4%)
Tip Size	50	52	54	57	62	66	68	70

After installation of the tip, titrate the mix to be sure the concentration is correct. Leaner dilutions will be needed for makeup.

For concentration analysis, use the Total Alkalinity Titration Procedure, the CIMCOOL "Mini" Kit III, or a refractometer.

TYPICAL PHYSICAL AND CHEMICAL PROPERTIES

Physical state .... liquid
Appearance and odor .... clear; sassafras
Color .... undyed, pink, blue
Solubility in water 100%
Weight, lb/gal, 60°F (15.6°C) 8.78
Specific gravity ($H_2O = 1$) 1.053
Boiling point, °F (°C) 212 (100)
Flash point, COC, °F (°C) .... none; self-extinguishing
Fire point, COC, °F (°C) .... none; self-extinguishing
Extinguishing media .... no fire hazard
Unusual fire & explosion hazards .... none
Freezing point (or pour point), °F (°C) .... -2 (-18.9)
If frozen, product separates. Thaw completely at room temperature and stir thoroughly.
pH, concentrate 9.6
pH, 1:25 (4.0%) mix, typical operating conditions 9.0
Total chlorine/chloride, wt %, calculated none/0.0030
Total sulfur, wt %, calculated .... none
Diethanolnitrosamine, mercurials, mineral oil, nitrites, phenols, PCB's, phosphates, PTBBA, silicones .... none

PACKAGING

Available in 5-gallon pails, 55-gallon drums, Liqua Bins, and in tank truck quantities.

Do not reuse container before cleaning. After cleaning, remove label and relabel appropriately.

DOT LABELING REQUIREMENTS

Hazardous Materials Description and Proper Shipping Name (49 CFR 172.101):	Hazard Class (49 CFR 172.101):
Not a hazardous material	Not applicable

Ask your local Cincinnati Milacron Distributor for a trial today.

CUTTING FLUIDS

Cincinnati Milacron Marketing Company
P.O. Box 9013
Cincinnati, OH 45209
(513) 841-8100

SEMI-SYNTHETIC METALWORKING FLUID FOR MODERATE-DUTY APPLICATIONS

CIMCOOL FIVE STAR® 40

DESCRIPTION
CIMCOOL FIVE STAR 40 metalworking fluid is a clear, water-soluble, nitrite-free, semi-synthetic product having good rust control and excellent rancidity control.

CIMCOOL FIVE STAR 40B has the same characteristics as CIMCOOL FIVE STAR 40, but is formulated to minimize foaming in soft water.

APPLICATION
For general-purpose grinding and machining operations where long cutting fluid life is needed, such as in large individual machine or central system reservoirs. Recommended for moderate-duty machining and grinding of ferrous metals, and some nonferrous metals.

When silicones are prohibited, always use CIMCOOL FIVE STAR 40 rather than CIMCOOL FIVE STAR 40B.

CINCINNATI
MILACRON

Cincinnati Milacron Marketing Company
P.O. Box 9013
Cincinnati, OH 45209
(513) 841-8100

CIMCOOL FIVE STAR 40

RECOMMENDED STARTING DILUTIONS

Use at 1:20 (5.0%) to 1:30 (3.3%) for machining and grinding.

CIMCOOL FIVE STAR 40 is to be mixed with water for use (add concentrate to water). Add no other substances to the concentrate or mix unless approved by Cimcool Technical Services (1-800-On 2 Time, or in Ohio 1-800-582-7522).

RECOMMENDED CIMCOOL MIX MASTER PROPORTIONER TIP SIZES

Dilution	1:20 (5.0%)	1:25 (4.0%)	1:30 (3.3%)	1:35 (2.9%)	1:40 (2.5%)	1:50 (2.0%)
Tip Size	52	54	56	58	60	62

After installation of the tip, titrate the mix to be sure the concentration is correct. Leaner dilutions may be needed for makeup.

For concentration analysis, use the Mixed Indicator (MI) Titration Procedure, the CIMCOOL "Mini" Kit II, or a refractometer.

TYPICAL PHYSICAL AND CHEMICAL PROPERTIES

Physical state .... liquid
Appearance and odor .... clear; chemical
Colors available .... pink, undyed, blue
Solubility in water 100%
Weight, lb/gal, 60°F (15.6°C) 8.84
Specific gravity (H_2O = 1) 1.060
Boiling point, °F (°C) 212 (100)
Flash point, COC, °F (°C) .... none; self-extinguishing
Fire point, COC, °F (°C) .... none; self-extinguishing
  Extinguishing media .... no fire hazard
  Unusual fire & explosion hazards .... none
Freezing point (or pour point), °F, (°C) .... 20 (-6.7)
  If frozen, thaw completely at room temperature and stir thoroughly.
pH, concentrate 10.3
pH, 1:30 (3.3%) mix, typical operating conditions 9.0
Total chlorine/chloride, wt %, calculated none/0.0033
Total sulfur, wt %, calculated 0.012
Phenols, phosphates, mercurials, PCB's, nitrites,
  PTBBA, diethanolnitrosamine, silicones .... none

PACKAGING

Available in 5-gallon pails, 55-gallon drums, Liqua Bins, and in tank truck quantities.

Do not reuse container before cleaning. After cleaning, remove label and relabel appropriately.

DOT LABELING REQUIREMENTS

Hazardous Materials Description and Proper Shipping Name (49 CFR 172.101):	Hazard Class (49 CFR 172.101):
Not a hazardous material	Not applicable

Cincinnati Milacron Marketing Company
P.O. Box 9013
Cincinnati, OH 45209
(513) 841-8100

GENERAL PURPOSE SEMISYNTHETIC METALWORKING FLUID FOR MODERATE-DUTY MACHINING AND GRINDING APPLICATIONS

CIMCOOL FIVE STAR® 50

(Tested as CIMCOOL K-50; CIMCOOL K-50B)

DESCRIPTION

CIMCOOL FIVE STAR 50 metalworking fluid uniquely blends the newest technological advantages of modern-day semisynthetics. CIMCOOL FIVE STAR 50 metalworking fluid is the newest addition to the popular CIMCOOL FIVE STAR "family" of metalworking fluids.

CIMCOOL FIVE STAR 50 metalworking fluid is a completely water-soluble, *nonnitrite, nonphenolic, nonsilicone,* semisynthetic product.

CIMCOOL FIVE STAR 50 metalworking fluid is an ideal product for individual machine sump and cental system applications due to its broad concentration range,1:15 (6.6%)–1:30 (3.3%), and performance capabilities.

CIMCOOL FIVE STAR 50 metalworking fluid is easy to maintain and control. This product can be recycled for reuse by using the CIMCOOL Full Cycle™ Module, or can easily be treated for disposal.

APPLICATION

Recommended for general-purpose machining and grinding of ferrous metals including carbon/cast steel, cast/nodular/malleable/gray iron, stainless steel, and high alloy steels. Can also be used on some light- to moderate-duty applications on nonferrous metals.

CIMCOOL FIVE STAR 50B metalworking fluid has the same characteristics as CIMCOOL FIVE STAR 50 metalworking fluid, but is formulated to minimize foaming in soft water. When silicones are prohibited, always use CIMCOOL FIVE STAR 50 metalworking fluid rather than CIMCOOL FIVE STAR 50B metalworking fluid.

CINCINNATI
MILACRON

Cincinnati Milacron Marketing Company
P.O. Box 9013
Cincinnati, OH 45209
(513) 841-8100

CIMCOOL FIVE STAR 50

RECOMMENDED STARTING DILUTIONS

Use at 1:15 (6.6%) to 1:30 (3.3%) for machining and grinding of ferrous and some non-ferrous metals.

CIMCOOL FIVE STAR 50 metalworking fluid is to be mixed with water for use (add concentrate to water). Add no other substances to the concentrate or mix unless approved by CIMCOOL Technical Services **(1-800-On 2 time, or in Ohio 1-800-582-7522).**

RECOMMENDED CIMCOOL MIX MASTER PROPORTIONER TIP SIZES

Dilution	1:10 (10.0%)	1:15 (6.7%)	1:20 (5.0%)	1:25 (4.0%)	1:30 (3.3%)	1:35 (2.9%)	1:40 (2.5%)
Tip Size	50	52	55	56	59	62	64

After installation of the tip, titrate the mix to be sure the concentration is correct. Leaner dilutions may be needed for makeup.

For concentration analysis, use the CIMCOOL "Mini" Kit I, Mixed Indicator (MI) Titration Procedure, or a refractometer.

TYPICAL PHYSICAL AND CHEMICAL PROPERTIES

Physical state .... liquid
Appearance and odor .... clear; chemical
Colors available .... pink, undyed, blue
Solubility in water 100%
Weight, lb/gal, 60°F (15.6°C) 8.38
Specific gravity (H_2O = 1) 1.005
Boiling point, °F (°C) 212 (100)
Flash point, COC, °F (°C) .... none; self-extinguishing
Fire point, COC, °F (°C) .... none; self-extinguishing
    Extinguishing media .... no fire hazard
    Unusual fire & explosion hazards .... none
Freezing point (or pour point), °F (°C) 32 (0)
    If frozen, product separates. Thaw completely at room temperature and stir thoroughly.
pH, concentrate 10.2
pH, 1:30 (3.3%) mix, typical operating conditions 9.1
Total chlorine/chloride, wt %, calculated none/0.0003
Total sulfur, wt %, calculated 0.167
Diethanolnitrosamine, mercurials, phenols, PCB's, nitrites, silicones,
    phosphates, PTBBA .... none

PACKAGING

Available in 5-gallon pails, 55-gallon drums, Liqua Bins, and in tank truck quantities.

Do not reuse container before cleaning. After cleaning, remove label and relabel appropriately.

DOT LABELING REQUIREMENTS

Hazardous Materials Description and Proper Shipping Name (49 CFR 172.101):	Hazard Class (49 CFR 172.101):
Not a hazardous material	Not applicable

Cincinnati Milacron Marketing Company
P.O. Box 9013
Cincinnati, OH 45209
(513) 841-8100

SEMISYNTHETIC METALWORKING FLUID FOR MODERATE-DUTY MACHINING AND GRINDING APPLICATIONS

CIMCOOL® QUAL STAR

(Tested as CX1002)

"A NEW-GENERATION SEMISYNTHETIC"

DESCRIPTION

As a new-generation semisynthetic fluid, CIMCOOL QUAL STAR metalworking fluid uniquely blends the technological advantages of modern day synthetics with the performance capabilities of soluble oils.

CIMCOOL QUAL STAR metalworking fluid is a clean, completely water-soluble, *nonnitrite, non-phenolic, nonsilicone,* semisynthetic product, having exceptional corrosion and rancidity control. CIMCOOL QUAL STAR metalworking fluid has a broad concentration range, 1:10 (10%)-1:30 (3.3%), which gives it the flexibility to fit many operations in the plant.

CIMCOOL QUAL STAR metalworking fluid is exceptionally mild, which makes it an ideal product for individual machine applications where mixes tend to concentrate. CIMCOOL QUAL STAR metal-working fluid can also be used in central system applications.

CIMCOOL QUAL STAR metalworking fluid's flexi-bility and broad operating range complements the new era of flexible manufacturing technology.

Easy to maintain and control. CIMCOOL QUAL STAR metalworking fluid can be recycled for reuse by the CIMCOOL Full Cycle™ Module, or can easily be treated for disposal.

APPLICATION

Recommended for general-purpose machining and grinding of ferrous metals including carbon/cast steel, cast/nodular/malleable/gray iron, stain-less steel, and high alloy steels. Can also be used on some light- to moderate-duty applications on nonferrous metals.

CINCINNATI
MILACRON

Cincinnati Milacron Marketing Company
P.O. Box 9013
Cincinnati, OH 45209
(513) 841-8100

CIMCOOL QUAL STAR

RECOMMENDED STARTING DILUTIONS

Use at 1:10 (10.0%) to 1:30 (3.3%) for machining and grinding of ferrous and some nonferrous metals.

CIMCOOL QUAL STAR metalworking fluid is to be mixed with water for use (add concentrate to water). Add no other substances to the concentrate or mix unless approved by CIMCOOL Technical Services **(1-800-On 2 time, or in Ohio 1-800-582-7522).**

RECOMMENDED CIMCOOL MIX MASTER PROPORTIONER TIP SIZES

Dilution	1:10 (10.0%)	1:20 (5.0%)	1:25 (4.0%)	1:30 (3.3%)
Tip Size	43	48	51	53

After installation of the tip, titrate the mix to be sure the concentration is correct. Leaner dilutions may be needed for makeup.

For concentration analysis, use the CIMCOOL "Mini" Kit I, Mixed Indicator (MI) Titration Procedure, or a refractometer.

TYPICAL PHYSICAL AND CHEMICAL PROPERTIES

Physical state .... liquid
Appearance and odor .... clear; chemical
Colors available .... pink, undyed
Solubility in water 100%
Weight, lb/gal, 60°F (15.6°C) 8.28
Specific gravity (H_2O = 1) 0.993
Boiling point, °F (°C) 212 (100)
Flash point, COC, °F (°C) .... none; self-extinguishing
Fire point, COC, °F (°C) .... none; self-extinguishing
    Extinguishing media .... no fire hazard
    Unusual fire & explosion hazards .... none
Freezing point (or pour point), °F (°C) .... 22 (−5.6)
    If frozen, thaw completely at room temperature.
pH, concentrate 9.5
pH, 1:30 (3.3%) mix, typical operating conditions 8.9
Total chlorine/chloride, wt %, calculated 0.143/0.002
Total sulfur, wt %, calculated 0.36
Diethanolnitrosamine, mercurials, phenols, PCB's, nitrites, silicones,
    phosphates, PTBBA .... none

PACKAGING

Available in 5-gallon pails, 55-gallon drums, Liqua Bins, and in tank truck quantities.

Do not reuse container before cleaning. After cleaning, remove label and relabel appropriately.

DOT LABELING REQUIREMENTS

Hazardous Materials Description and Proper Shipping Name (49 CFR 172.101):	Hazard Class (49 CFR 172.101):
Not a hazardous material	Not applicable

Cincinnati Milacron Marketing Company
P.O. Box 9013
Cincinnati, OH 45209
(513) 841-8100

METALWORKING FLUID FOR EXTREMELY HEAVY-DUTY MACHINING AND GRINDING

CIMPERIAL® 20

DESCRIPTION
CIMPERIAL 20 metalworking fluid is a premium soluble oil which forms a milky mix with water.

APPLICATION
Designed to replace straight cutting oils, CIMPERIAL 20 is recommended for extremely heavy-duty machining and grinding of ferrous metals. It is also nonstaining to many alloys of aluminum and copper.

A high degree of chemical extreme pressure (EP) lubricity makes it excellent for low-speed, low-clearance machining operations such as threading, tapping, broaching, reaming, gun drilling and gear hobbing.

Because CIMPERIAL 20 has grinding characteristics similar to sulfurized oil, it is highly recommended for diamond dressed form grinding operations such as ball track grinding of bearing races and bar grinding. These jobs and similar difficult grinding operations can be done more quickly, accurately, and economically with CIMPERIAL 20 than with straight cutting oils or conventional soluble oils.

CINCINNATI
MILACRON

Cincinnati Milacron Marketing Company
P.O. Box 9013
Cincinnati, OH 45209
(513) 841-8100

CIMPERIAL 20

RECOMMENDED STARTING DILUTIONS

Operation	Carbon Steels, Cast Steels	High Alloy Steels, Stainless Steels	Aircraft, Tool Steels
Heavy Duty Machining	1:20 (5.0%)	1:20 (5.0%)	1:15 (6.7%)
Reaming, Tapping, Broaching, Form Milling	1:20 (5.0%)	1:15 (6.7%)	1:15 (6.7%)
Grinding, Sawing, Gear Shaping, Hobbing	1:20 (5.0%)	1:15 (6.7%)	1:20 (5.0%)
Form and Centerless Grinding and Heavy Duty Grinding to Replace Oil	1:20 (5.0%)	1:20 (5.0%)	1:20 (5.0%)

CIMPERIAL 20 is to be mixed with water for use (add concentrate to water). Add no other substances to the concentrate or mix unless approved by Cimcool Technical Services (1-800-On 2 Time, or in Ohio 1-800-582-7522).

RECOMMENDED CIMCOOL MIX MASTER PROPORTIONER TIP SIZES

Dilution	1:10 (10.0%)	1:15 (6.7%)	1:20 (5.0%)	1:25 (4.0%)	1:30 (3.3%)
Tip Size	1/8	46	50	52	54

After installation of the tip, titrate the mix to be sure the concentration is correct.

For concentration analysis, use the CIMCOOL "Mini" Kit I, the CIMCOOL Mixed Indicator (MI) Titration Procedure, or a refractometer.

TYPICAL PHYSICAL AND CHEMICAL PROPERTIES

Physical state .... liquid
Appearance and odor .... clear; sassafras
Color .... undyed
Solubility in water .... (appreciable; emulsifiable) greater than 10%
Weight, lb/gal, 60°F (15.6°C) 8.30
Specific gravity (H_2O = 1) 0.996
Boiling point, °F (°C) .... not applicable
Flash point, COC, °F (°C) 390 (199)
Fire point, COC, °F (°C) 420 (215)
When mixed with water for use, the flash and fire points are raised beyond the range of the COC method.

Extinguishing media .... foam, carbon dioxide
Unusual fire & explosion hazards .... none
Freezing point (or pour point), °F (°C) .... −2 (−19)
If frozen, thaw completely at room temperature.
pH, concentrate .... not applicable
pH, 1:20 (5.0%) mix, typical operating conditions 9.1
Total chlorine/chloride, wt %, calculated 5.2/0.0001
Total sulfur, wt %, calculated 3.5
Phenols, phosphates, mercurials, PCB's, nitrites, diethanolnitrosamine, silicones .... none

PACKAGING

Available in 5-gallon pails, 55-gallon drums, Liqua Bins, and tank truck quantities.

Do not reuse container before cleaning. After cleaning, remove label and relabel appropriately.

DOT LABELING REQUIREMENTS

Hazardous Materials Description and Proper Shipping Name (49 CFR 172.101):	Hazard Class (49 CFR 172.101):
Not a hazardous material	Not applicable

Cincinnati Milacron Marketing Company
P.O. Box 9013
Cincinnati, OH 45209
(513) 841-8100

GENERAL-PURPOSE METALWORKING FLUID WITH HEAVY-DUTY CAPABILITY

CIMPERIAL® 1011

DESCRIPTION

CIMPERIAL 1011 metalworking fluid is a premium soluble oil, which forms a milky mix with water.

APPLICATION

CIMPERIAL 1011 metalworking fluid is recommended for many machining and grinding operations such as broaching, gear cutting, tapping, reaming, turning, drilling, internal grinding, and centerless grinding of low and high machinability ferrous and nonferrous metals.

An excellent product for cutoff and power sawing operations.

CINCINNATI
MILACRON

CIMPERIAL 1011

RECOMMENDED STARTING DILUTIONS

Operation	Carbon Steels, Cast Steels	High Alloy Steels, Stainless Steels	Aircraft, Tool Steels	Copper, Aluminum
Machining—Milling, Drilling, Turning, Cutoff, & Power Sawing	1:20 (5.0%)	1:20 (5.0%)	1:15 (6.7%)	1:20 (5.0%)
Reaming, Tapping, Broaching, Form Milling	1:10 (10.0%)	1:10 (10.0%)	1:10 (10.0%)	1:10 (10.0%)
Grinding	1:25 (4.0%)	1:25 (4.0%)	1:20 (5.0%)	1:25 (4.0%)

CIMPERIAL 1011 metalworking fluid performs well in these operations. It can be used, however, on other applications. For specific recommendations, contact your local CIMCOOL representative.

CIMPERIAL 1011 metalworking fluid is to be mixed with water for use (add concentrate to water). Add no other substances to the concentrate or mix unless approved by CIMCOOL Technical Services **(1-800-On 2 time, or in Ohio 1-800-582-7522).**

RECOMMENDED CIMCOOL MIX MASTER PROPORTIONER TIP SIZES

Dilution	1:10 (10.0%)	1:20 (5.0%)	1:25 (4.0%)	1:30 (3.3%)	1:40 (2.5%)
Tip Size	46	50	52	53	56

After installation of the tip, titrate the mix to be sure the concentration is correct. Leaner dilutions may be needed for makeup.

For concentration analysis, use the CIMCOOL "Mini" Kit I, Mixed Indicator (MI) Titration Procedure, or a refractometer.

TYPICAL PHYSICAL AND CHEMICAL PROPERTIES

Physical state .... liquid
Appearance and odor .... hazy; sassafras or evergreen
Colors available .... green, undyed
Solubility in water .... (appreciable; emulsifiable) greater than 10%
Weight, lb/gal, 60°F (15.6°C) 8.34
Specific gravity (H_2O = 1) 1.008
Boiling point, °F (°C) .... not determined
Flash point, COC, °F (°C) 335 (168)
Fire point, COC, °F (°C) 365 (185)
When mixed with water for use, the flash and fire points are raised beyond the range of the COC method.
Extinguishing media .... foam, carbon dioxide
Unusual fire & explosion hazards .... none
Freezing point (or pour point), °F (°C) .... 30 (-1.1)
If frozen, thaw completely at room temperature.
pH, concentrate .... not applicable
pH, 1:20 (5.0%) mix, typical operating conditions 8.6
Total chlorine/chloride, wt %, calculated 8.4/0.0002
Total sulfur, wt %, calculated 0.75
Diethanolnitrosamine, mercurials, PCB's, phosphates, nitrites, PTBBA .... none

PACKAGING

Available in 5-gallon pails, 55-gallon drums, Liqua Bins, and in tank truck quantities.

Do not reuse container before cleaning. After cleaning, remove label and relabel appropriately.

DOT LABELING REQUIREMENTS

Hazardous Materials Description and Proper Shipping Name (49 CFR 172.101):	Hazard Class (49 CFR 172.101):
Not a hazardous material	Not applicable

Cincinnati Milacron Marketing Company
P.O. Box 9013
Cincinnati, OH 45209
(513) 841-8100

CIMCLEAN®

Industrial Cleaners

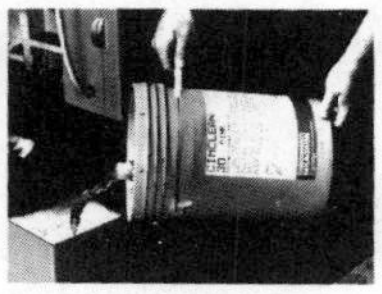

CIMCLEAN 30

CIMCLEAN 30 industrial cleaner is a general purpose, synthetic product.

Application

Recommended for cleaning individual machine and central system reservoirs, fluid lines, premix tanks, machines, and parts. It can also be used in parts washers, steam cleaners, floor washing equipment, and for brush cleaning by adjusting the concentration of CIMCLEAN 30 and water.

Features And Benefits

- Handles the toughest cleaning jobs, such as decontaminating and removing insoluble soaps, caused by hard water.
- Thoroughly removes chip deposits and tramp oil.
- Improves rancidity control.
- Contains corrosion-inhibiting ingredients to protect machines and parts.
- Saves money: CIMCLEAN 30 is highly concentrated cleaner. It can be used from 1:50 (2%) to 1:150 (0.6%) depending on the severity of the cleaning job.
- Eliminates fire hazards. CIMCLEAN 30 is a water-based formulation with no petroleum, no so-called "safety solvents." No solvent odor.

Packaging

Available in 5-gallon pails, 55-gallon drums, and tank trucks.

CIMCLEAN 10

CIMCLEAN 10 industrial cleaner, when diluted with water at 1:25 (4%), removes grit, dirt, oil and grease from individual machine and central filtration reservoirs.

Application

CIMCLEAN 10 cleans fluid reservoirs, pumps and piping, and central filtration system components. It can be mixed with the metalworking fluid before dumping or used after the fluid is dumped for more thorough cleaning.

Features And Benefits

- Excellent cleaner to use when converting from cutting oil to water-soluble metalworking fluid. Removes oily, greasy deposits and restores bright finish to machine tool exteriors.
- Improves rancidity control by removing sludge and dirt - breeding grounds for bacteria.
- Noncorrosive - will not rust exposed parts of machines or remove factory-applied machine paints.
- Has excellent nonferrous corrosion control which makes it an effective cleaner for aluminum and copper parts.

Packaging

Available in 5-gallon pails, 55-gallon drums, and tank trucks.

CIMTAP®

Tapping Compounds

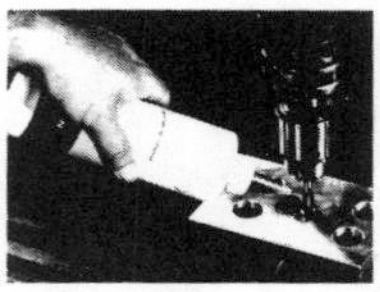

CIMTAP II

CIMTAP II tapping compound is a water-soluble liquid.

Application

CIMTAP II is designed for blind holes, deep vertical holes and high production operations requiring quick application. Can be used on both ferrous and nonferrous metals.

Features And Benefits

- CIMTAP II comes in an easy application squirt bottle.
- Maximum tap life - CIMTAP II contains extreme-pressure additives that form a solid lubricant between chip and tool.
- Accurate threads on steels, cast irons, stainless and nickel alloys, copper and die castings.
- Clean and water-soluble, leaving no deposits on the machine.
- Compatible with and readily soluble in water-based metalworking fluids.
- Nontoxic, pleasant to use, with no smoke or fumes.

Packaging

Available in cases of 12 reusable 1-pint squirt bottles, 5-gallon pails, 55-gallon drums, and tank trucks.

CIMTAP

CIMTAP tapping compound is a water-soluble paste that contains conventional and extreme pressure (EP) lubricants.

Application

Recommended for a wide range of tapping operations, from hand tapping to high production automatic jobs, on ferrous and nonferrous metals. Excellent for small partial assemblies.

Can be used in three ways: 1) Applied to tap directly;
2) Diluted with water and applied with a squirt can; or
3) Mixed with CIMCOOL metalworking fluids for high production, multiple operations.

Features And Benefits

- Maximum tap life - good lubricity eliminates chip welding, tap burning and resultant tap breakage.
- Accurate threads on steels, cast irons, stainless and nickel alloys, as well as on aluminum, brass and copper alloys.
- Clean, water-soluble, does not leave deposits on machine.
- Pleasant to use - doesn't smoke or smell. It is nontoxic, and doesn't produce fumes as do some chlorinated solvent tapping compounds.

Packaging

Available in cases of 24 1-pint cans, 40-lb. pails, and 250-lb. drums.

Crown Industrial Products Company, Incorporated
100 State Line Road/P.O. Box 326
Hebron, IL 60034
(815) 648-2424

Cutting Oil

Crown's #7020 Cutting Oil is designed for use with non-ferrous metals that are easily stained by the presence of active sulfur additives in cutting oils. The product contains no chlorine and approximately 2.0% inactive sulfur which will not stain up to 250°F.

Essentially, cutting fluids have two basic functions:

1. As a coolant - cool the tool so as to reduce loss of hardness and cool the work to prevent any distortions of the substrate.

2. As a lubricant - improve surface finish; also prevent welding of metal to tool point and subsequent rough surfaces due to material build-up on the finished face of the work.

In addition to these functions there are numerous secondary but important requirements which must be satisfied.

They must protect the finished surfaces against rust. They must not promote discoloration of the finished surface. This could result from chemical action between the fluid and the metal. For example, active sulfur will promote a black discoloration when machining copper alloys or brass. Surface discoloration may result in overheating of the surface if to much friction is built up during the machining operation.

Sulfur in cutting oils is frequently classified as being either active or nonactive. It is assumed that sulfur must be classified as active in order to be effective. The terms "active" or "inactive" only refer to how sulfur affects a copper strip at a specified temperature. For example, a copper strip is immersed into the oil at 212°F and if the strip is not discolored, or is only slightly tarnished, the sulfur is termed "inactive." If the strip is badly tarnished the term active is used. Normally these temperatures are below those temperatures encountered in machining operations. It is not recommended for cutting oils with "active" sulfur to be used in the machining of copper alloys or similar materials, since staining might occur.

Crown's Cutting Oil is considered to be an all purpose lubricant and coolant. It is specially blended with a combination of various additives to produce a cutting fluid to handle most machining operations.

DoALL Company
254 North Laurel Avenue
Des Plaines, IL 60016-4398
(312) 824-1122

NEW!

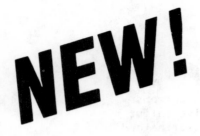

EXTRA HEAVY DUTY FOR TOUGHEST METALS

Make mixtures more concentrated for greater lubricity, more dilute for greater cooling effect.

POWER-CUT® 390 EHD

The Super-Versatile Cutting Fluid

A SUPERIOR SOLUBLE OIL FOR TOUGHER OPERATIONS

- ***Best Tool Life***
- ***Highest Lubricity***
- ***Toughest Metals***
- ***Best Finishes***

Recommended, where a soluble is desired, for all machining operations from broaching to grinding on all ferrous metals, high alloys, stainless steels, and exotic metals that are considered to be especially difficult to machine.

In sawing: Best where the greater portion of metals being cut is moderately to very tough or so plastic that they are prone to weld to the saw teeth. Chip welding and the resulting blade breakage and tooth stripping is reduced. Band life is extended, sawing rates are increased, down time is reduced, and profitability improved. Best band velocities 125-300 fpm.

Increased tool life and improved finishes are commonly experienced on these metals as a result of POWER-CUT 390 EHD's high level of lubricity, EP characteristic, and measured cooling effect.

Used at recommended mixtures, POWER-CUT 390 EHD forms a stable emulsion that will replace many cutting oils on most operations. Has better "keep" in the machine than many ordinary soluble oils over extended service.

POWER-CUT 390 EHD is a finely controlled mixture of petroleum oil, emulsifiers, and sulfurized and fatty additives. This combination allows maximum production rates with the toughest metals on tougher machining operations.

A valuable tool in the grinding of metals which are very prone to load the wheel, preventing maximum grinding rates and accurate sizing. The interval between wheel dressing is extended, production rates are increased, and sizing accuracy is improved.

Copper-bearing alloys, which are subject to active sulfur staining or corrosion, should be test cut before full production is undertaken.

DoALL®

DoALL Company
254 North Laurel Avenue
Des Plaines, IL 60016-4398
(312) 824-1122

POWER-CUT 390 EHD

The Super-Versatile Cutting Fluid

DILUTIONS:

NOTE: To assure complete mixing add POWER-CUT 390 EHD to at least three times the amount of water while stirring the mixture.

Operations	Parts Concentrate To Parts Water
Sawing, Cut-Off, tougher metals	1:5
Sawing, Cut-Off, higher band speeds	1:7-1:10
Sawing, Contour	1:7-1:10
Tapping, threading, broaching, gear hobbing	1:5-15
Turning, milling, drilling	1:10-20
General grinding	1:20-40
Form and centerless grinding	1:20

AVAILABLE IN:	CATALOG NO.
1-gallon cans (1 per box, 4 boxes per case)	452-041213
5-gallon pails	452-041411
55-gallon drums	452-042013

ALL DoALL CUTTING FLUIDS ARE AVAILABLE AT ALL LOCAL DoALL INDUSTRIAL SUPPLY CENTERS

For proper method to use to prepare your machines to use mixtures of POWER-CUT 390 EHD and other cutting fluids, ask your local DoALL Industrial Supply Center for a free copy of DoALL's Bulletin, Cutting Fluid System Care, Catalog No. 986-104406 11/79.

968-104513 10/81

Printed in U.S.A.

DoALL Company
254 North Laurel Avenue
Des Plaines, IL 60016-4398
(312) 824-1122

Synthetic cutting fluids for high tool

Milling woodcutter teeth in SAE 1018, 140 BHN, at 100 fpm, 0.001 in./tooth with POWER-CUT HD-600.

POWER-CUT HD-600

HEAVY-DUTY SYNTHETIC CONCENTRATE

- Increased production rates
- Best synthetic lubrication
- Cutting of tougher metals than other synthetic fluids
- Ferrous and nonferrous machining
- Clean parts for sound welds
- Clean machines and tools

*Heavy-duty—for use with a fully loaded tool generating the maximum chip for the material, operation and tool geometry.

POWER-CUT HD-600 is a heavy-duty* synthetic cutting fluid concentrate combining superior cooling ability and significant lubricity which make it possible to machine tougher materials than with other synthetic fluids. Extended tool life or increased production rates are routine with this superb fluid.

This concentrate is adaptable to many machining conditions. More concentrated mixtures are used on tougher operations and materials requiring greater lubrication and less concentrated mixtures on less severe operations.

POWER-CUT HD-600 concentrate may be used to machine nonferrous as well as ferrous metals. It is especially useful for high speed sawing of aluminums. Prolonged contact with alloys of copper may result in staining. This can be prevented by rinsing the part in water after it is machined.

And it is safe to use. No oily vapors are produced. Neither the concentrate nor the residue film is combustible. Piece parts become dry to the touch—are not slippery. Machine cleaning is easy and less frequently needed. It does not gum up slides, ways or controls. The natural cleaning action of this fluid assures an open, clean, trouble-free cutting fluid system.

Less solution is carried off on chips and piece parts. Less fluid concentrate is needed for the same working period. Controlled settling rate clears chips and grit from the work area, yet allows them to settle out correctly to prevent clogging the fluid system.

Users report that they are usually able to perform welding operations without removing the transparent, microscopic, residue film remaining on parts.

GENERAL MACHINING TIPS

When replacing straight oils with HD-600, increase tool speed 25 to 35%.

When replacing soluble oils with HD-600, increase tool speed 10 to 25%.

Workpiece materials: Ferrous and nonferrous except magnesium. HD-600 performs best on materials in 50% and over machinability area.

Best band tool velocities are above 250 fpm.

Available in 1, 5 and 55 gal containers.

STARTING DILUTIONS FOR HD-600

Broaching	1:4—1:15
Drilling	1:10—1:20
Deep hole drilling	1:5—1:15
Milling	1:15—1:20
Turning and Boring	1:15—1:20
Screw Machines**	1:5—1:20
Tapping Chasing Reaming Gear cutting	1:4—1:10
Sawing, cut-off	1:5-1:10
Sawing, vertical	1:5-1:15

***Designed to use water-soluble fluids.*

DoALL Company
254 North Laurel Avenue
Des Plaines, IL 60016-4398
(312) 824-1122

speeds, moderate-to-easy machining metals

Initially formulated specifically as a grinding concentrate to give superior results in precision surface grinding, Kleen-Kool concentrate has also proven successful on moderate to light-duty chip producing operations as well. A synthetic coolant, it combines exceptional cooling, wetting and rust inhibiting qualities to effectively cool the tool and workpiece for more accurate size control and extended tool life. Kleen-Kool does not attack copper or aluminum as do certain other synthetic fluids.

Kleen-Kool concentrate is a true solution type which is readily soluble in soft and hard waters. It produces a crystal-clear solution with a pleasant blue tint. The solution has a natural cleaning action which maintains a clean machine. Parts are clean and easy to handle.

It flows freely, through grinding wheels and into minute crevices between the tool and chip. Grinding wheels are kept open and free cutting. Tool, workpiece and chips remain cool.

Chips and metal fines settle out rapidly helping to maintain a clean solution and coolant system. Clean coolant results in finer finishes and lower coolant system maintenance costs, since pumps and other moving parts are exposed to less abrasion.

Kleen-Kool is an ideal coolant to use with the Cool Grinding™ cutting fluid attachment. With this coolant system, the concentrate is introduced into the grinding wheel at each side near the hub and flows freely through the wheel. As it emerges, it creates a fine mist at the point of contact with the work. Evaporative cooling effectively removes heat as it is generated to assure close maintenance of tolerances. Heavier cuts may be made without incurring excessive surface stressing.

Surface grinding SAE 1045 steel on DoALL high-speed grinder with Kleen-Kool synthetic coolant.

KLEEN-KOOL

LIGHT- TO MEDIUM-DUTY SYNTHETIC CONCENTRATE

Concentrate provides:

- **Super cooling**
- **Dimensional control**
- **Crystal clear solution**
- **Free-cutting grinding wheels**
- **Controlled chip settle out**

STARTING DILUTIONS FOR KLEEN-KOOL CONCENTRATE

Drilling	1:10—1:30
Deep hole drilling	1:10—1:30
Grinding	1:10—1:30
Milling	1:10—1:30
Turning and boring	1:10—1:30
Tapping	
Chasing	1:10—1:30
Reaming	
Sawing, cut-off	1:10—1:30
Sawing, vertical	1:10—1:30

GENERAL MACHINING TIPS

Use a harder wheel when changing from a soluble oil to Kleen-Kool concentrate. Use where cooling of work and tool is most important. Use richer concentrations for greater rust control in hard water areas.

Workpiece materials: Ferrous and nonferrous except magnesium. Kleen-Kool concentrate performs best on materials in 50% and over machinability area.

Best band tool velocities are above 250 fpm.

Available in 1, 5 and 55 gal containers.

DoALL Company
254 North Laurel Avenue
Des Plaines, IL 60016-4398
(312) 824-1122

Soluble oils for moderate-to-rapid tool speeds,

A soluble concentrate that lets you increase tool speed 10 to 25% over straight cutting oils.

POWER-CUT 390 EHD

EXTRA HEAVY-DUTY CONCENTRATE FOR TOUGHEST METALS

- **Highest lubricity**
- **Best tool life**
- **Tougher operations**
- **Better finishes on high-alloy and stainless steels, ferrous and exotic metals.**

POWER-CUT 390 EHD concentrate is recommended for all machining operations from broaching to grinding, whenever the application calls for a soluble oil. It has a high lubricity, EP characteristic and measured cooling effect. As a result, it can help produce better finishes and increase tool life in heavy duty machining of tough metals like high-alloy and stainless steels, ferrous and exotic metals.

Used at recommended mixtures, POWER-CUT 390 EHD concentrate forms a stable emulsion that will replace many cutting oils on most operations. Has better "keep" in the machine than many ordinary soluble oils over extended service.

POWER-CUT 390 EHD is a finely-controlled mixture of petroleum oil, emulsifiers and sulfurized and fatty additives. This combination allows maximum production rates on tough machining jobs:

In grinding metals which are prone to load the wheel, preventing maximum grinding rates and accurate sizing, POWER-CUT 390 EHD is a valuable tool. The interval between wheel dressing is extended and production rates improved.

In sawing, POWER-CUT 390 EHD concentrate smooths the way when the greater part of the metals being cut is moderately to very tough or so plastic that they are prone to weld to the saw teeth. Chip welding and the resulting blade breakage and tooth stripping is reduced. As a result, band life is extended, sawing rates increased, downtime reduced and overall profitability improved.

GENERAL MACHINING TIPS

When replacing straight cutting oils with 390 EHD, increase tool speed 10 to 25% and use mixture of 1:3 to 1:15.

Workpiece material: All ferrous metals. Not recommended for alloys containing major amounts of copper or magnesium.

Best band tool velocities are 150 to 300 fpm.

Available in 1, 5 and 55 gal containers.

STARTING DILUTIONS FOR 390 EHD

Milling	1:10—1:15
Turning and boring	1:10—1:15
Broaching	1:5—1:10
Grinding	1:10—1:40
Drilling	1:10—1:15
Deep hole drilling	1:5—1:10
Tapping Chasing Reaming Gear cutting	1:5—1:20
Screw machines*	1:5—1:15
Sawing, cut-off	1:5
Sawing, vertical	1:5—1:10

**Designed to use water-soluble fluids.*

difficult-to-moderate machining metals

POWER-CUT No. 360 is a controlled chemical mixture of emulsifiers, mineral oil and fortified additives. This water-soluble type fluid provides the added capacity to reduce the amount of heat generated as well as to effectively remove heat to allow high machining rates and faster chip removal.

This soluble oil delivers good performance on alloy steels as well as on mild steels. In a concentrated mixture with water, POWER-CUT No. 360 concentrate possesses the lubricity of many cutting oils. An operation which requires taking a heavy chip is the ideal spot to use this fine cutting fluid. The combination of lubricity, low-temperature lubrication and efficient heat removing capacity makes POWER-CUT No. 360 concentrate an extremely valuable tool for those operations which are characterized by deep tool penetration into the work. Tapping, drilling, form grinding and deep milling are all ideal operations for this quality product. Also specifically recommended for grinding with Borazon® CBN wheels.

POWER-CUT No. 360 concentrate is not recommended for use on copper-bearing alloys which are highly sensitive to sulfur staining or corrosion.

POWER-CUT No. 360 is easy to handle. It shows a willingness to disperse in and form stable emulsions with water with a hardness of 20 grains and less. The ease of mixing with water is almost equal to that found in chemical or synthetic cutting fluid concentrates. Therefore, the formation of invert emulsions or gels is not experienced as frequently as with other emulsifiable oils.

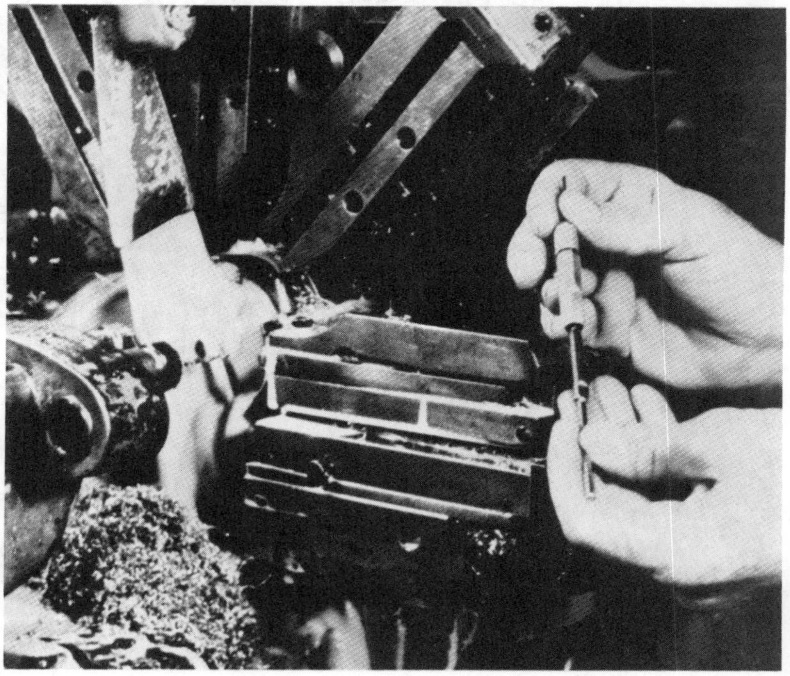

Deep hole drilling 303 stainless with No. 360 at 105 fpm. Hole is .109 in. dia by 1.270 in. deep.

POWER-CUT No. 360 concentrate

RUGGED SOLUBLE FOR TOUGHER METALS

- **Extended tool life**
- **High lubricity**
- **Improved finishes**
- **Reduced chipwelding**

STARTING DILUTIONS FOR NO. 360

Milling	1:15—1:40
Turning and boring	1:15—1:40
Grinding	1:10—1:40
Drilling	1:10—1:40
Deep hole drilling	1:10—1:20
Tapping Chasing Reaming Gear cutting	1:3—1:30
Screw machines*	1:5—1:15
Sawing, cut-off	1:3—1:15
Sawing, vertical	1:5—1:10

**Designed to use water-soluble fluids.*

GENERAL MACHINING TIPS

When replacing straight cutting oils with 360, increase tool speed 10 to 25% and use mixture of 1:3 to 1:15.

Workpiece material: All ferrous metals. Not recommended for alloys containing major amounts of copper or magnesium.

Best band tool velocities are 150 to 300 fpm.

Available in 1, 5 and 55 gal containers.

DoALL Company
254 North Laurel Avenue
Des Plaines, IL 60016-4398
(312) 824-1122

For moderate-to-rapid tool speeds, difficult-to-easy machining metals.

No. 470 Soluble Oil used to machine bronze bushings on Warner & Swasey chucking machine.

No. 470 SOLUBLE OIL

- Broad range of metals and operations
- Nonstaining type
- Long service life
- Better ground finishes on high-alloy steels
- Wide dilution range

No. 470 Soluble Oil is formulated with sufficient emulsifiers and rust inhibitors to make it adaptable across a broad range of operations and dilutions (1:5 to 1:60). Lubricity is achieved without resorting to active chemical additives which sometimes cause staining of copper, brass, bronze and other nonferrous metals.

These factors make it the ideal compromise cutting fluid for light- to medium-duty work on both ferrous and nonferrous materials. It may be used to machine all metals except magnesium.

No. 470 had excellent "keep" in the machine, remaining clean and sweet over extended service.

Many machining operations may be performed with this one fluid, thus reducing cutting fluid inventory. These qualities alone make No. 470 an exceptionally economical machining aid.

Recommended for grinding high-alloy steels where its balanced lubrication helps to prevent premature wheel breakdown and metallic wheel loading or where better surface finishes are desired.

The mixture ratio of No. 470 may be quickly measured with the Industrial Fluids Tester (see page 9 of this brochure). Ratio control at the machine tool can be monitored by production personnel with this simple-to-use instrument. To adjust the emulsion ratio to the proper range, see Figure 6 on page 6 of DoALL's Cutting Fluid Selector, Form No. 968-104554 7/79.

Although No. 470 does not contain nitrates, phenols, chromates, mercury, lead, arsenic, halogens, or other materials considered to be toxic, a mist collector should be employed when it is used on grinders or in mist-generating units due to its petroleum oil content.

GENERAL MACHINING TIPS

Use richer concentrations for finer finishes and greater rust control.

Workpiece materials: All metals except magnesium.

Best band tool speeds are 150 to 300 fpm.

Available in 1, 5 and 55 gal containers.

STARTING DILUTIONS FOR NO. 470

Milling	1:10–1:50
Turning and boring	1:10–1:50
Grinding	1:10–1:60
Drilling	1:10–1:40
Deep hole drilling	1:5–1:30
Chasing Reaming Tapping Gear cutting	1:5–1:20
Screw machines*	1:5–1:20
Sawing, cut-off	1:5
Sawing, vertical	1:5–1:10

*Designed to use water-soluble cutting fluids

DoALL Company
254 North Laurel Avenue
Des Plaines, IL 60016-4398
(312) 824-1122

CUTTING OILS For slower tool speeds and very difficult-to-machine materials.

No. 240 CUTTING OIL

For very slow speeds, tough stainless steels, high alloys

No. 240 Cutting Oil is formulated from a mineral oil base to which are added fatty oils for lubrication, sulfur to prevent chip welding and chlorine to add greater wear resistance. No. 240 is a medium viscosity, dark-colored stable product, free from suspended particles. It has sufficient body to cling to the tool, preventing fluid starvation deep in the cut. Intended for machining the toughest materials, No. 240 will allow longer tool life and finer finishes than straight sulfurized or sulfurized fat-type oils. It may be cut with No. 80 Cutting Oil or 40-second mineral seal oil where conditions are less severe. Use it on tough stainless steels, high alloy steels and titanium. Use at band speeds less than 175 fpm.

Available in 5 and 55 gal containers.

High pressure forged ell being drilled with No. 240 Cutting Oil.

No. 150 CUTTING OIL

For multipurpose operations, faster machining of mild alloys, grinding high-chrome, high-carbon alloys

A cutting oil with fatty and chlorinated additives, No. 150 is a multi-purpose oil for both ferrous and nonferrous machining. A unique combination of medium-low viscosity, outstanding wetting and high film strength permits faster machining rates on mild alloy steels of machinability groups I and II and aluminums.

High-chrome, high-carbon steels may be ground with reduced wheel loading using No. 150. Use it also to get better finishes when grinding metals with machinability in areas III and IV.

Especially effective when used with DoALL's Dart band tool at band speeds less than 175 fpm.

Available in 1, 5 and 55 gal containers.

Drilling half-hard, free-cutting brass with No. 150 Cutting Oil.

CUTTING FLUIDS

No. 120 CUTTING OIL

For nonferrous metals, automatic screw machines

This medium viscosity cutting oil contains a sulfurized fatty additive with a non-staining form of sulfur. Its lower viscosity allows greater machining speeds than higher viscosity oils. No. 120 Cutting Oil is especially adapted to use on automatic screw machines where lubricating oils and cutting oils frequently can intermingle. It assures complete wetting of the tool and controls chip build-up on the cutting edge resulting in improved tool life. It is recommended for the machining of aluminum, copper-bearing alloys, other nonferrous metals and mild steels.

Band speeds less than 175 fpm.

Available in 5 and 55 gal containers.

Drilling aluminum clamping device with No. 120 Cutting Oil.

No. 80 CUTTING OIL

For sawing at faster rates than No. 240 Cutting Oil in a wide range of steels

A sawing oil providing antiweld and heavy-duty lubricity in a broad range of band speeds below 200 fpm. Combines lower viscosity, penetrating ability, greater heat removal and high lubricity for heavy-duty machining at faster speeds than normal for cutting oils. Also useful for deep cavity machining where cutting fluid accessibility is limited. Use on all metals except those susceptible to sulfur staining.

Available in 5 and 55 gal containers.

No. 40 CUTTING OIL

For light duty machines

Developed to fill a need for a low-cost cutting fluid for low-cost, cut-off band machines in small shops doing one or two of a kind sawing or light production work. A lower viscosity cutting oil from active, sulfur-bearing mineral oil with added chlorinated hydrocarbons. Use on all ferrous metals which are not sensitive to sulfur staining. Band speeds less than 175 fpm.

Available in 5 and 55 gal containers.

BN-30 GRINDING OIL

Grinding fluid for Borazon* CBN wheels

A specially-blended, sulfo-chlorinated oil recommended for use with Borazon CBN wheels when grinding aerospace alloys, nickel base alloys and austenitic stainless steels. Use with adequate ventilation and safety precautions to meet modern workpiace standards for grinding operations.

Available in 5 and 55 gal containers.

*T.M. General Electric Company

DoALL Company
254 North Laurel Avenue
Des Plaines, IL 60016-4398
(312) 824-1122

DoALL Cutting Fluid Selector

TO SELECT THE CUTTING OR GRINDING FLUID BEST SUITED TO THE METAL AND MACHINING OPERATION FOLLOW STEPS INDICATED.

STEP 1. Locate the metal to be machined in the Machinability Rating Table. Its comparable machinability rating will be found to the right in the machinability percent column.

This machinability rating table is arranged in order of SAE-AISI identification numbers. Machinability rating percentage numbers are not to be taken as exact values from which there are no variations. However, for comparative purposes they serve as a convenient point to begin the analysis of a machining and cutting fluid application.

Ratings shown are comparisons of machining ease in terms of tool life and cutting speeds, using high-speed steel tools, with AISI B1112 cold drawn bars serving as the base of 100%. As an aid to using the selector we will use C-1045 as an example. It is outlined in the first column to the left in fig. 1 below. When C-1045 has a hardness number of 199 on the Brinell scale is is considered to have a 60% Machinability rating.

Fig. 1. MACHINABILITY RATING TABLE

Indicated percentages of alloying elements are maximums as commonly found in each steel category. The actual quantities may vary considerably.

SAE-AISI Numbers	BHN	Machinability %
Plain Carbon Steels		
B-1006	147	78
B-1010	147	78
C-1008	175	66
C-1010	172	65
C-1015	160	72
C-1016	148	78
C-1017	163	72
C-1019	146	78
C-1020	162	72
C-1022	147	78
C-1023	154	75
C-1025	162	72
C-1030	164	70
C-1035	162	70
C-1040	179	64
C-1043	178	64
1 [redacted]		
C-1046	203	57
C-1050	210	55
C-1054	217	53
C-1055	221	52
C-1059	222	52
C-1060	223	51
C-1064	224	50
C-1C65	229	50
C-1069	231	48
C-1070	230	49
C-1075	238	48
C-1080	271	42
C-1085	269	42
C-1090	273	42
C-1095	274	42
Resulphurized Carbon Steels Bessemer FCC		
C-1106	150	79
C-1108	149	80
C-1109	152	81
C-1110	148	83
B-1111	131	94
B-1112	122	100
B-1113	101	132
C-1113	120	100
C-1115	150	81
C-1116	139	94
C-1118	139	91
C-1119	120	100
C-1125	152	81
C-1126	150	81
C-1137	169	72
C-1138	164	75
C-1140	171	72
C-1146	167	76
C-1151	180	70
Manganese Steels Mn 1.75%		
1320	210	57
1321	212	59
NE 1330	210	60
1335	211	60
NE 1340	216	57
Nickel Steels Ni 3.50%		
2317	185	66
2330	220	55
2335	242	51
2340	210	57
2345	231	51

SAE-AISI Numbers	BHN	Machinability %
Nickel Steels Ni 5.00%		
2512	210	51
2515	212	52
NE 2517	215	51
Nickel-Chrome Steels Ni 1.25% Cr 0.65% or 0.80%		
3115	191	66
3120	190	66
3130	213	57
3135	225	53
3140	282	44
3145	192	64
3150	201	60
Nickel-Chrome Steels Ni 3.50% Cr 1.55%		
E 3310	241	51
E 3316	250	55
Molybdenum Steels Mo 0.25%		
4017	185	78
4023	182	78
4024	182	78
4027	212	66
4028	191	72
4032	184	76
4037	189	73
4042	198	70
4047	204	65
4053	261	53
4063	153	52
Chrome-Moly Steels Cr 0.95% Mo 0.20%		
4130	181	72
E 4132	190	72
4135	189	70
4137	209	65
E 4137	205	67
4140	212	62
4142	227	59
4145	221	60
4147	219	60
4150	242	59
Nickel-Chrome-Moly Steels Ni 1.80% Cr 0.50% Mo 0.25%		
4317	215	60
4320	201	63
E 4337	243	54
4340	240	57
E 4340	239	57
Nickel-Moly Steels Ni 1.80% Mo 0.25%		
4608	242	58
E 4617	201	66
4615	192	66
4620	198	64
X 4620	193	66
E 4620	202	64
4621	199	66
4640	198	66
E 4640	245	51

SAE-AISI Numbers	BHN	Machinability %
Nickel-Moly Steels Ni 3.50% Mo 0.25%		
4812	249	51
4815	256	51
4817	251	51
4820	248	53
Chrome Steels Cr 0.30% or 0.60%		
5045	188	70
5046	186	70
Chrome Steels Cr 0.80%, 0.95% or 1.05%		
5120	187	75
5130	241	57
5132	189	72
5135	188	72
5140	192	70
5145	210	65
5147	211	66
5150	215	64
5152	216	64
Carbon-Chrome Steels C 1.00% Cr 0.50%, 1.00% or 1.45%		
E 50100	211	45
E 51100	221	40
E 52100	220	40
Chrome-Vanadium Steels Cr 0.85% or 0.95% V 0.10% or 0.15%		
6102	202	57
6145	182	66
6150	192	60
6152	195	60
Nickel-Chrome-Moly Steels Ni 0.55% Cr 0.50% Mo 0.20%		
8617	182	66
8620	183	66
8622	185	65
8625	189	62
8627	188	64
8630	161	72
8635	165	70
8637	164	70
8640	172	66
8642	177	65
8645	182	64
8647	194	60
8650	195	60
8653	203	56
8655	205	57
8660	215	54
Nickel-Chrome-Moly Steels Ni 0.55% Cr 0.50% Mo 0.259%		
8719	175	67
8720	178	66
8735	171	70
8740	183	66
8742	185	64
8747	192	60
8750	194	60
Manganese-Silicon Steels Mn 0.55% Si 2.00%		
9255	122	54
9260	238	51
9262	235	49

SAE-AISI Numbers	BHN	Machinability %
Nickel-Chrome-Moly Steels Ni 3.25% Cr 1.20% Mo 0.12%		
E 9310	243	48
E 9315	238	50
E 9317	239	49
Manganese-Nickel-Chrome-Moly Steels Mn 1.00% Ni 0.45% Cr 0.40% Mo 0.12%		
9437	182	66
9440	183	66
9442	179	66
9445	181	64
Nickel-Chrome-Moly Steels Ni 0.55% Cr 0.17% Mo 0.20%		
9747	187	64
9763	215	54
Nickel-Chrome-Moly Steels Ni 1.00% Cr 0.80% Mo 0.25%		
9840	232	50
9845	238	49
9850	242	45
Stainless Steels		
302		45
303*		60
304		45
308+		27
309+		28
314+		32
317+		29
321		36
330+		27
347		36
403		39
410		54
416*		72
420		57
420 F*		79
430		54
430 F**		91
440		37
440 A		45
440 B		42
440 C		40
440 F*		59

\+ Poorest Machining Properties.
* Fairly Good Machining—Contain Sulfur and Selenium.
** Best Machining Properties

SAE-AISI Numbers	BHN	Machinability %
Cast Iron		
Soft	130	81
Medium	168	64
Hard	243	47
Malleable Iron		
Malleable Iron	115	106
Malleable Iron	135	80
Cast Steel		
Cast Steel	121	85
Cast Steel	219	50
Cast Steel	245	44

SAE-AISI Numbers	BHN ROC	Machinability %
Nonferrous Metals Aluminum		
2 S	23-24	300-1500
11 S-T3	95	500-2000
17 S-T	100	300-1500
Brass		
Yellow	58	80
Red	55	60
Leaded	B-70	280
Red, Leaded	55	180
Bronze		
Aluminum Bronze	B-102	60
Manganese Bronze	B- 90	60
Phosphor Bronze	B-140	40
Phosphor Bronze Leaded	60	140
Silicon Bronze		60
Copper & Alloys		
Copper, Cast	30	70
Copper, Rolled	80	60
Everdur	200	60
Everdur, Leaded	200	120
Gun Metal, Cast	65	60
Magnesium		
Dow H	50	500-2000
Dow J	58	500-2000
Dow R		1150
Nickel & Alloys		
Nickel	135	26
Monel, Cast		45
Monel, Rolled		55
Monel, "K"		35
Inconel (B)		35
Zinc		200

Tool steels are usually compared with type W. These ratings have been converted to compare with B1112 using the W group as a base of 30.

Group	Machinability %
W	30
L	27
O	27
A	26
S	26
F	23
P	23 to 30
H (Cr)	23
H (W or Mo)	15 to 18
M	13 to 18
T	12 to 17
D	12 to 15

STEP 2. Refer to the Cutting Fluid Rapid Selection Chart, fig. 2. Locate the machinability rating number for the material in question on the white SAE-AISI STEELS scale, or any of the other scales of the chart for which you have a value. Following the example, 60% machinability is located at "A" and circled. The color appearing in that area will indicate the cutting fluids which should be considered first. Point "A" falls in an area where two colors overlap, making it possible to consider any one of five cutting fluids if both heavy-duty and light-duty machining are included.

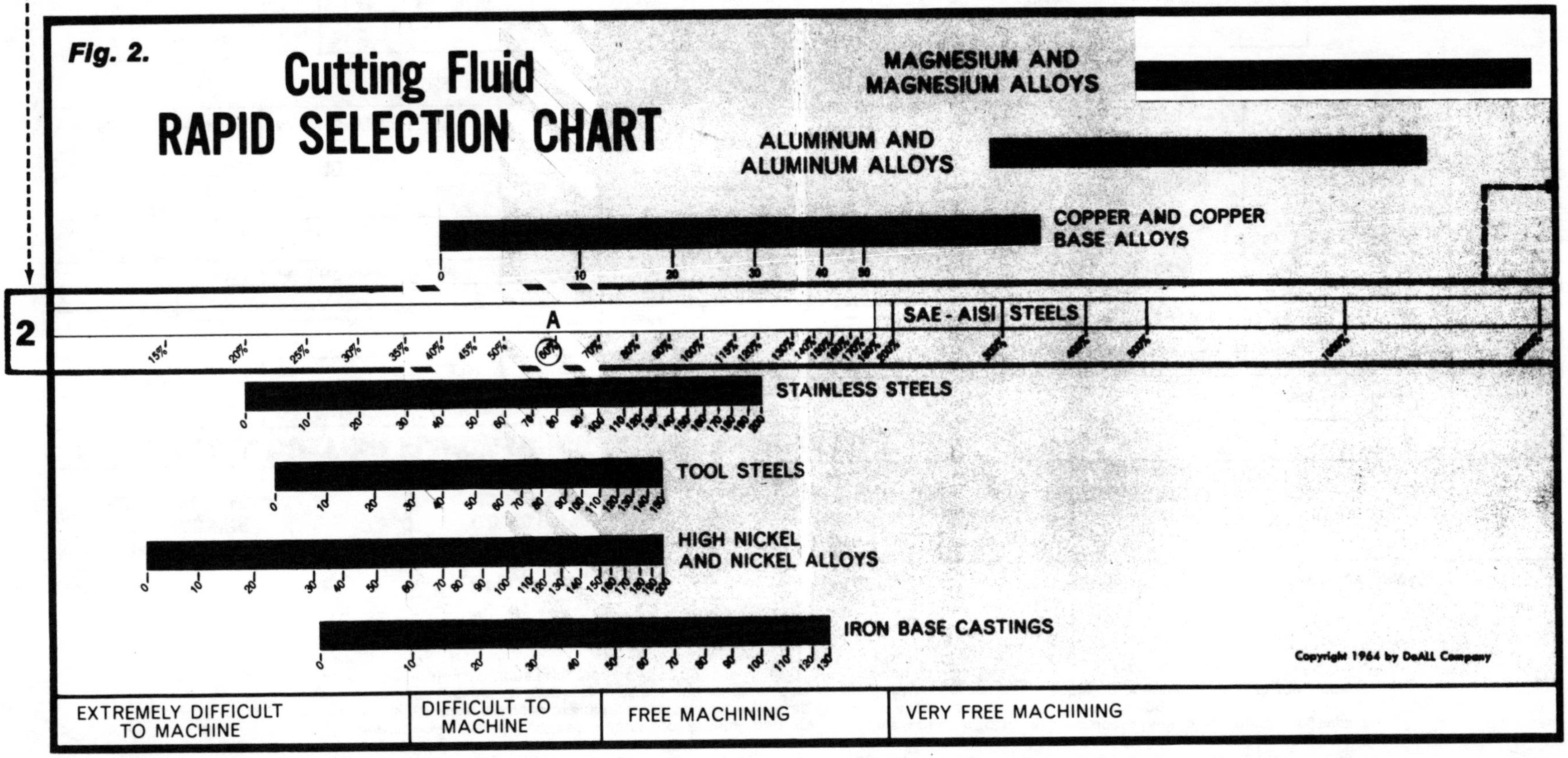

Fig. 2.

DoALL Company
254 North Laurel Avenue
Des Plaines, IL 60016-4398
(312) 824-1122

CUTTING FLUIDS

STEP 3. Refer to the color coded list of cutting fluids. Any of the fluids outlined in red theoretically can be used to machine metals with a 60% machinability rating. In most instances the heavy-duty fluids, with the notation (HD), are chosen. Be certain to read the notations below the cutting fluid list.

STEP 4. Machinability ratings are more directly related to turning than to other operations. For other operations, follow these general rules: If the operation is more difficult (see fig. 8, Machining Operation Difficulty; or finer finishes are desired or drill, tap, or reamer is cutting undersized; or rake angles become more negative, use *more* concentrated dilutions *first*, then move to left of "A".

If the operation is *less* difficult; or greater tool speeds than are normal for H.S.S. tooling are used (example: carbide tools); or drill, tap, or reamer is cutting oversize; or rake angles become more positive, use *less* concentrated dilutions first, then move to right of "A".

STEP 5. For band tool recommendations, refer to General Band Machine Recommendations Chart, fig. 3. Figures shown here were compiled from field experience with cut-off saws. However, they are an accurate guide for all sawing applications.

Note: When machining magnesium or its alloys, cutting fluids containing water usually are not recommended because of the fire hazard associated with fine chips of these metals. Cutting oils are indicated as shown on the Cutting Fluid Rapid Selection Chart.

STEP 6. The information shown in the DoALL Cutting Fluid Quick Comparison Chart, fig. 4, will help you decide which cutting fluids are more suitable for a particular application. For example, fluid having active sulfur are not usually considered suitable for machining copper or its alloys because of resultant staining.

STEP 7. For various machining operations refer to the General Machining Selection Chart.

Fig. 3.

GENERAL BAND MACHINE RECOMMENDATIONS

Band Speed	Class of Fluid	Carbon Alloy, H.S.S, and Tungsten Carbide
LESS THAN 175 SFPM	CUTTING OIL	#240, #80
150-300 SFPM	SOLUBLE-OIL	POWER-CUT 390 EHD, POWER-CUT No. 360
MORE THAN 250 SFPM	SYNTHETIC FLUID	POWER-CUT HD-600

***Fig. 4.* DoALL CUTTING FLUIDS QUICK COMPARISON CHART**

Cutting Fluid	Duty	Metals to Use on	Cutting Speed	Class	Lubricity	Heat Transfer	Active Sulfur	Chlorine	Trace Natural Inclusions	Clarity	Color	Dilution	Miscibility
# 80	**Lt/Med**	**FE**	**Lo/Int**	**Cut. Oil**	**xxx**	**x**	**Y**	**Y**	**—**	**Opaque**	**Brn**	**None**	**—**
#150	Lt/Med	FE & NONFE	Lo/Int	Cut. Oil	xxx	x	—	Y	—	Opaque	Lt. Brn	None	—
#120	**Lt/Med**	**FE & NONFE**	**Lo**	**Cut. Oil**	**xxx**	**x**	**—**	**—**	**S (1)**	**Opaque**	**Dk. Brn**	**None**	**—**
#240	Hvy	FE	Lo	Cut. Oil	xxxx	x	Y	Y	—	Opaque	Brn	Vary (3)	—
#470	**Lt/Med**	**FE & NONFE(2)**	**Int**	**Sol. Oil**	**xx**	**xx**	**—**	**—**	**S (1)**	**Opaque**	**Wht**	**1:5/1:60**	**Gd.**
POWER-CUT No. 360	Med/Hvy	FE (2)	Int.	Sol. Oil	xx	xx	Y	Y	—	Opaque	Grn	1:5/1:40	Gd.
POWER-CUT 390 EHD	**Hvy**	**FE (2)**	**Int/Lo**	**Sol. Oil**	**xx**	**xx**	**Y**			**Opaque**	**Tan**	**1:5/1:20**	**Gd.**
Kleen-Kool	Lt/Med	FE & NONFE (2)	Hi	Syn.	x	xxx	—	—	—	Transparent	Blue	To 1:30	Inst.
POWER-CUT HD-600	**Hvy**	**FE & NONFE (2)**	**Hi**	**Syn.**	**x**	**xxx**	**—**	**—**	**—**	**Transparent**	**Clr**	**1:5/1:20**	**Inst.**

3

POWER-CUT HD-600	**(HD)***
Kleen-Kool	**(LD)**
POWER-CUT 390 EHD	**(HD)**
POWER-CUT No. 360	**(HD)**
No. 470 Soluble Oil	**(LD)****

No. 240 Cutting Oil	**(HD)*****
No. 150 Cutting Oil	**(LD)**
No. 80 Cutting Oil	**(HD/LD)**
No. 150 Cutting Oil	**(HD)**
No. 120 Cutting Oil	**(LD)**

*Heavy duty meaning a fully-loaded tool generating the maximum chip normal for the operation and tooling.

**Light duty meaning a lightly-loaded tool generating a less than maximum chip normal for the operation and tooling.

***Work hardening is more important than per cent machinability when selecting a cutting fluid for materials in this area. Therefore a soluble-oil type, with its heat removal capacity, may be better than a cutting oil having higher lubricity. Go to dry machining only as a last resort.

(1) Inactive

(2) Do not use on magnesium or its alloys.

(3) Vary to meet conditions with either mineral seal oil or 100/100 paraffin oil.

DoALL Company
254 North Laurel Avenue
Des Plaines, IL 60016-4398
(312) 824-1122

Fig. 5 GENERAL SELECTION CHART Machinability Ratings of Materials **

Machine Operation	I Greater than 70%	II 50%-60%	III 40%-50%	IV Less than 40%	V Greater than 100%		VI Less than 100%	
	Malleable Iron Cast Steels Free Cutting Steels Carbon Steels	Alloy Steels Nickel Steel Nickel-Chromium Steel	Ingot-Iron Wrought-Iron Stainless Steels (Free Cutting)	Stainless (aust.) High-Speed Steel Tool Steels Manganese Steels Hi-Temp Alloys	Magnesium Alloys	Aluminum Alloys Leaded Brass Phosphor Bronze Zinc	Gun Metal Brass	Monel Inconel Nickel
Power Sawing	HD-600(X) #360 # 80	HD-600 #360 390 EHD # 80	#360 390 EHD # 80 #240(2)	#360(1) 390 EHD # 80 #240 Dry	# 80 #150 #120	HD-600 #150 #120	HD-600 #150 #120	#360(1) 390 EHD # 80(4) #240(2) Dry
Contour Sawing	HD-600 #360 # 80 Tool-Saver	HD-600 #360 390 EHD # 80 Tool-Saver	#360 390 EHD # 80 #150 #240 Tool-Saver	#360 390 EHD # 80 #240 Dry Tool-Saver	# 80 #150 #120 Tool-Saver Dry	HD-600 #470 #150 #120 Tool-Saver	HD-600 #470 #150 Tool-Saver	#360 # 80 #240 Dry Tool-Saver
Grinding (Surface Form, Cylindrical Centerless)	Kleen-Kool #470(6) Tool-Saver	Kleen-Kool #470(6) 390 EHD Tool-Saver	Kleen-Kool #470(6) 390 EHD #150(6)	#470 390 EHD #150	#150 #120 Tool-Saver Dry	Kleen-Kool #470 #150	Kleen-Kool #470 #150	#470 #360 390 EHD #150
Broaching Tapping	Reducer HD-600 #360 #150 #240	HD-600 #360 390 EHD #150 #240	#360 390 EHD #150 #240 Tapping Cmpd.	390 EHD #150 #240 Tapping Cmpd.	# 80 #150 #120 Dry	HD-600 #150 Tapping Cream	HD-600 #150 #120 Tapping Cream	#360 390 EHD # 80 #240 Tapping Cmpd
Threading Gear Shaving Gear Cutting	HD-600 #360 #150	#360 390 EHD #150 #240	390 EHD # 80 #150 #240	390 EHD # 80 #150 #240	# 80 #150 #120 Dry	#150 #120	HD-600 #150 #120	390 EHD # 80 #150 #240
Deep Drilling Milling Boring	HD-600 #470 #360	HD-600 #470 #360 390 EHD	HD-600 #360 390 EHD #150	#360 390 EHD #150 #240 Dry	# 80 #150 #120 Dry	HD-600 #150 #120	HD-600 #470 #150	#470 #360 390 EHD # 80 #150 #240
Multiple Spindle Turret Lathes	HD-600 #470 #360 #150	HD-600 #470 #360 #150	#470 #360 #150 #240	#360 390 EHD #150 #240	#150 #120	HD-600 #470 #150 #120	HD-600 #470 #150 #120	#360 390 EHD #150 #240
Automatic Screw Machine High Speed Light Feed	#470(5) #360 #150 #120	#470(5) #360 #150 #120	#470(5) #360 #150 #120	#360(5) 390 EHD #150 #120	#150 #120 Dry	HD-600(5) #470 #150 #120	HD-600(5) #470 #150 #120	#360(5) 390 EHD(5) #150 #240
Drilling Planing Shaping Turning	HD-600 #470 #360 #150 #240	HD-600 #470 #360 390 EHD #240	#470 #360 390 EHD #150 #240	#360 390 EHD #150 #240 Dry	# 80 #150 #120 Dry	HD-600 #470 #150	HD-600 #470 #150 #120	#360 390 EHD # 80 #240 Dry

X—Listings in each column are in order of tool surface speed or ease of machining. The fluids at the top are for the greatest speeds and most easily machined material.

1—Lower limit approximately 35%

2—Upper limit approximately 35%

3—Upper limit approximately 70%

4—Generates smoke above 10 sq. in. per min.

5—Do not use water soluble fluids in machine, which employ the cutting fluid as a machine lubricant.

6—Use #470 and #150 to improve finish and to prevent heat checking, burning and wheel loading.

**Compiled from METALS HANDBOOK.

DoALL Company
254 North Laurel Avenue
Des Plaines, IL 60016-4398
(312) 824-1122

Fig. 6. SOLUBLE OIL EMULSION ADJUSTMENT CHART

DESIRED RATIO \ EMULSION AS TESTED	5 to 1	8 to 1	10 to 1	15 to 1	20 to 1	25 to 1	30 to 1	35 to 1	40 to 1	50 to 1
5 to 1		.67	.91	1.25	1.43	1.54	1.61	1.67	1.71	1.76
8 to 1	5.0		.23	.55	.71	.82	.89	.94	.98	1.03
10 to 1	8.33	2.23		.31	.48	.58	.65	.70	.73	.78
15 to 1	16.67	7.78	4.55		.16	.26	.32	.37	.41	.46
20 to 1	25.00	13.34	9.10	3.13		.096	.16	.21	.24	.29
25 to 1	33.33	18.89	13.64	6.25	2.39		.06	.11	.15	.20
30 to 1	41.67	24.45	18.19	9.38	4.77	1.93		.05	.08	.13
35 to 1	50.00	30.00	22.73	12.50	7.15	3.85	1.62		.03	.08
40 to 1	58.33	35.56	27.28	15.63	9.53	5.77	3.22	1.39		.05
50 to 1	75.00	46.67	36.37	21.88	14.29	9.62	6.46	4.17	2.44	

ADD CONCENTRATE

ADD WATER

Note: Figures above the diagonal line show gallons of soluble-oil concentrate to be added for every 10 gal. of tested emulsion. Gallons of water to be added are found below the diagonal line.

Fig. 7. FLUID CAPABILITY

Dilution Ratio	Percentage of Concentrate in Finished Emulsion or Solution	Per Cent Additional Concentrate Compared to Next Lower Dilution
1:2	33.33%	33.2% more than 1:3
1:3	25.00%	49.7% more than 1:5
1:5	16.70%	83.5% more than 1:10
1:10	9.10%	91.2% more than 1:20
1:20	4.76%	

Fig. 8. MACHINING OPERATION DIFFICULTY

.30 Internal Broaching
.35 External Broaching
.35 Pipe Threading
.40 Tapping
.65 Threading
.65 Generating Gear Teeth
.70 Deep Drilling
.75 Milling, Multiple
.75 Boring
.75 Form Tools
.80 Hi-Speed, Lit-Feed
.80 Drilling
1.00 Shaping
1.00 Planning
1.00 Turning
1.10 Grinding

Note: The values shown are presented for comparison purposes only. An operation's relative difficulty is greatly influenced by cutting tool and machine tool selection, use, and maintenance.

ETNA PRODUCTS, INC.

PHONE:
(216) 543-9845
TELEX:
980131 WDMR

16824 PARK CIRCLE DRIVE
CHAGRIN FALLS, OHIO 44022 U. S. A.

MAILING ADDRESS:
P.O. BOX 630
CHAGRIN FALLS, O.
44022-0630 U.S.A.

MASTER DRAW #WS-1177-C

HEAVY DUTY SYNTHETIC METALWORKING FLUID

Description

WS-1177-C is a new heavy duty synthetic emulsifiable fluid that has been formulated to provide excellent performance on difficult metalworking and metal machining operations. It was developed by blending a unique EP agent and a group of synthetic additives to prevent corrosion, staining, foaming and bacterial or fungal attack. This special combination of items provides excellent tool life and trouble free performance on both ferrous and non-ferrous metals.

Application

WS-1177-C is specifically formulated to handle tough machining applications such as bar turning, centerless grinding machining and other general machining and grinding operations on all types of metals.

Outstanding Qualities

Excellent tool life. Extends tool life and increases productivity.

- -Economical to use, resists depletion.
- -Exceptional rancidity control.
- -Protects against corrosion and staining.
- -No nitrates, phenols, chlorine or sulphur.
- -Long lasting in machine.
- -Maintains emulsion stability in hard or soft water.
- -The high cooling rate of water allows increased production speeds.
- -Excellent foam control.
- -Dilution monitored by titration or refractometer.
- -Leaves a light film on part, yet rinses off easily.
- -Operator compatible and easy to mix.

Instructions for Use
Applications/Dilutions

Before adding the WS-1177-C, the reservoir and machines should be thoroughly cleaned and rinsed to minimize bacterial contamination. The system can then be charged with water and the proper amount of WS-1177-C added to the water while agitating to insure dispersion of the product. Laboratory service provided for start-up and maintenance of the system.

Etna Products, Incorporated
16824 Park Circle Drive/P.O. Box 630
Chagrin Falls, OH 44022-0630
(216) 543-9845

MASTER DRAW #WS-1177-C
Recommended Dilutions

	Cast Iron & Malleable Iron	Cast Steel & Carbon Steel	Alloy Steel & Stainless Steel	Nonferrous Copper Brass & Aluminum
All operations in multi-spindle screw machines & automatic chucking machines	19:1 (5%)	19:1 (5%)	19:1 (5%)	29:1 (3.5%)
Grinding, O.D., I.D. and Centerless	39:1 (2.5%)	29:1 (3.5%)	29:1 (3.5%)	49:1 (2.0%)
Milling, drilling and turning	24:1 (4%)	19:1 (5%)	19:1 (5%)	39:1 (2.5%)
Reaming, Boring & Sawing	19:1 (5%)	14:1 (6.5%)	14:1 (6.5%)	29:1 (3.5%)

TYPICAL SPECIFICATIONS

Base Value	95-115mg KOH/g
Pounds per Gallon	8.8
Flash Point	None
PH (Neat)	9.5 typical
PH @ 5% in Tap Water	8.9 typical
Specific Gravity	1.06 typical

For further technical information or to place an order, please call at (216) 543-9845.

PHONE:
(216) 543-9845
TELEX:
980131 WDMR

ETNA PRODUCTS, INC.

16824 PARK CIRCLE DRIVE
CHAGRIN FALLS, OHIO 44022 U.S.A.

MAILING ADDRESS:
P.O. BOX 630
CHAGRIN FALLS, O.
44022-0630 U.S.A.

MASTER DRAW #WS1740

EMULSIFIABLE METALWORKING FLUID

Description

#WS1740 is a new heavy duty emulsifiable fluid that has been formulated to provide excellent performance on difficult metalworking operations that previously demanded the use of neat cutting oil. It was developed by blending a mineral oil with a synthetic chlorinated frictional modifier and a group of additives to prevent corrosion, staining, foaming and bacterial or fungal attack. This special combination of items provides excellent tool life and trouble free performance on both ferrous and non-ferrous metals.

Application

#WS1740 is specifically formulated to handle tough machining applications such as gear hobbing, gear shaping, spline cutting, broaching and other general machining and grinding operations on all types of metals. It can also be used neat as a headstock lubricant.

Outstanding Qualities

Excellent tool life. Extends tool life and increases productivity.

- Economical to use resists depletion.
- Exceptional rancidity control.
- Protects against corrosion and staining.
- Resistant to rancidity.
- No nitrates, phenols or added sulfur.
- Long lasting in machine.
- Maintains emulsion stability in hard or soft water.
- The high cooling rate of water allows increased production speeds.
- Excellent foam control.
- Dilution monitored by titration or refractometer.
- Leaves a light film on part yet rinses off easily.
- Operator compatible and easy to mix.

Instructions for Use
Applications/Dilutions

Before adding the #WS1740, the reservoir and machines should be thoroughly cleaned and rinsed to minimize bacterial contamination. The system can then be charged with water and the proper amount of #WS1740 added to the water while agitating to insure dispersion of the neat oil. Laboratory service provided for start-up and maintenance of the system.

Etna Products, Incorporated
16824 Park Circle Drive/P.O. Box 630
Chagrin Falls, OH 44022-0630
(216) 543-9845

MASTER DRAW #WS1740
Recommended Dilutions

	Cast Iron & Malleable Iron	Cast Steel & Carbon Steel	Alloy Steel & Stainless Steel	Nonferrous Copper Brass & Aluminum
Gear hobbing, gear shaping broaching, spline milling heavy tapping, thread rolling and gun drilling	14:1 (6.5)	9:1 (10%)	9:1 (10%)	19:1 (5%)
All operations in multi-spindle screw machines & automatic chucking machines	24:1 (4%)	19:1 (5%)	19:1 (5%)	29:1 (3.5%)
Grinding, O.D., I.D. and Centerless	39:1 (2.5%)	29:1 (3.5%)	29:1 (3.5%)	49:1 (2.0%)
Milling, drilling and turning	29:1 (3.5%)	24:1 (4%)	19:1 (5%)	39:1 (2.5%)
Deep Hole Drilling	19:1 (5%)	9:1 (10%)	9:1 (10%)	29:1 (3.5%)
Reaming, Boring, Form Milling & Sawing	19:1 (5%)	14:1 (6.5%)	14:1 (6.5%)	29:1 (3.5%)

TYPICAL SPECIFICATIONS

Viscosity @ 100°F	500-550 SSU
Viscosity @ 210°F	90-120 SSU
Acid Number	14-16mg KOH/g typical
Base Value	30-35mg KOH/g
Pounds per Gallon	8.49
Flash Point	320°F (COC)
Specific Gravity	1.02 typical
5% Tap Water PH	9.3 typical

Etna Products, Incorporated
16824 Park Circle Drive/P.O. Box 630
Chagrin Falls, OH 44022-0630
(216) 543-9845

PHONE:
(216) 543-9845
TELEX:
980131 WDMR

ETNA PRODUCTS, INC.

16824 PARK CIRCLE DRIVE
CHAGRIN FALLS, OHIO 44022 U. S. A.

MAILING ADDRESS:
P.O. BOX 630
CHAGRIN FALLS, O.
44022-0630 U.S.A.

DAPHNE CUT HS-4M

Technical Data Sheet
Special High Performance Cutting Fluid
(For High Speed Gear Hobbing)

INTRODUCTION: With the evolution of new high speed gear hobbers equipped with carbide tooling, the conventional gear hobbing oils compounded with sulfur or chlorine fatty compounds began to fail to provide adequate tool life. To answer the stringent demands of the new equipment, DAPHNE CUT HS-4M was developed by blending a special organic extreme pressure agent and various frictional modifiers.

DAPHNE CUT HS-4M represents a new concept in the technology of fluids for gear hobbing. The product is a non-active and non-water based gear cutting fluid that provides excellent tool life whether one is using solid carbide, carbide tipped, or conventional high-speed steel hobs. DAPHNE CUT HS-4M was developed in conjunction with a leading manufacturer of gear hobbing machines. The manufacturer developed a new drive system that has eliminated the chipping of carbide when cutting carbon steel gears at high speeds and feeds. The lubricity provided by DAPHNE CUT HS-4M has aided in eliminating the chipping and greatly extended tool life.

DAPHNE CUT HS-4M is compatible with the elastomers normally employed in machine tool lubrication systems. The product is inhibited to protect against staining and corrosion of the non-ferrous and ferrous parts in the lubrication reservoir and on the actual metal machining application operation. This permits its use on ferrous as well as non-ferrous machining operations. Other advantages include:

1. Excellent resistance to oxidation and chemical breakdown.
2. Will not separate with normal use and storage.
3. Does not contain sulfur or chlorine-fatty type EP additives, which can cause problems when they leak into machinery lubrication systems.
4. Transparent green color allows the operator to view the work piece.
5. While the DAPHNE CUT HS-4M was developed for gear hobbing, the fluid can also be used on a wide variety of metal cutting operations such as turning, drilling, milling, grinding, etc.

Typical Physical Properties:

Appearance	Low viscosity, transparent green oil
Viscosity Kin.	
at 100°F	120-140 SSU
at 210°F	35-45 SSU
Specific Gravity	.879
Pounds per gallon	7.32
Flash Point (COC)	356°F
Total Acid No. ASTM D-974	3.96
Copper Corrosion (210°F x 1 hr.)	1 (1A)

DAPHNE CUT HS-4M
Manufactured in the U.S.A. by ETNA PRODUCTS, INC. of Chagrin Falls, Ohio under license from Idemitsu Apollo Corporation of New York, New York
MASTER DRAW LUBRICANTS®

Etna Products, Incorporated
16824 Park Circle Drive/P.O. Box 630
Chagrin Falls, OH 44022-0630
(216) 543-9845

PHONE:
(216) 543-9845
TELEX:
980131 WDMR

ETNA PRODUCTS, INC.

16824 PARK CIRCLE DRIVE
CHAGRIN FALLS, OHIO 44022 U.S.A.

MAILING ADDRESS:
P.O. BOX 630
CHAGRIN FALLS, O.
44022-0630 U.S.A.

FIELD TESTS: One of the more important properties of a gear hobbing oil is its ability to provide good tool life. Additional data was generated on two field tests which compared a conventional chlorine fatty gear hobbing oil to DAPHNE CUT HS-4M. This data available upon request.

STORAGE: DAPHNE CUT HS-4M is not affected by freezing and thawing but indoor storage is recommended. If the product has been stored at below freezing temperatures for prolonged periods, it is important to allow the oil to return to ambient temperature prior to use.

SHELF LIFE: Under normal conditions the shelf life for DAPHNE CUT HS-4M is twelve months.

PACKAGING AND PRICING: DAPHNE CUT HS-4M is packaged in 55 U.S. gallon steel non-returnable drums. Bulk shipments by over the road transport, 5,500 gallon minimum, are also available. Product shipped F.O.B. Chagrin Falls, Ohio.

Prices are available upon request with quantity discounts offered.

Terms: Net 30 Days

For further technical information or to place an order, please call at:

(216) 543-9845

DAPHNE CUT HS-4M
Manufactured in the U.S.A. by ETNA PRODUCTS, INC. of Chagrin Galls, Ohio
under license from Idemitsu Apollo Corporation of New York, New York

PHONE:
(216) 543-9845
TELEX:
980131 WDMR

ETNA PRODUCTS, INC.

16824 PARK CIRCLE DRIVE
CHAGRIN FALLS, OHIO 44022 U. S. A.

MAILING ADDRESS:
P.O. BOX 630
CHAGRIN FALLS, O.
44022-0630 U.S.A.

DAPHNE CUT(TM) HS-8M

Technical Data Sheet

Description: DAPHNE CUT(TM) HS-8M represents a new concept in the technology of cutting and grinding oils. To answer the stringent demands for improved tool life, grinding wheel life, and increased productivity, DAPHNE CUT(TM) HS-8M was developed by blending a special organic extreme-pressure agent with various frictional modifiers. This unique combination of components provides improved tool life and better surface finishes.

DAPHNE CUT(TM) HS-8M is compatible with the elastomers normally employed in machine tool lubrication systems. The product is inhibited to protect against staining and corrosion of the non-ferrous and ferrous components of the machine tool and on the actual metalworking operation. This permits its use on ferrous as well as non-ferrous metals. Other advantages include:

1) Excellent resistance to oxidation and chemical breakdown.

2) Provides exceptional wheel life and tool life.

3) Does not contain chlorine EP additives, which can cause staining.

4) Transparent green color allows the operator to view the work piece.

5) Light viscosity allows fines to drop out rapidly and allows for easy filtration.

DAPHNE CUT(TM) HS-8M has been formulated for high-speed machining operations such as turning, threading, gear hobbing, gear shaping, spline cutting, and flute grinding. It is extensively used on bar machines and Swiss-type automatics equipped with carbide tooling.

Typical Physical Properties:

Appearance	Low-viscosity, transparent green oil
Viscosity, Kinematic	
100° F SSU	110-130
210° F SSU	40-50
Specific Gravity	.877 typical
Pounds per Gallon	7.31 typical
Flash Point (COC)	355° F
Sulphur	Positive
Chlorine	Negative
Copper Corrosion	
(210° F x 1 hour)	1 (1A)

DAPHNE CUT(TM) HS-8M
Manufactured in the USA by ETNA PRODUCTS, INC. of Chagrin Falls, Ohio, under license from Idemitsu Apollo Corporation, of New York, New York

Etna Products, Incorporated
16824 Park Circle Drive/P.O. Box 630
Chagrin Falls, OH 44022-0630
(216) 543-9845

PHONE:
(216) 543-9845
TELEX:
980131 WDMR

ETNA PRODUCTS, INC.

16824 PARK CIRCLE DRIVE
CHAGRIN FALLS, OHIO 44022 U. S. A.

MAILING ADDRESS:
P.O. BOX 630
CHAGRIN FALLS, O.
44022-0630 U.S.A.

-2-

Handling/Storage: DAPHNE CUT(TM) HS-8M is not affected by freezing and thawing, but indoor storage is recommended. If the product has been stored outdoors at below freezing temperatures, allow to warm to room temperature prior to use.

Disposal: DAPHNE CUT(TM) HS-8M can be reclaimed for reuse by ETNA PRODUCTS, INC., or it can be incinerated to meet EPA requirements.

Packaging and Pricing: DAPHNE CUT(TM) HS-8M is packaged in 55-gallon U.S. steel non-returnable drums. Bulk shipments by over-the-road transports, 5,500 gallon minimum is also available. Product is shipped F.O.B. Chagrin Falls, Ohio. Prices are available upon request, with quantity discounts offered.

Terms: Net 30 Days.

FOR FURTHER TECHNICAL INFORMATION OR TO PLACE AN ORDER, PLEASE CALL AT:

216/543-9845

DAPHNE CUT(TM) HS-8M
Manufactured in the USA by ETNA PRODUCTS, INC. of Chagrin Falls, Ohio, under license from Idemitsu Apollo Corporation, of New York, New York

FarBest Corporation
24551 Raymond Way, Suite 110
El Toro, CA 92630
(714) 855-3881

FARBEST TECHNICAL BULLETIN

PRODUCT:

KING K-53
E.T. TAPPING
COMPOUND

KING K-53 E.T. TAPPING COMPOUND

DESCRIPTION:

KING K-53 E.T. TAPPING COMPOUND is a heavy duty lubricant designed to provide the carefully balanced properties required for tapping various ferrous metals, and contains extreme pressure additives to prolong tool life.

BENEFITS:

- Excellent lubricity
- Load-carrying properties
- Long tool life

TYPICAL PROPERTIES:*

Viscosity @ 100 deg. F (SUS)	500
Specific Gravity, @ 60 deg. F	1.096
Pounds per Gallon	9.13
Flash Point (deg. F, COC)	250

* Average values subject to minor manufacturing variances which do not affect performance.

SALES OFFICES:

24551 Raymond Way
El Toro, CA 92630
714/855-3881

6715 McKinley Avenue
Los Angeles, CA 90001
213/758-3181

1401 Greenleaf Avenue
Elk Grove Village, IL 60007
312/437-1450

Plum Street
Verona, PA 15147
412/828-5880

418 Tango Street
San Antonio, TX 78216
512/349-0321

Chemical Road
Plymouth Meeting, PA 19462
215/825-5050

FF-3227 AA020885

CUTTING FLUIDS

FarBest Corporation
24551 Raymond Way, Suite 110
El Toro, CA 92630
(714) 855-3881

FARBEST **TECHNICAL BULLETIN**

PRODUCT:

MIRROR BASE

SALES OFFICES:

24551 Raymond Way
El Toro, CA 92630
714/855-3881

6715 McKinley Avenue
Los Angeles, CA 90001
213/758-3181

1401 Greenleaf Avenue
Elk Grove Village, IL 60007
312/437-1450

Plum Street
Verona, PA 15147
412/828-5880

418 Tango Street
San Antonio, TX 78216
512/349-0321

Chemical Road
Plymouth Meeting, PA 19462
215/825-5050

CUTTING FLUIDS

MIRROR BASE

DESCRIPTION:

MIRROR BASE is a synthetic coolant designed for use in a wide range of machining operations on ferrous and non-ferrous metals.

MIRROR BASE has been compounded using the latest in corrosion inhibitors, wetting agents and lubricity additives to provide you with a metal-working fluid capable of producing finely finished corrosion-free parts that are easily cleaned.

BENEFITS:

- Performs a wide range of cutting operations on both ferrous and non-ferrous metals.
- Excellent wetting properties promote efficient chip removal from workpiece and prolong tool life.
- Rust and corrosion additives protect both your equipment and production parts.
- Clear working solution permits direct view of production process.

TYPICAL PROPERTIES:*

Specific Gravity	1.018
Pounds/Gallon	8.5
Flash Point, C.O.C. deg. F	None
Boiling Point, deg. F	212
pH (as is)	9.1
pH (20:1 dilution)	8.2
Color	Dark Blue

* Average values subject to minor manufacturing variances which do not affect performance.

FF-3421 AA013185

FarBest Corporation
24551 Raymond Way, Suite 110
El Toro, CA 92630
(714) 855-3881

FARBEST TECHNICAL BULLETIN

PRODUCT:

FARBEST NOSUL CUTTING OIL

FARBEST NOSUL CUTTING OIL

DESCRIPTION:

FARBEST NOSUL CUTTING OIL has been specially developed to meet a need in the electronics industry for a quality cutting oil capable of producing parts out of high alloy and stainless steels where the use of sulfur, chlorine or phosphorus is prohibited. In addition to lubricity and metal wetting additives, it contains a film strength improver to prevent rupture of the lubricant film.

BENEFITS:

- Performs difficult jobs without the use of sulfur, chlorine or phosphorus.
- Good service life - chemically stable and compatible with most machine lubricants.

APPLICATIONS:

FARBEST NOSUL CUTTING OIL is recommended for general machine shop use on all metals including magnesium.

TYPICAL PROPERTIES:*

Viscosity @ 100 deg. F (SUS)	110-120
Specific Gravity, @ 60 deg. F	0.8800
Pounds per Gallon	7.33
Flash Point (deg. F, COC)	380

* Average values subject to minor manufacturing variances which do not affect performance.

CUTTING FLUIDS

SALES OFFICES:

24551 Raymond Way
El Toro, CA 92630
714/855-3881

6715 McKinley Avenue
Los Angeles, CA 90001
213/758-3181

1401 Greenleaf Avenue
Elk Grove Village, IL 60007
312/437-1450

Plum Street
Verona, PA 15147
412/828-5880

418 Tango Street
San Antonio, TX 78216
512/349-0321

Chemical Road
Plymouth Meeting, PA 19462
215/825-5050

FF-6871 AA020885

FarBest Corporation
24551 Raymond Way, Suite 110
El Toro, CA 92630
(714) 855-3881

FARBEST TECHNICAL BULLETIN

PRODUCT:

FARBEST SOLUBLE OIL NO. 472

SALES OFFICES:

24551 Raymond Way
El Toro, CA 92630
714/855-3881

6715 McKinley Avenue
Los Angeles, CA 90001
213/758-3181

1401 Greenleaf Avenue
Elk Grove Village, IL 60007
312/437-1450

Plum Street
Verona, PA 15147
412/828-5880

418 Tango Street
San Antonio, TX 78216
512/349-0321

Chemical Road
Plymouth Meeting, PA 19462
215/825-5050

FARBEST SOLUBLE OIL NO. 472

DESCRIPTION:

FARBEST SOLUBLE OIL NO. 472 is a multi-purpose product recommended for light or heavy duty cutting, milling, turning, drilling, broaching and grinding.

FARBEST SOLUBLE OIL NO. 472 has been compounded with especially designed extreme pressure additives that produce excellent results through a range of machining operations without the disadvantages that may follow use of a chlorinated product. This lack of chlorine also makes the product most suitable for titanium work.

BENEFITS:

- Versatile over a wide range of operations and metals.
- Reputation for not causing skin irritations.
- Excellent choice for applications requiring halogen-free (no active sulfur or chlorine) coolants.

APPLICATIONS:

Cutting, milling, drilling, etc.	20-30:1
Grinding	40:1

TYPICAL PROPERTIES:*

Viscosity @ 100 deg. F (SUS)	1500
Specific Gravity, @ 60 deg. F	0.9212
Pounds per Gallon	7.67
Flash Point (deg. F, COC)	415

* Average values subject to minor manufacturing variances which do not affect performance.

FF-6584 AA020885

FarBest Corporation
24551 Raymond Way, Suite 110
El Toro, CA 92630
(714) 855-3881

FARBEST TECHNICAL BULLETIN

PRODUCT:

FARBEST LASERS PLUS CUTTING OIL NO. 7281

FARBEST LASERS PLUS CUTTING OIL NO. 7281

DESCRIPTION:

FARBEST LASERS PLUS CUTTING OIL NO. 7281 is a heavy duty, light colored, non-foaming, transparent cutting oil. It contains a high concentration of Extreme Pressure sulfur and chlorine-bearing compounds. Fatty derivatives contribute to its excellent lubricity properties.

FARBEST LASERS PLUS CUTTING OIL NO. 7281 has excellent anti-weld properties due to its ability to form a hard film between the work and tool. In addition, Extreme Pressure additives provide load-carrying properties for hard metal chips to slide freely over the tool, giving a high finish and long tool life.

FARBEST LASERS PLUS CUTTING OIL NO. 7281 is a versatile, multi-purpose cutting oil that will eliminate many other products used in your plant. It is ideally suited for use with all ferrous metals, stainless steel and high alloy steel.

BENEFITS:

- Versatile - performs a variety of cutting operations on a wide range of metals.
- Excellent anti-weld and lubricity properties.

APPLICATION:

May be used straight or cut back with pale oil for less severe operations.

TYPICAL PROPERTIES:*

Viscosity @ 100 deg. F (SUS)	200
Specific Gravity, @ 60 deg. F	0.8973
Pounds per Gallon	7.47
Flash Point (deg. F, COC)	365

* Average values subject to minor manufacturing variances which do not affect performance.

SALES OFFICES:

24551 Raymond Way
El Toro, CA 92630
714/855-3881

6715 McKinley Avenue
Los Angeles, CA 90001
213/758-3181

1401 Greenleaf Avenue
Elk Grove Village, IL 60007
312/437-1450

Plum Street
Verona, PA 15147
412/828-5880

418 Tango Street
San Antonio, TX 78216
512/349-0321

Chemical Road
Plymouth Meeting, PA 19462
215/825-5050

Fiske Brothers Refining Company
129 Lockwood Street
Newark, NJ 07105
(201) 589-9150

FISKE'S CUT-N-COOL

HEAVY DUTY SOLUBLE OIL

COMBINES THE MACHINING EFFICIENCY OF CUTTING OIL WITH THE COOLING ABILITY OF WATER IN ONE PRODUCT

WHAT MAKES CUT-N-COOL SUPERIOR TO ORDINARY SOLUBLE OILS?

The conventional soluble oil available in the per gallon price range is a blend of mineral oil and emulsifiers. General purpose products of this type form milky emulsions when added to water and are sold under many trade names. In all cases, it is the water in the emulsion (not the soluble oil) that keeps the tool and work cool. The soluble oil serves principally as a rust inhibitor to the coolant water. While emulsions of general purpose soluble oils provide cooling, they cannot effectively prevent metal "build-up" on the cutting edges of the tool because they have no extreme pressure additives such as sulfurized and chlorinated compounds. Their lubricating value is no better than the mineral oil from which they are made. They provide cooling . . . but they can do no more.

HERE IS WHY CUT-N-COOL IS SUPERIOR

- **SULFURIZED AND CHLORINATED COMPOUNDS HAVE BEEN ADDED** — The same effective bases have been added to CUT-N-COOL that normally are added only to straight cutting oils. The supply of sulfurized and chlorinated additives present in an emulsion of CUT-N-COOL is so high that machining efficiency is comparable to a general purpose straight cutting oil.
- **GIVES SUBSTANTIALLY LONGER TOOL LIFE, IMPROVED FINISH AND MORE ACCURATE CUTS** — The extreme pressure components (sulfur and chlorine) in CUT-N-COOL provide performance characteristics that are far better than general purpose soluble oils. Metal "build-up" at the cutting tool edge is minimized or eliminated, resulting in machining efficiencies associated with more costly undiluted oils.
- **KEEPS WORK COOLER** — There is less friction with corresponding less heat to carry away. The heat that is generated is cooled by water, a material twice as effective as cutting oil in cooling ability.
- **PERMITS INCREASED PRODUCTION** — The greater efficiency of CUT-N-COOL frequently permits feeds and speeds of the machine to be increased. Greater output reduces costs, increases profits.
- **CURTAILS TRADITIONAL PROBLEMS** — A generous supply of germicides have been added to CUT-N-COOL to retard rancidity and dermatitis. Rust inhibitors have been added to provide maximum protection against ferrous corrosion. More expensive emulsifiers have been used that resist oxidation, the cause of gummy deposits . . . moving parts of your machine do not stick.
- **ECONOMICAL** — Use cost is usually less than per gallon. CUT-N-COOL represents the real economy when you consider increased production afforded, longer tool life, better looking finished work, longer emulsion life, reduced labor expense by eliminating unproductive work, etc. "Low priced" soluble oils are much more expensive when all costs are calculated.

Fiske Brothers Refining Company
129 Lockwood Street
Newark, NJ 07105
(201) 589-9150

PRODUCT USE

1. For all machining and grinding operations where a water soluble product is currently in service on ferrous materials, steel alloys or aluminum.

   Note: CUT-N-COOL will tarnish copper, brass or bronze and certain other stain sensitive non-ferrous metals unless work is cleaned shortly after use.

2. As a replacement for cutting oil on ferrous materials and steel alloys when operating conditions are not severe. Under such conditions, benefits will include low use cost, less mess, much easier cleaning of work and a cooler operation without smoke.

   Note: Not recommended for machines where there is leakage into the lubricating oil system, brush applications or machining magnesium.

PRODUCT DESCRIPTION

Free flowing, brown oil
Forms a light tan emulsion in water
Emulsion is stable in soft or hard water
Contains a rich supply of chlorine and active sulfur

DIRECTIONS FOR USE

MACHINING — Add 1 part of product to 10 parts of water. Normal operating ratios are 1-5 to 1-20, concentration depending on severity of conditions.

GRINDING — Add 1 part of product to 40 parts of water.

Fiske Brothers Refining Company
129 Lockwood Street
Newark, NJ 07105
(201) 589-9150

THE ECONOMICS OF CUTTING OILS: PRICE VERSUS PERFORMANCE

Fiske's heavy duty cutting oils have been formulated to provide the significant savings in the overall expenses involved in metal fabrication.

It is false economy to consider the price of the cutting oil as the primary reason for its selection. While cost is the most obvious comparison, it represents only a minor part of the total expenses. When you find a better performing cutting oil capable of increasing production, reducing tool replacement costs, save labor expense by eliminating unproductive tool dressing and gives you a better looking finished product . . . then, and only then, do you have ***significant savings.***

A test of Fiske's Darl or Clorsul Oils will convince you that performance is much more important than the price of the cutting oil ***when all costs are considered.***

PRODUCT CHARACTERISTICS

	Darl Cutting Oil No. 2	Clorsul Cutting Oil No. 2
Type	Sulfurized Oil	Sulfurized and Chlorinated Oil
Color	Brown	Brown
Viscosity	SAE 10	SAE 10
Soluble in	Kerosene, mineral spirits, petroleum oil or chlorinated solvents	Kerosene, mineral spirits, petroleum oil or chlorinated solvents

DIRECTIONS FOR USE

Use either product straight in a circulating system or by brush application.

Fiske Brothers Refining Company
129 Lockwood Street
Newark, NJ 07105
(201) 589-9150

FISKE BROTHERS
REFINING COMPANY

NEWARK, NEW JERSEY 07105 • TOLEDO, OHIO 43605

1230 CUTTING OIL

Description	Light-colored oil
Used	Formulated to meet industry's need for a transparent cutting oil, permitting the operator to see his work while machining. Especially recommended for ferrous metals. Suitable for all machining operations including threading, tapping, turning, milling, drilling, etc.
Properties	· Contains both chlorine and active sulfur · Operator can see through the oil · Prevents metal "build-up" on tools · Gives superior finish to the work · Holds dimensions true to size · Avoids dermatitis
Application	Use as received in a circulating system or brush to the tool.

Gulf Oil Products Company
P.O. Box 2001
Houston, TX 77252
(713) 754-1565

Gulfcut® Soluble Oil

for machining and grinding

General Qualities and Characteristics

Gulfcut Soluble Oil is a specially compounded emulsifying product composed of a petroleum oil and special emulsifiers. It mixes readily with hot or cold water. It forms homogeneous and exceptionally stable emulsions which show no apparent separation after long periods of usage or storage.

Application Requirements

Metal cutting and grinding operations generate heat. The prime requirement, therefore, is to select a fluid which will reduce generation and increase dissipation of heat. The effectiveness with which this is accomplished has a direct bearing on increasing feed, speed, tool life, and improving surface finish. An efficient and economical way to control heat is to use the excellent cooling properties of water combined with some of the lubricating properties of oil. Additionally, an effective cutting and grinding fluid must contain anticorrosion properties to protect the workpiece during the manufacturing process. Another characteristic which enhances its usefulness is stability, whether in use or in storage.

Recommendations

Gulfcut Soluble Oil emulsions are the preferred choice for grinding and for machining operations where cooling is of the utmost importance. It meets the basic application requirements described above in a highly satisfactory and economical manner.

Gulfcut Soluble Oil is usually diluted in oil/water ratios ranging from 1:10 to 1:60. Because there are so many variables to be considered from job-to-job, it is impractical to make a specific recommendation of optimum oil to water ratios. However, in general, the richer emulsions are used for the more difficult jobs, the intermediate for the free-machining metals and the lean mixtures are used for such operations as grinding, where massive cooling is required.

Gulfcut Soluble Oil is specially inhibited to give it effective anticorrosion properties. It will prevent corrosion of both ferrous and nonferrous machined metal parts. Workpieces machined with such emulsions ordinarlly need no further corrosion protection during subsequent handling or manufacturing processes because the thin film protects for a suitable period.

The properties of Gulfcut Soluble Oil permit production of clean work, excellent surface finish and minimum tool wear. When used as a grinding coolant, emulsions will contribute to long wheel life, accurate tolerances and superior finish.

When preparing emulsions of Gulfcut Soluble Oil and water, always add oil to water. Stability of the emulsion is enhanced when it is put through a homogenization process.

Gulf Oil Products Company
P.O. Box 2001
Houston, TX 77252
(713) 754-1565

Typical Properties

Gravity: °API	21.5
Viscosity	
Kin: cSt	
40°C (104°F)	38.43
100°C (212°F)	5.16
SUV: Sec	
37.8°C (100°F)	200
98.9°C (210°F)	43.6
Flash, OC: F (C)	320 (160)
Fire, OC: F (C)	350 (177)
Pour: F (C)	– 20 (-28.9)
Color, ASTM D 1500	4.0
Sulfur: %	0.69
Copper Strip Test	
212 F, 3 Hr.	1
Corrosion Test (x)	
77 F and 100 F, 168 Hr.	passes
Neutralization No.	
ASTM D 974	
Total Acid No.	0.53
pH Value (x)	8.8
Emulsion Test (x)	
1 Part Oil; 9 Parts Synthetic	
Hard Water	
Froth, 15 Min.: MI.	nil
Separated Oil, 72 Hr.: %	trace
1 Part Oil; 9 Parts Synthetic	
Hard Water; 10 Parts Methyl Alcohol	
Froth, 15 Min.: MI.	nil
Separated Oil, 72 Hr.: %	1.8

(x) Method described in the latest issue of MIL-C-4339 Specification

Gulf Oil Corporation
P.O. Box 1563
Houston, Texas 77251

Gulf Oil Products Company
P.O. Box 2001
Houston, TX 77252
(713) 754-1565

Gulfcut® Heavy Duty Soluble Oil
for machining and grinding

General Qualities and Characteristics

Gulfcut Heavy Duty Soluble Oil combines the good features of a compounded cutting oil with those of regular emulsions to produce a fluid with maximum cooling ability, excellent lubricity and antiweld properties. It offers a high degree of emulsion stability, corrosion protection, control of rancidity, resistance to foaming and protection from serious staining of nonferrous metals.

Application Requirements

Modern tooling and high machining speeds demand the maximum cooling effect of water. At the same time, they require the good features of petroleum cutting oils: lubricity, antiweld properties and corrosion protection. By combining these qualities, it is possible to obtain longer tool life, higher production and better finishes even on tough-to-machine materials. In addition to these basic qualities, preferred soluble oils are also fortified with an effective germicide to control bacteria growth, rancidity and odor. Other additives are used to control corrosion and foaming.

Recommendations

Gulfcut Heavy Duty Soluble Oil offers six specific advantages which make it exceptionally suitable for tough machining jobs.

1. It combines maximum cooling ability with excellent lubricity and antiweld properties. These qualities lead to greater accuracy because tools and workpieces are cooled efficiently.
2. It includes an effective germicide which controls bacteria growth, rancidity and odor.
3. It provides exceptional corrosion control.
4. Gulfcut H. D. Soluble can be diluted to a much greater extent than can general purpose emulsifying oils, thus lowering cutting fluid costs.
5. At all dilutions, its emulsions remain homogeneous, even in hard water, and an effective antifoam agent controls foaming when soft water is used.
6. It can be used for machining nonferrous metals without serious staining or corrosion resulting.

Each specific combination of metal machinability, tool set-up, feed, and speed dictates its own optimum dilution ration. In general the richest dilutions range from (oil to water) 1:15 to 1:30 for most difficult jobs. Free machining steels and other materials with high machinability ratings commonly use emulsions of 1:30 to 1:50. Because grinding usually requires maximum cooling with less emphasis on lubrication, dilution ratios of 1:60 to as high as 1:150 yield excellent results in both finish quality and length of wheel life.

Gulf Oil Products Company
P.O. Box 2001
Houston, TX 77252
(713) 754-1565

Typical Properties

Gravity: °API	19.1
Viscosity, Kin: cSt	
40°C (104°F)	66.6
100°C (212°F)	7.82
SUV: Sec	
37.8°C (100°F)	346
98.9°C (210°F)	52.5
Pour: F (C)	−25 (-32)
Sulfur: %	3.3
Rust Preventive Test	
ASTM D 665 Procedure A&B	
Procedure A, 24 Hr.	passes
Procedure B, 24 Hr.	passes
Neutralization Number	
ASTM D 664	
Total Acid No.	1.0
Emulsion Test, 70-90 F	
FTMS 791-3205	
Distilled Water	passes
Synthetic Hard Water	passes
Corrosion Test, MIL-C-4339	
1 Pt. Oil; 4 Pt. Distilled Water, 77 F	
168 Hr., Aluminum, Steel, Tin-Lead	passes
Falex Wear Test, 10% Emulsion, 70-80 F	
500 Lb. Gauge Load, 15 Min.	
Wear: No. of Teeth	2
Gauge Load at Seizure: Lb.	4,500+

Summary Application Table

Operating Situation	Dilution Range—oil to water
Difficult machining jobs involving high speeds or feeds or low machinability metal	1:15 to 1:30
Mild blanking and forming on strip	1:10
Free machining steels and nonferrous materials with high machinability ratings	1:30 to 1:50
Grinding	1:60 to 1:150
Deep drawing stainless steel	1:5 to 1:10

Note: Always add oil to water when preparing emulsions.

Gulf Oil Products Company
P.O. Box 2001
Houston, TX 77252
(713) 754-1565

Quick Guide to Gulfcut® Oils

Ranges of Duty

Several Gulfcut oils have the same rating as to machinability.
The ranking below from light to heavy duty within each rating.

	General Ranking	Severity
Group I	Free-machining steels	43B 41B 41M 41D 41E
Group II	Intermediate steels	31A 31C 45A
Groups III, IV	Difficult steels	21D 45B 45B/44A blends
Group V	Free-machining nonferrous	11D 11A
Group VI	Difficult nonferrous	11A 41B 41M 41D 41E 45A

Classifications by Machinability

Ferrous	Machinability Rating*	Gulfcut Product													
		11A	11D**	21D**	31A**	31C	41B**	41D	41E**	41H**	41M**	43B**	44A	45A	45B
Group I	>70						x	x	x	x	x	x			
Group II	>50-<70				x	x				x				x	
Groups III, IV	<50			x						x			(x	blend	x)
Nonferrous															
Group V	>100	x	x												
Group VI	<100	x					x	x	x		x			x	

* Machinabilities: Ferrous ratings based on 100% for 1112 screw stock.
Nonferrous ratings based on 100% for leaded yellow brass.

** Antimist oils

Gulfcut Specifications

	11A	11D	21D	31A	31C	41B	41D	41E	41H	41M	43B	44A	45A	45B
Gravity: API	23.1	39.2	22.1	27.3	23.0	24.5	23.3	26.3	21.7	26.1	24.4	26.7	21.5	19.7
Viscosity,														
Kin: cSt														
40°C (104°F)	40.0	4.3	31.4	20.96	34.79	22.83	29.82	31.34	44.44	18.87	29.26	11.56	38.27	67.1
100°C (212°F)	6.6	—	4.69	4.24	5.26	3.95	4.74	5.54	6.27	3.95	4.79	2.99	5.92	7.83
SUV: Sec														
37.8°C (100°F)	202	40.7	163.6	109.9	180.4	120	155.0	161.3	230	101	151.9	67.5	197.5	350
98.9°C (210°F)	47.8	—	42.0	40.5	43.9	39.6	42.2	44.8	47.3	39.6	42.4	36.3	46.1	52.6
Flash, OC: °C	179	129	177	179	171	163	171	188	193	163	168	157	163	182
°F	355	265	350	355	340	325	340	370	380	325	335	315	325	360
Fire, OC: °C	196	146	182	204	182	177	188	210	216	182	179	168	177	210
°F	385	295	360	400	360	350	370	410	420	360	355	335	350	410
Pour: °C	−15	−7	−48	−34	−34.4	−46	−37	−15	−34	−21	−40	−21	−32	−21
°F	+5	+20	−55	−30	−30	−50	−35	+5	−30	−5	−40	−5	−25	−5
Color, ASTM D 1500	L 4.5	L 1.5	L 6.5	5.5 dil	5.5 dil	5.0	L5.5 dil	L 2.0	L 3.5	L5.5 dil	6.0	5.5 dil	L6.5 dil	6.5 dil
Neutralization No., Total Acid No.	7.73	0.13	0.08	0.36	0.46	0.11	0.36	0.35	3.2	0.38	0.07	0.51	1.5	0.53
Sulfur, Total: %	—	—	2.66	1.90	1.80	0.25	0.83	0.38	*0.07	0.90	0.14	3.44	2.02	3.38
Sulfur, Active ASTM D 1662	—	—	1.02	0.75	0.75	Nil	Nil	Nil	Nil	Nil	Nil	1.35	Nil	1.55
Chlorine: %	—	—	1.00	—	—	0.11	0.41	1.69	2.35	0.41	0.22	1.98	0.98	2.00
Fatty Oil: %	24	1.8	1.00	5.4	5.5	1.81	7.25	1.0	**	7.25	1.0	8.85	18.15	9.45

*—Phosphorus, 0.13%
**—Saponification No. 6.4

Gulf Oil Products Company
P.O. Box 2001
Houston, TX 77252
(713) 754-1565

Gulfcut® Mineral-Lard Oils

General Qualities and Characteristics

Gulfcut 11A and 11D are mineral-lard oils differing mainly in the concentration of fatty oils they contain. Both are compounded for machining of nonferrous metals where staining from sulfurized oils is objectionable.

The higher concentration of fatty oil in Gulfcut 11A makes it more desirable where depth of cut, feed and/or speed is greater. By mixing the two, you can obtain the exact concentration of fatty oil needed for a wide range of machining.

Gulfcut 11D is also widely used for blending with other cutting oils to fit the requirements of special jobs. It contains an antimist additive.

Application Requirements

Modern feeds, speeds and cuts in machining of nonferrous metals demand cutting oils that give better finish, longer tool life and better antiweld properties than straight mineral oils can give. The addition of lard oils to straight mineral oils has provided these properties. Gulfcut mineral-lard oils go beyond these basic requirements through careful selection and blending of highly-refined base stocks. They keep inventory down by providing you — in just two cutting oils that are compatible for blending — with a complete range of cutting fluids for most every need in nonferrous metalworking.

Recommendations

Each specific combination of metal machinability, tool setup, feed and speed dictates its own optimum concentration of fatty oils. The most difficult jobs will call for Gulfcut 11A undiluted. With less difficult jobs, Gulfcut 11A can be diluted with Gulfcut 11D; in general, the less difficult the job, the more dilution with 11D. Gulfcut 11D is the preferred recommendation for machining aluminum and magnesium and their alloys.

Properties

	Gulfcut 11A	Gulfcut 11D*
Gravity, °API	23.1	39.2
Viscosity,		
Kin: cST 40°C (104°F)	40.0	4.3
100°C (212°F)	6.6	—
SUV: Sec 37.8°C (100°F)	202.0	40.7
98.9°C (210°F)	47.8	—
Flash, OC: °C	179	129
°F	355	265
Fire, OC: °C	196	146
°F	385	295
Pour: °C	−15	−7
°F	+5	+20
Color, ASTM D 1500	L4.5	L1.5
Neutralization No. ASTM D 974		
Total Acid No.	7.73	0.13
Sulfur: %	none	none
Chlorine: %	none	none
Fatty Oil: %	24	1.8

*Antimist

Ask the pro from Gulf

Gulf Oil Corporation
P.O. Box 1563
Houston, Texas 77001

Gulf Oil Products Company
P.O. Box 2001
Houston, TX 77252
(713) 754-1565

Gulfcut® 21D Broad-Spectrum • Heavy Duty Oil for Tough Alloy Machining

General Qualities and Characteristics

Gulfcut 21D is a mineral oil fortified with sulfur, chlorine, and synthesized sulfurized fatty oil. This new synthetic fatty oil gives improved oxidation and lead corrosion resistance over previously used sulfurized lard oil. The blend of chlorine, sulfur, and fatty oil additives in Gulfcut 21D confer anti-weld and extreme pressure characteristics that produce long tool life and good finish when working in a wide variety of tough steels. The oil is light in color and transparent to permit close inspection by machine operators, and contains antimist for effective mist control.

Applications

The extra-heavy-duty character of Gulfcut 21D makes it suitable for turning, threading, tapping, and broaching operations on a wide variety of metals. It is particularly recommended for use in automatic screw machines handling a variety of tough alloy steels including stainless steels.

Recommendations

The flexibility of Gulfcut 21D makes it ideal for shops working with a variety of tough steels. The oil can be used instead of two or three limited heavy duty oils and result in much simplification of oil storage and maintenance systems.

Properties

	21D*
Gravity: °API	22.1
Viscosity,	
Kin: cSt 40°C (104°F)	31.4
100°C (212°F)	4.69
SUV: Sec 37.8°C (100°F)	163.6
98.9°C (210°F)	42.0
Flash, OC: °C	177
°F	350
Fire, OC: °C	182
°F	360
Pour: °C	−48
°F	−55
Color, ASTM D 1500	L 6.5
Neutralization No., ASTM D 664 Total Acid No.	0.08
Sulfur, Total: %	2.66
Sulfur, Active ASTM D 1662	1.02
Chlorine: %	1.00
Fatty Oil: %	1.00

**Antimist*

CUTTING FLUIDS

Ask the pro from Gulf

Gulf Oil Corporation
P.O. Box 1563
Houston, Texas 77001

Gulf Oil Products Company
P.O. Box 2001
Houston, TX 77252
(713) 754-1565

Gulfcut® Sulfurized-Mineral-Fatty Oils

General Qualities and Characteristics

Gulfcut 31A and 31C are compounded sulfurized-mineral-fatty oils differing from each other mainly in viscosity. Both contain sulfur in three forms: free sulfur, active sulfurized-mineral oil and sulfurized-fatty oil. Gulfcut 31A also contains an antimist additive.

The outstanding load-carrying and antiweld properties of Gulfcut 31A and 31C make them ideal for tough machining jobs — even stainless steel work. Yet the inherent flexibility of these oils makes both ideally suited for multi-purpose duty. They are regularly used on bronze, brass and aluminum in addition to many types of steels. Nonferrous parts may require a solvent wash immediately following machining to prevent surface stains.

Application Requirements

Where cutting pressures are high and where tool feeds are likely to be excessive, a cutting oil is needed that will provide excellent lubricity and antiweld characteristics. The addition of sulfur in various forms to straight mineral oil will produce these properties. The further addition of specially formulated fatty oils makes possible a better finish on the workpiece, and also reduces the tendency to stain.

The compounding of Gulfcut sulfurized-mineral fatty oils begins with the careful selection and blending of highly refined base stocks, and continues through blending-in of additives. The end result is a series of oils which provides effective lubrication, heat dissipation and antiweld action through a broad range of machining operations. These oils control build-up edge for long tool life, fine finish and high machining efficiency.

Recommendations

Each specific combination of metal machinability, tool setup, feed and speed dictates its own optimum in viscosity of a sulfurized-mineral-fatty oil. In general, the slower and tougher jobs require a higher-viscosity oil. Conversely, for easier jobs, a lower-viscosity oil can be used. Consult your Gulf Sales Engineer for a specific recommendation for your type of job.

Properties

	31A*	31C
Gravity: API	27.3	23.0
Viscosity,		
Kin: cSt		
40°C (104°F)	20.96	34.79
100°C (212°F)	4.24	5.26
SUV: Sec		
37.8°C (100°F)	109.9	180.4
98.9°C (210°F)	40.5	43.9
Flash, OC: °C	179	171
°F	355	340
Fire, OC: °C	204	182
°F	400	360
Pour: °C	−34	−34.4
°F	−30	−30
Color, ASTM D 1500	5.5 dil	5.5 dil
Neutralization No.,		
ASTM D 974		
Total Acid No.	0.36	0.46
Sulfur, Total: %	1.90	1.80
Sulfur, Active		
ASTM D 1662	0.75	0.75
Fatty Oil: %	5.4	5.5

**Antimist*

Gulf Oil Corporation
P.O. Box 1563
Houston, Texas 77001

Gulf Oil Products Company
P.O. Box 2001
Houston, TX 77252
(713) 754-1565

Gulfcut® Moderate-to Heavy-Duty Sulfo-Chlorinated Fatty Oils

Gulfcut Grades 41B, 41D, 41M, 45A

General Qualities and Characteristics

Gulfcut grades 41B, 41D, 41M, and 45A cover the range of moderate- to heavy-duty machining of ferrous alloys and moderate- to difficult-to-machine nonferrous metals. All are sulfo-chlorinated fatty oils formulated without free sulfur and tailored for a variety of machining operations. Newly formulated Gulfcut Oils now contain synthesized sulfurized fatty oil for improved oxidation and lead corrosion resistance. Gulfcut 41B and 41M contain an antimist additive.

Applications

Gulfcut 41B—This oil is specially suited for high production automatic screw machines working on free-machining steels or free to moderately difficult-to-machine nickel and copper alloys. A special antioxidant counteracts fine lead dispersions making this a preferred oil for leaded alloy steels.

Gulfcut 41D and 41M—Gulfcut 41D and 41M are general purpose cutting oils having a higher level of compounding than Gulfcut 41B. They are best suited for large, diversified machine shops such as engine manufacturing plants where free-machining and intermediate steels and difficult-to-machine nonferrous alloys are worked. Turning, form cutting, gear cutting and shaving, deep hole drilling, and broaching are typical uses of these oils. Gulfcut 41M is blended to a light viscosity for higher speeds and moderate feeds while Gulfcut 41D is formulated for heavier cuts and slower speeds. A special antioxidant counteracts the lead in leaded alloy steels.

Gulfcut 45A—Gulfcut 45A is the most highly compounded of Gulfcut nonstaining oils and suitable for very tough alloys of nickel and copper as well as ferrous alloys of moderate difficulty. A very high content of sulfurized fatty oil yields superior EP performance and smooth finished machined parts.

Recommendations

This Gulfcut series of sulfo-chlorinated fatty oils falls into three general areas: the lighter duty oil is Gulfcut 41B for free machining steels and moderately difficult nonferrous alloys. Gulfcut 41D and 41M will handle steels of intermediate difficulty and difficult-to-machine nonferrous alloys. Gulfcut 45A covers heavy duty nonferrous applications. These are general ratings only. A shop with mostly light duty machining might do well with Gulfcut 41B. Another possibility is Gulfcut 45A blended back with Gulfcut 11D for most work and 45A alone for difficult nonferrous alloys. Such applications are difficult to generalize; selection of a cutting oil for a particular plant is best made through consultation with a Gulf pro.

Properties

	41B*	41D	41M*	45A
Gravity:°API	24.5	23.3	26.1	21.5
Viscosity,				
Kin: cSt 40°C (104°F)	22.83	29.82	18.87	38.27
100°C (212°F)	3.95	4.74	3.95	5.92
SUV: Sec 37.8°C (100°F)	120	155.0	101	197.5
98.9°C (210°F)	39.6	42.2	39.6	46.1
Flash, OC: °C	163	171	163	163
°F	325	340	325	325
Fire, OC: °C	177	188	182	177
°F	350	370	360	350
Pour: °C	−46	−37	−21	−32
°F	−50	−35	−5	−25
Color, ASTM D 1500	5.0	L5.5 dil	L5.5 dil	L6.5 dil
Neutralization No.,				
Total Acid No.	0.11[a]	0.36[a]	0.38[b]	1.5[b]
Sulfur, Total: %	0.25	0.83	0.90	2.02
Sulfur, Active ASTM D 1662	Nil	Nil	Nil	Nil
Chlorine: %	0.11	0.41	0.41	0.98
Fatty Oil: %	1.81	7.25	7.25	18.15

*Antimist

a) ASTM D 974
b) ASTM D 664

CUTTING FLUIDS

Ask the pro from Gulf

Gulf Oil Corporation
P.O. Box 1563
Houston, Texas 77001

Gulf Oil Products Company
P.O. Box 2001
Houston, TX 77252
(713) 754-1565

Gulfcut® Oils for Grinding, Deep Hole or Gun Drilling, and Honing

Gulfcut grades 41H, 44A, 45A and 45B

General Qualities and Characteristics

Many forming, grinding and deep hole drilling operations impose special requirements. The heavy duty Gulfcut sulfo-chlorinated fatty oil formulations have a synthesized sulfurized fatty oil additive for improved performance in these operations (Gulfcut 44A, 45A and 45B). Gulfcut 41H, a chlorinated mineral oil containing phosphorus, antioxidants, and antifoam agents is specially formulated for form grinding without burning or staining, and contains antimist for effective mist control.

Application

Gulfcut 41H—Gulfcut 41H is a chlorine and phosphorous compounded mineral oil formulated specifically for grinding hardened tool steels. It is effectively inhibited against oxidation and provided with high air release and antifoam character for grinding of flutes on drills, taps, and reamers as performed on Hertlein type grinders. The formulation leaves surfaces free from burning or staining. The oil is light in color, free from odor, and additives are not diminished by diatomaceous earth filtration. Gulfcut 41H is also widely applicable to thread grinding operations.

Gulfcut 44A and 45B—Gulfcut 44A and 45B are heavy duty sulfochlorinated fatty oils containing free sulfur for maximum EP performance and antiweld character. Gulfcut 44A is useful in many types of gundrilling operations and can be blended with Gulfcut 45B in various percentages to meet exact viscosity requirements.

Gulfcut 45A—Except for Gulfcut 41H, Gulfcut 45A is the most highly compounded Gulfcut oil that does not contain free sulfur. In many honing operations where staining is a problem, the high quantities of sulfochlorinated fatty oils in Gulfcut 45A will yield excellent results when blended with a light viscosity honing oil such as Gulfcut 11D.

Recommendations

The severe operations of form grinding, thread grinding, deep hole drilling, and honing of hardened tool steels impose special requirements. The Gulfcut grades 41H, 44A, 45A and 45B are most useful in these difficult operations. In addition, these special purpose oils can be blended with many lighter duty oils to handle a broad range of cutting requirements.

Properties

	41H***	44A	45A	45B
Gravity: °API	21.7	26.7	21.5	19.7
Viscosity,				
Kin: cSt 40°C (104°F)	44.44	11.56	38.27	67.1
100°C (212°F)	6.27	2.99	5.92	7.83
SUV: Sec 37.8°C (100°F)	230	67.5	197.5	350
98.9°C (210°F)	47.3	36.3	46.1	52.6
Flash, OC: °C	193	157	163	182
°F	380	315	325	360
Fire, OC: °C	216	168	177	210
°F	420	335	350	410
Pour: °C	−34	−21	−32	−21
°F	−30	−5	−25	−5
Color, ASTM D 1500	L 3.5	5.5 dil	L 6.5 dil	6.5 dil
Neutralization No.,				
Total Acid No.	3.2[a]	0.51[a]	1.5[b]	0.53[b]
Sulfur, Total: %	*0.07	3.44	2.02	3.38
Sulfur, Active ASTM D 1662	Nil	1.35	Nil	1.55
Chlorine: %	2.35	1.98	0.98	2.00
Fatty Oil: %	**	8.85	18.15	9.45

**Phosphorus, 0.13%* ***Saponification No. 6.4* ****Antimist* *a) ASTM D 974* *b) ASTM D 664*

Gulf Oil Corporation
P.O. Box 1563
Houston, Texas 77001

SP15275-781

Gulf Oil Products Company
P.O. Box 2001
Houston, TX 77252
(713) 754-1565

Gulfcut(R) Multi-purpose Antimist Oils

Gulfcut 43B for free-machining steels
Gulfcut 41E for free-machining steels and tough nonferrous alloys

General Qualities and Characteristics

Oil mist in the air which operators breathe in a machine shop should not exceed 5 milligrams per cubic meter under provisions of the Occupational Safety and Health Act—OSHA. Outstanding resistance to misting has been incorporated into two multi-purpose Gulfcut oils—43B and 41E. In addition, both oils can function as a machine tool lubricant and Gulfcut 41E may be used as a hydraulic fluid. The resistance to misting in both oils combined with these lubricating properties permits economical solution of difficult misting problems.

Applications

Gulfcut 43B— Gulfcut 43B is an ideal fluid for high production automatic screw machine plants working with free-machining steels. The effective mist suppressant in the oil effectively minimizes misting in all high speed machining operations. Both the viscosity of Gulfcut 43B and the nonstaining characteristics of the oil permit use as a spindle lubricant as well as a cutting fluid.

Gulfcut 41E—Gulfcut 41E is a premium-quality, dual-purpose cutting oil with additives blended for superior performance on steels and tough nonferrous alloys. Its selective additive package provides high cutting efficiency on both types of metal. High resistance to oxidation and corrosion allows the fluid to function as lubricating and hydraulic oils as well. A light color and transparency permit viewing workpieces during use.

Recommendations

Gulfcut 43B is recommended for plants working with free-machining steels on high speed equipment, particularly automatic screw machines where misting is a problem. Viscosity of the fluid permits its use as a spindle lubricant as well as cutting fluid.

Outstanding mist resistance makes Gulfcut 41E an ideal anti-mist oil for plants working both ferrous and nonferrous stocks on high speed machinery. The multi-purpose use of the oil for lubricating and hydraulics promises economy as well as mist suppression.

Properties

	41E*	43B*
Gravity: API	26.3	24.4
Viscosity,		
Kin: cSt 40°C (104°F)	31.34	29.26
100°C (212°F)	5.54	4.79
SUV: Sec 37.8°C (100°F)	161.3	151.9
98.9°C (210°F)	44.8	42.4
Flash, OC: °C	188	168
°F	370	335
Fire, OC: °C	210	179
°F	410	355
Pour: °C	−15	−40
°F	+5	−40
Color, ASTM D 1500	L 2.0	6.0
Neutralization No., Total Acid No.	0.35[a]	0.07[b]
Sulfur, Total: %	0.38	0.14
Sulfur, Active ASTM D 1662	Nil	Nil
Chlorine: %	1.69	0.22
Fatty Oil: %	1.0	1.0

*Antimist

a) ASTM D 974
b) ASTM D 664

Gulf Oil Corporation
P.O. Box 1563
Houston, Texas 77001

SP15367-781

Gulf Oil Products Company
P.O. Box 2001
Houston, TX 77252
(713) 754-1565

Gulfcut® Heavy Duty Active Oils For Tough Alloy Steels and Difficult Machining Operations

Gulfcut grades 44A and 45B

General Qualities and Characteristics

Gulfcut 44A and 45B are formulated with high quantities of combined sulfur in a synthesized fatty oil, and chlorine and free sulfur for maximum load carrying ability and antiweld performance.

Application

Gulfcut 44A—Gulfcut 44A is formulated especially for gun-drilling and other deep hole drilling, reaming, and trepanning operations on tough alloy steels. Its low viscosity makes it ideal for automatic screw machines working tough alloys.

Gulfcut 45B—Gulfcut 45B is highly compounded to give excellent results in severe machining operations on tough ferrous metals. It may be used alone for tapping, threading, and broaching of tough alloys.

Blends

Gulfcut 44A and 45B may be mixed to intermediate viscosities for operations requiring lighter oils. For example, a 50-50 mixture of Gulfcut 44A and 45B yields a fluid with a viscosity of 135 SSU at 100°F, a favorable viscosity for automatic machines working very difficult-to-machine steels.

Gulfcut 45B may be blended back with Gulfcut 11D or other light viscosity oils to reduce compounding and handle a wide variety of machining jobs of intermediate difficulty.

Recommendations

Gulfcut 44A and 45B may be used in the toughest machining operations either alone or blended to intermediate viscosities. Both oils may function as concentrates for blending with lighter duty oils to provide cutting fluids for steels of intermediate difficulty.

Properties

	44A	45B
Gravity: °API	26.7	19.7
Viscosity,		
Kin: cSt40°C (104°F)	11.56	67.1
100°C (212°F)	2.99	7.83
SUV: Sec 37.8°C (100°F)	67.5	350
98.9°C (210°F)	36.3	52.6
Flash, OC: °C	157	182
°F	315	360
Fire, OC: °C	168	210
°F	335	410
Pour: °C	−21	−21
°F	−5	−5
Color, ASTM D 1500	5.5 dil	6.5 dil
Neutralization No., ASTM D 974 Total Acid No.	0.51	0.53
Sulfur, Total: %	3.44	3.38
Sulfur, Active ASTM D 1662	1.35	1.55
Chlorine: %	1.98	2.00
Fatty Oil: %	8.85	9.45

Gulf Oil Corporation
P.O. Box 1563
Houston, Texas 77001

SP15280-781

LPS® TAP-ALL™ TAPPING FLUID

TECHNICAL DATA

- **For use on ferrous and non-ferrous metals, including aluminum.**
- Extends tool life.
- Provides clean cuts and finer finishes.
- Has staying power to prevent tool seizure.
- Non-flammable, safe to use.
- Does not contain Sulfur or Chlorine, which can stain metal.

LPS® TAP-ALL™ is available in the following sizes:

Product	Part No.
4 oz. Squeeze	01204
16 oz.	01216
1 gal.	01228
5 gal.	01205

Types of Industries that use LPS® Tap-All™:

* Industrial Maintenance
* Automotive Repair
* Metal Working
* Plumbing and Heating
* Aviation and Aerospace
* Marine
* Mining

General Information:

LPS® Tap-All™ is a machining aid designed for use on steel, aluminum, and virtually all other metal surfaces. It combines one of the most effective coolants known to man with lubricity and machining aids to reduce friction, heat, and tool failure. Independent laboratory tests have shown that LPS® Tap-All™ dramatically extends the life of machine tools. It does not contain any chlorine or sulfur which can stain or corrode metal surfaces. LPS® Tap-All™ is also fortified with corrosion inhibitors to prevent rusting of fresh machined surfaces.

Form #2024

Formulation:

LPS® Tap-All™ is an engineered blend of coolants, lubricity aids and corrosion inhibitors for use on all metals except Magnesium.

Application Procedure:

For best results apply LPS® Tap-All™ as tool comes in contact with work. Use as needed. Do not use on Magnesium.

Properties:

Weight per Gallon:	8.34 Lbs.
Flash Point:	Non-flammable
Appearance:	Clear, colorless liquid
Odor:	Cinnamon
Sulfur/Chlorine Content:	None
Four Ball Scar Diameter:	0.64mm
Drilling Tests:*	100 Holes without tool failure
Tapping Tests:*	100 Holes without tool failure

*Tests conducted using 4340 steel quenched and tempered to 341 BHN, at a cutting speed of 30 ft. per minute, with 0.015 inch wear required for tool failure.

Specific Uses:

- Drilling - Tapping - Sawing
- Reaming - Threading - Broaching
- Engraving - Turning - Facing
- Boring - Milling of ferrous and non-ferrous metals.

Material Safety Data Sheets (OSHA 20 Form) are available upon request.

LPS®

Holt Lloyd Corporation
4647 Hugh Howell Road
Tucker, GA 30084
(404) 934-7800 1-800-241-8334
TWX 810-766-4920

ASIA
MEMA
APAA

NIDA
SIDA
ASMMA

Keystone/Pennwalt Corporation
21st and Lippincott Streets
Philadelphia, PA 19132
(215) 225-7473

KEYSTONE® KEYCUT NO. 101 CUTTING OIL

Features

Translucent Red-Brown
Contains moderate treatment with extreme pressure additives
Homogeneous; no sulphur settling, excellent stability
Lubricates and carries high loads
Prolongs tool life - Reduces heat
Excellent cooling on high speed operations
Good wetting ability
Oil and solvent soluble
Performs well on all metals including stainless (Not recommended for use on aluminum without dilution with light oil).

Uses

Threading and tapping
General purpose machine shop operations
Automatic screw machine work

Method of Application

Brush
Recirculation - Reservoir
Oil Can (for hand operations)

NOTE: Keycut 101 can be used straight or diluted with blending oil (approximately 100 SUS @ 100°F), fuel oil or kerosene. The amount of dilution is determined by the severity of the job.

TYPICAL PROPERTIES

Gravity: API	23.5
Viscosity	
SUS @ 100°F	180
SUS @ 210°F	44
Viscosity Index	83
Color-ASTM-D-1500	Dark Brown
Flash - °F	370
Pour Point - °F	-25
ASTM Viscosity Grade No.	S150

Keystone Division of Pennwalt Corporation, with a policy of continuous improvement, reserves the right to change specifications as our technology progresses. We are not responsible for misuse and/or misapplication of our products.

Keystone/Pennwalt Corporation
21st and Lippincott Streets
Philadelphia, PA 19132
(215) 225-7473

PRODUCT INFORMATION

KEYSTONE PENNWALT
PRECISION LUBRICANTS
21st and LIPPINCOTT STREETS PHILADELPHIA, PA 19132 U.S.A.

KEYSTONE® KEYCUT NO. 106 SYNTHETIC COOLANT AND CUTTING FLUID

Features

Extreme pressure treatment
Anti-weld additives
Excellent rust protection at recommended dilutions
Excellent solution stability
Contains additives to prevent oxidation and rancidity
Recommended for use in mist coolant units
Sulphur and chlorine free
Transparent green color for visibility of work piece and lubricant

Uses

General machining operations - to replace straight oil or water soluble cutting fluids on all metals except Magnesium and Magnesium alloys
Especially recommended for machining of stainless steel and titanium when the speed is increased 15 to 20 percent and depth of cut is reduced

Method of Application

Recirculation - Reservoir
Brush (for hand operations)
Mist system

NOTE: Recommended Dilution:

Grinding - 50 to 150 parts water to 1 part 106
Cutting - 10 to 15 parts water to 1 part 106

TYPICAL PROPERTIES

Gravity: API	1.0
Viscosity Index	NA
Color	Green
Flash - °F	None
Sulfur	0
Chlorine	0
Heavy Metals	0

Keystone Division of Pennwalt Corporation, with a policy of continuous improvement, reserves the right to change specifications as our technology progresses. We are not responsible for misuse and/or misapplication of our products.

Keystone/Pennwalt Corporation
21st and Lippincott Streets
Philadelphia, PA 19132
(215) 225-7473

PRODUCT INFORMATION

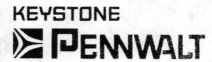

KEYSTONE
PENNWALT
PRECISION LUBRICANTS
21st and LIPPINCOTT STREETS, PHILADELPHIA, PA. 19132 U.S.A.

CUTTING FLUIDS

KEYSTONE® KEYCUT SOLUBLE OIL

Features

Forms milky emulsion in water
Reduces heat from friction
Lubricates tool and work piece
Good wetting ability on metal surfaces
Keycut film cleans readily in alkali or degreasing operations
Does not contain chlorinated hydrocarbons
Soluble in oil or kerosene

Uses

General purpose machine shop work, including milling, drilling, cutting, boring, sawing and grinding
Stamping and drawing steel and aluminum parts

NOTE: Recommended Dilution:

General purpose - 10 to 40 parts water to 1 part Keycut
Most Grinding - 50 parts water to 1 part Keycut
Grinding Cast Iron - 30 parts water to 1 part Keycut

TYPICAL PROPERTIES

Gravity: API	18.7
Viscosity SUS @ 100°F	235
Flash - °F	380
ASTM Viscosity Grade No.	S215
Sulphur, active %	0
Chlorine %	0

Keystone Division of Pennwalt Corporation, with a policy of continuous improvement, reserves the right to change specifications as our technology progresses. We are not responsible for misuse and/or misapplication of our products.

KoolMist Division
400 Roosevelt Avenue
Montebello, CA 90640
(213) 723-1165

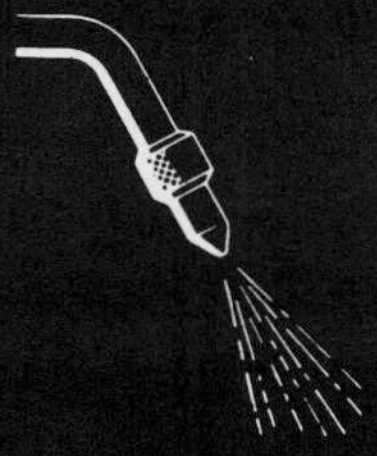

TWO-WAY (MIST OR FLOOD) COOLANT WITH HIGH LUBRICITY

KOOLMIST FORMULA "77"

Kool Mist "77" provides the high lubricity needed when metals are unusually tough and abrasive, or when tool pressures are exceptionally high. In other respects it has essentially the same properties as Kool Mist "78". It may be used in either mist or flood coolant systems, eliminating the trouble and cost of maintaining double inventories. It is also OSHA-safe and biodegradable, and may be discharged into any waste stream without injuring the environment.

REDUCES CUTTING EDGE ABUSE

Where extreme tool pressure or abrasive materials cause premature breakdown of cutting edges. Formula "77" is indispensable for controlling destructive friction and shock. Highly effective lubricating components penetrate the interface between tool and cut. where they minimize chipping and galling of the cutting edge.

NOT JUST A SOLUBLE OIL

Although it disperses in water, Kool Mist "77" is a product of advanced chemical technology with no similarity to ordinary soluble oils. It contains no customary halogens, will not turn rancid. and will not gum machine surfaces. Instead, its lubricating components protect and lubricate all machine parts contacted.

A UNIVERSAL COOLANT

Use Kool Mist "77" for mist or flood cooling of any metal or alloy, ferrous or non-ferrous . . . even Monel, Inconel, magnesium. titanium and all the new exotic materials. In some instances it will prove to be the only practical way to machine a difficult metal. The ability of "77" to reduce friction and keep tools cutting under heavy load cannot be equalled by conventional coolants.

OSHA-SAFE, BIODEGRADABLE

Kool Mist "77" is non-flammable, non-toxic, and non-irritating. Powerful bactericides prevent dermatitis and rancidity, and there is no objectionable odor. The formulation is biodegradable, and may be discharged into any waste stream with no ill effects on the environment.

ENDS COMMON PROBLEMS

Kool Mist "77" contains no sulpher, chlorine or other reactive chemicals that lead to staining and stress cracking. Highly effective inhibitors prevent rust and corrosion at dilutions up to 32:1, and none of the scientifically balanced formula's ingredients can precipitate to cause scale and buildup in coolant lines and valves. Machines and workpieces stay clean and bright.

READILY AVAILABLE

Kool Mist "77" is available throughout the United States and in many other countries. Standard packaging includes one-gallon plastic jugs, singly or four per case; five-gallon cans; and 55-gallon drums.

FREE SAMPLES AVAILABLE FOR TESTING IN YOUR OWN PLANT AND LABORATORY

AVAILABLE THROUGH	MANUFACTURED BY
YOUR INDUSTRIAL DISTRIBUTOR	**KOOLMIST** DIVISION OF ALL-POWER MFG. CO. **400 ROOSEVELT AVE., MONTEBELLO, CALIF. 90640** **(213) 723-1165**

FORM KM-2

KoolMist Division
400 Roosevelt Avenue
Montebello, CA 90640
(213) 723-1165

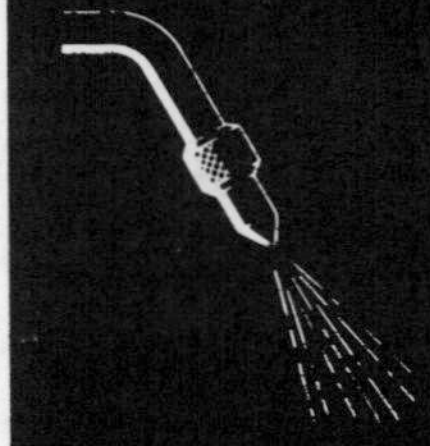

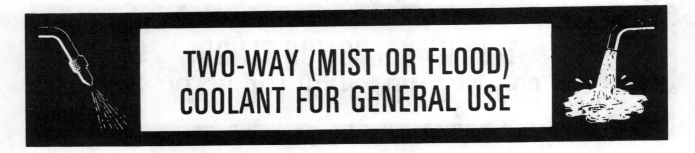

TWO-WAY (MIST OR FLOOD) COOLANT FOR GENERAL USE

KOOLMIST FORMULA "78"

Why stock two different concentrates for your mist and flood coolant systems? Kool Mist Formula "78" concentrate may be used for either purpose, and gives outstanding results. Formula "78" eliminates double inventory, reduces costs, cuts storage space requirements. It also prevents accidental use of the wrong coolant through human error. Whether used for mist or flood, this updated and improved formulation is OSHA-safe and biodegradable.

FOR ANY MATERIAL OR PROCESS

Kool Mist "78" is the most universal light duty coolant available. It may be used in mist or flood form on any metal or alloy, ferrous or non-ferrous. The formula is also highly efficient for cutting Monel, Inconel, magnesium, titanium and all the new exotic materials. It will perform on any cutting or grinding operation in which the use of coolant is feasible.

SAFE FOR THE ENVIRONMENT

Kool Mist "78" is now biodegradable and meets prevailing regulations and guidelines. The effluent may be discharged into any waste stream without special treatment and will have no adverse effect on the environment.

SAFE FOR METALS

Because it contains no sulphur, chlorine or other reactive chemicals, Kool Mist "78" cannot produce stains or induce latent stress cracks in metals. Neither will any residual coolant film cause subsequent chemical reactions or health hazards.

SAFE FOR PERSONNEL

Kool Mist "78" is non-toxic, and gives off no objectionable odor. Multiple bactericides prevent dermatitis and rancidity, extending the tank life of the coolant. There are no harsh components to irritate the operator's skin. The formula is non-flammable in either concentrated or dilute form.

SAFE FOR MACHINES

Machine surfaces remain free of rust and corrosion when Kool Mist "78" is used in dilutions up to 32:1 because the formulation features non-reactive components and highly effective rust inhibitors. Coolant lines and valves remain free-flowing since there is nothing in the coolant to precipitate and cause buildup.

CONVENIENTLY PACKAGED

No need to buy a lot when you need a little. You may order Kool Mist in one-gallon plastic jugs, singly or four per case; in five-gallon cans; or in 55-gallon drums. Available throughout the United States and in many other countries.

FREE SAMPLES AVAILABLE FOR TESTING IN YOUR OWN PLANT AND LABORATORY

AVAILABLE THROUGH	MANUFACTURED BY
YOUR INDUSTRIAL DISTRIBUTOR	**KOOLMIST** DIVISION OF ALL-POWER MFG. CO. **400 ROOSEVELT AVE., MONTEBELLO, CALIF. 90640** **(213) 723-1165**

FORM 8175

KoolMist Division
400 Roosevelt Avenue
Montebello, CA 90640
(213) 723-1165

KOOLMIST FORMULA "300"

ENVIRONMENTALLY *Clean* COOLANT

MEETS OSHA, NASA, AEC and EPA REQUIREMENTS

- Cuts machining time by about 30% on all metals including exotics.
- Biodegradable and user-safe. Meets current environmental regulations.
- Eliminates heat. Lets cutter peel chips instead of bulldozing.
- Ends frequent resharpening. Can triple, even quadruple, tool life.
- Dramatically reduces coolant cost per year.
- Equally effective for flood coolant systems and mist generators.

ECOLOGICALLY SAFE

KoolMist "300" effluent is biodegradable and may be disposed of in any primary sewage treatment plant. Tramp oils will not mix with the coolant and will eventually float to the top of the discharge holding tank, where they may be removed by vacuum skimmer and sold for salvage value.

SAFE FOR PERSONNEL

Multiple bactericides prevent dermatitis, rancidity and unpleasant odors. Independent laboratory tests show KoolMist "300" to be virtually free of problems relating to eye and skin irritation, ingestion, inhalation, absorption and nitrosamine effects.

SAFE FOR PARTS, EQUIPMENT

Non-reactive components plus highly effective rust inhibitors assure total absence of rust, corrosion and buildup when KoolMist "300" is used in dilutions up to 42:1. Workpieces stay bright, and coolant lines and valves do not become restricted. The formula contains no sulfur, chlorine or other ingredients that lead to stress cracking and stains, and any residual coolant film will not cause chemical reactions or health hazards.

SAVES TIME AND MONEY

Although KoolMist "300" is not initially costly (a 42:1 mix with water costs about 33¢ a gallon), much of this cost can be recovered during use. By eliminating heat and friction it saves about 30% of machining time, lowers the rate of rejects, multiplies tool life and minimizes costly downtime for resharpenings. When kept free of trash and waste, KoolMist "300" may be recirculated indefinitely. This saves untold hours of shutdowns normally required to drain, clean and recharge sumps; and drastically reduces the amount of coolant used per year. In the long run, this superior formulation is actually far more economical to use than many "economy" coolants.

FREE SAMPLES AVAILABLE FOR TESTING IN YOUR OWN PLANT AND LABORATORY

AVAILABLE THROUGH	MANUFACTURED BY
YOUR INDUSTRIAL DISTRIBUTOR	**KOOLMIST** SUBSIDIARY OF ALL-POWER MFG. CO. 13141 MOLETTE ST., SANTA FE SPRINGS, CA. 90670 (213) 802-2640

KM-10680

Marston Bentley Incorporated
1810 Star Batt Drive, Building 5
Rochester, MI 48063
(313) 852-6730

ZEROLENE range of cutting fluids are highly chlorinated and sulphurised compounded oils.

Originally developed for service machining operations, they are formulated to give excellent tool life combined with consistently high surface finish. They are particularly effective on difficult metals, giving good chip removal, and are non staining.

Zerolene SW4A/4D

Main Applications Broaching, Gun Drilling, Gear Cutting of high quality steels, low alloy steels, high nickel alloys.

Typical properties

Kinematic Viscosity at 25°C	4.5 c/s
Chlorine Content	23% minimum
Sulphur Content	1% w/w
Saponifiable Matter	7% w/w

Zerolene SW4A contains molybedenum disulphide for applications where additional lubricating properties are desirable.

Zerolene 65

Main Applications Similar to Zerolene SW4A/4D, Zerolene 65 has the advantage of being a better lubricant, is more viscous, and has a higher flashpoint.

Typical Properties

Kinematic Viscosity at 25°C	16 c/s
Chlorine Content	23% minimum
Sulphur Content	0.9% w/w
Saponifiable Matter	7% w/w

Zerolene 97

Main Applications Broaching, Gun Drilling, reaming of a wide range of materials from low alloy steels to high nickel alloys. Heavy duty applications.

Typical Properties

Kinematic Viscosity at 25°C	12 c/s
Chlorine Content	16.5% w/w
Sulphur Content	0.8% w/w
Saponifiable Matter	4.5% w/w

The information given herein is based on general experiences and, where applicable, on the results of tests on samples of typical manufacture. However because of many factors which are outside our control, which can affect the use of these products, we cannot accept any liability for any injury, loss or damage resulting from reliance upon such information. The product is sold subject to our standard conditions of sale, copies of which are available on request.

Master Chemical Corporation
501 West Boundary Street
Perrysburg, OH 43551
(419) 874-7902

TRIM® CLEAR

DATA AND INFORMATION

GENERAL DESCRIPTION:

TRIM® CLEAR is a WATER CLEAR fluid concentrate designed PRIMARILY for use in GRINDING STEEL and CAST IRON. This product is especially RECOMMENDED for operations in which HIGH HEAT transfer is necessary—such as SURFACE grinding, DOUBLE DISC grinding, WET ABRASIVE grinding and ABRASIVE CUT-OFF.

ADVANTAGES:

- WATER CLEAR—to SEE THE JOB IN PROCESS.
- RUST INHIBITING—grind with the COOLNESS of water, still leave a RUST INHIBITING FILM.
- Designed to be NON-FOAMING as a grinding solution with MAXIMUM HEAT-TRANSFER.
- SETTLES CHIPS RAPIDLY.
- EFFECTIVE IN HARD WATER.
- CONTAINS NO NITRITES.

Available in 5 gallon and 54 gallon Drums and Tank Wagon Lots.

PHYSICAL PROPERTIES: (TYPICAL DATA)

Form	Clear fluid
Color	Colorless
Specific Gravity	1.10
Flash Point	None
Viscosity	40 SUS @ 100 °F
Pour Point	−10 °F
Residue	Fluid, semi-crystalline
Odor	Mild, pleasant
pH	9.4 at 1% conc. 9.5 at 2% conc.
Refractometer Factor	1.8
Titration Factor	0.22

CUTTING FLUIDS

Master Chemical Corporation
501 West Boundary Street
Perrysburg, OH 43551
(419) 874-7902

TRIM® 9106 CS

DATA AND INFORMATION

GENERAL DESCRIPTION:

TRIM® 9106 CS is a CLEAR, WATER MISCIBLE FLUID concentrate specifically formulated for use in CENTRAL SYSTEMS. This product is particularly suitable for HIGH SPEED TURNING AND GRINDING of ferrous and non-ferrous metals. As manufactured, TRIM® 9106 CS contains NO SULFUR, PHOSPHORUS, CHLORINE OR OTHER HALOGENS.

ADVANTAGES:

- WILL NOT form AMMONIA or other gases when grinding or machining cast iron.
- RECOMMENDED for HIGH HEAT OPERATIONS such as: all types of grinding (except form grinding), especially Blanchard type, double disc grinding, wet abrasive belt, abrasive cut off operations and high speed turning with carbide cutting tools.
- FLOATS CARBON PARTICLES.
- INHIBITS CHIP PACKING.
- EXCELLENT WATER SOFTENING properties.
- WILL NOT ATTACK BRASS SERRATIONS OF MAGNETIC CHUCKS.
- CONTAINS NO NITRITES.

PHYSICAL PROPERTIES: (TYPICAL DATA)

Form	Liquid
Color	Straw
Specific Gravity	1.12
Flash Point	None
Viscosity	30 SUS @ 100°F (37.8°C)
Pour Point	−10°F
Residue	Crystalline-redissolvable in its own solution or water
Odor	Mild, pleasant
pH	8.6 at 10% conc.
Refractometer Factor	2.5
Titration Factor	0.38

MASTER CHEMICAL CORPORATION
501 West Boundary Street • Perrysburg, Ohio 43551 • 419-874-7902

TRIM® HM

Pat. 2,999,064

DATA AND INFORMATION

GENERAL DESCRIPTION:

TRIM® HM is a WATER CLEAR fluid concentrate designed for ABRASIVE OPERATIONS and is PRIMARILY RECOMMENDED FOR HARD MATERIALS such as CAST ALLOYS, CARBIDES, CERAMICS, GLASS and CAST IRON. This product has strong RUST INHIBITING qualities and is especially effective in operations where HIGH HEAT transfer is necessary—such as SURFACE grinding.

ADVANTAGES:

- WATER CLEAR—to SEE THE JOB IN PROCESS.
- RUST INHIBITING—grind with the COOLNESS of water, still leave a RUST INHIBITING FILM.
- Designed to be LOW-FOAMING as a grinding solution with MAXIMUM HEAT-TRANSFER. Particularly effective with diamond wheels.
- SETTLES CHIPS RAPIDLY.
- LOW-MISTING properties provide CLEANER OPERATING CONDITIONS.
- RESISTS emulsification of TRAMP OIL.
- CONTAINS NO NITRITES.

Available in 5 gallon and 54 gallon Drums and Tank Wagon Lots.

PHYSICAL PROPERTIES: (TYPICAL DATA)

Form	Clear, thin liquid	**Residue**	Slightly Crystalline
Color	Greenish Amber	**Odor**	Mild, pleasant
Specific Gravity	1.10	**pH**	8.8 at 1% conc. 8.9 at 2% conc.
Flash Point	None		
Viscosity	35 SUS @ 100 °F (37.8 °C)	**Refractometer Factor**	1.6
Pour Point	−13 °F	**Titration Factor**	0.28

CUTTING FLUIDS

Master Chemical Corporation
501 West Boundary Street
Perrysburg, OH 43551
(419) 874-7902

TRIM® ep extreme pressure

DATA AND INFORMATION

GENERAL DESCRIPTION:

TRIM® EP is a SEMI-CHEMICAL FLUID concentrate containing STABLE friction reducing addititves. This product is designed for EXTREME PRESSURE applications (tough alloys and severe operations) but is also excellent for GENERAL MACHINING assignments. TRIM® EP is very efficient due to its combination of strong cooling and high lubricating characteristics.

ADVANTAGES:

- HIGH LUBRICITY—due to lubricity ADDITIVES acting as ANTI-WELD AGENTS.
- WILL NOT cause CORROSION.
- WILL NOT cause STAINING.
- CLEAN—TRIM® EP is not MESSY and will not SMOKE under normal operating conditions.
- EFFICIENT—due to a UNIQUE COMBINATION of SUPERIOR COOLING and HIGH LUBRICITY characteristics.
- EXTREME PRESSURE applications include outstanding performance on TOUGH ALLOYS, SEVERE OPERATIONS and SAWING.

Available in 5 gallon and 54 gallon drums and tank wagon lots.

PHYSICAL PROPERTIES: (TYPICAL DATA)

Form . . . . Thick Liquid
Color . . . . Brown
Specific Gravity . . . . 1.04 (8.7#/gal)
Flash Point . . . . None
Viscosity . . . . . . 230 SUS @ 100°F (37.8°C)

Residue . . . . Soft, readily resoluble in its own solution or in water.
Odor . . . . Mild, pleasant
pH . . . . 8.4 at 5% conc.
8.3 at 10% conc.
Refractometer Factor 1.66

Master Chemical Corporation
501 West Boundary Street
Perrysburg, OH 43551
(419) 874-7902

TRIM® REGULAR

PAT. 2,999,064

DATA AND INFORMATION

GENERAL DESCRIPTION:

TRIM® REGULAR is an ALL CHEMICAL cutting and grinding fluid concentrate designed for general machining and grinding of CAST IRON and STEEL (including stainless). TRIM® REGULAR as manufactured contains NO SULFUR, PHOSPHORUS, CHLORINE OR OTHER HALOGENS.

ADVANTAGES:

- ALL CHEMICAL properties IMPROVE COOLING and LUBRICATION.
- Proven results with INCREASED SPEEDS and FEEDS.
- MILD, PLEASANT ODOR.
- COMPENSATES for MINERAL effects of most hard waters.
- Keeps MACHINES and PARTS CLEAN.
- EXCEPTIONAL for machining CAST IRON.
- WILL NOT SMOKE.
- Offers more pleasant working conditions for the operator.
- Allows machining of TOUGHER METALS with CLOSER SIZE CONTROL.

Available in 5 gallon and 54 gallon Drums and Tank Wagon Lots.

PHYSICAL PROPERTIES: (TYPICAL DATA)

Form	Soft Jel	**Specific Gravity**	1.04
Color	Straw	**Flash Point**	None
Odor	Mild, pine	**Residue**	Soft liquid
Viscosity	12,000 cps (#3 spindle @ 6 rpm)	**pH**	8.3 at 5% con. 8.4 at 10% conc.
(Brookfield)	1,880 cps (#3 spindle @ 60 rpm)		

MASTER CHEMICAL CORPORATION
501 West Boundary Street • Perrysburg, Ohio 43551 • 419-874-7902

Master Chemical Corporation
501 West Boundary Street
Perrysburg, OH 43551
(419) 874-7902

TRIM® HD HEAVY DUTY

DATA AND INFORMATION

GENERAL DESCRIPTION:

TRIM® HD is a HEAVY DUTY, WATER MISCIBLE fluid concentrate designed for GENERAL as well as HEAVY DUTY applications (tough alloys and severe operations). This product contains a STABLE CHLORINE additive which acts as an effective FRICTION REDUCING AGENT. TRIM® HD is extremely effective for the machining of STAINLESS STEELS and other HEAT RESISTANT ALLOYS.

ADVANTAGES:

- DESIGNED for HEAVY DUTY applications. TOUGH ALLOYS and SEVERE OPERATIONS are made easier with an overall UPGRADING OF EFFICIENCY.
- CHEMICAL COOLANT providing a water system with the HIGH LUBRICITY characteristics of CHLORINE.
- Reduced forces LOWER HEAT GENERATION—deliver BETTER FINISHES—CLOSER SIZE CONTROL—and UNUSUALLY LONGER TOOL LIFE.
- TRIM® HD approaches the LUBRICITY of STRAIGHT OIL with DOUBLE the COOLING CAPACITY of oil.
- INCREASES production RATES—LOWERS unit COST.

Available in 5 gallon and 54 gallon Drums and Tank Wagon Lots.

PHYSICAL PROPERTIES: (TYPICAL DATA)

Form	Soft Jel	**Odor**	Mild, pleasant
Color	Cream colored	**Flash Point**	None
Specific Gravity	1.04	**Residue**	Liquid
Viscosity	12,000 cps (#3 spindle @ 6 rpm)	**pH**	8.3 at 5% conc. 8.4 at 10% conc.
(Brookfield)	1,880 cps (#3 spindle @ 60 rpm)		

Master Chemical Corporation
501 West Boundary Street
Perrysburg, OH 43551
(419) 874-7902

TRIM®

PAT. 2,999,064

70/30

DATA AND INFORMATION

GENERAL DESCRIPTION:

TRIM® 7030 is an ALL CHEMICAL cutting and grinding fluid concentrate designed for general machining and grinding of CAST IRON and STEEL. As manufactured, TRIM® 7030 contains NO SULFUR, PHOSPHORUS, CHLORINE OR OTHER HALOGENS.

ADVANTAGES:

- ALL CHEMICAL properties IMPROVE COOLING and LUBRICATION.
- Proven results with INCREASED SPEEDS and FEEDS.
- MILD, PLEASANT ODOR.
- COMPENSATES for MINERAL effects of most hard waters.
- Keeps MACHINES and PARTS CLEAN.
- EXCEPTIONAL for machining CAST IRON.
- WILL NOT SMOKE.
- Offers more pleasant working conditions for the operator.
- Allows machining of TOUGHER METALS with CLOSER SIZE CONTROL.

Available in 5 gallon and 54 gallon Drums and Tank Wagon Lots.

PHYSICAL PROPERTIES: (TYPICAL DATA)

Form	Heavy liquid
Color	Light straw
Specific Gravity	1.08
Flash Point	None
Viscosity	5870 cps, (#3 spindle @ 6 rpm)
(Brookfield)	985 cps, (#3 spindle @ 60 rpm)
Residue	Moist crystal—redissolvable in its own solution or water.
Odor	Mild, pine
pH	8.3 at 5% conc. 8.4 at 10% conc.

CUTTING FLUIDS

Master Chemical Corporation
501 West Boundary Street
Perrysburg, OH 43551
(419) 874-7902

TRIM® CE

CHEMICAL EMULSION

DATA AND INFORMATION

GENERAL DESCRIPTION:

TRIM® CE is a CHEMICAL EMULSION CONCENTRATE containing an extremely effective FRICTION REDUCING LUBRICANT. TRIM® CE is designed specifically for use as an INITIAL CHARGE for CENTRAL SYSTEMS and INDIVIDUAL MACHINES or as MAKEUP for INDIVIDUAL MACHINES. This product is recommended for a wide variety of metal removal operations on most FERROUS METALS and many nonferrous metals such as BRASS, COPPER, AND ALUMINUM.

ADVANTAGES:

- WIDE RANGE JOB APPLICATION from tough assignments such as gear hobbing and broaching to lighter duties like turning. TRIM® CE has proven EQUALLY EFFECTIVE.
- EASILY ADAPTABLE to NONFERROUS and FERROUS METALS including tough STAINLESS STEEL ALLOYS; even delivers good finish on soft materials such as ALUMINUM ALLOYS.
- STABILITY. TRIM® CE forms an extremely tight emulsion of fine particle size. This tight, stable emulsion allows TRIM® CE to be run for extended periods without pump-outs.
- HOUSEKEEPING IS EASY with TRIM® CE'S built-in CLEANING ACTION—metal chips and dirt will not build up, therefore machines stay CLEAN.
- FLUID RESIDUE prohibits sticky ways and slides—a most important consideration in the operation of AUTOMATIC and NUMERICALLY CONTROLLED MACHINES.

Available in 5 gallon and 54 gallon drums and tank wagon lots.

PHYSICAL PROPERTIES: (TYPICAL DATA)

Form	Fluid	**Flash Point**	(Cleveland Open Cup) 345 °F (174 °C)
Color	Dark Green	**Residue**	Liquid
Specific Gravity	.99	**pH**	9.2 at 2% conc. 9.4 at 5% conc.
Odor	Mild	**Refractometer Factor**	1.0
Pour Point	−27 °F (-33 °C)		
Viscosity	301 SUS @ 100 °F (37.8 °C)		

MASTER CHEMICAL CORPORATION
501 West Boundary Street • Perrysburg, Ohio 43551 • 419-874-7902

Master Chemical Corporation
501 West Boundary Street
Perrysburg, OH 43551
(419) 874-7902

TRIM® SOL

DATA AND INFORMATION

GENERAL DESCRIPTION:

TRIM® SOL is a CHEMICAL EMULSION CONCENTRATE containing an extremely effective FRICTION REDUCING LUBRICANT. This product is designed for a wide variety of metal removal operations on most FERROUS METALS and many nonferrous metals such as BRASS, COPPER and ALUMINUM.

ADVANTAGES:

- WIDE RANGE JOB APPLICATION from tough assignments such as gear hobbing and broaching to lighter duties like turning, TRIM® SOL has proven EQUALLY EFFECTIVE.
- EASILY ADAPTABLE to NONFERROUS and FERROUS METALS including tough STAINLESS STEEL ALLOYS; offers excellent finish on soft materials such as ALUMINUM ALLOYS with NO STAINING.
- STABILITY. TRIM® SOL forms an extremely tight emulsion of fine particle size. This tight, stable emulsion allows TRIM® SOL to be run for extended periods without pump-outs.
- HOUSEKEEPING IS EASY with TRIM® SOL'S built-in CLEANING ACTION—metal chips and dirt will not build-up, therefore machines stay CLEAN.
- FLUID RESIDUE prohibits sticky ways and slides—a most important consideration in the operation of AUTOMATIC and NUMERICALLY CONTROLLED MACHINES.

Available in 5 gallon and 54 gallon Drums and Tank Wagon Lots.

PHYSICAL PROPERTIES: (TYPICAL DATA)

Form .... Fluid
Color .... Dark Green
Specific Gravity .... .99
Odor .... Mild
Pour Point .... −20°F (−28.9°C)
Flash Point .... (Cleveland Open Cup) 305°F (151.7°C)
Fire Point .... (Cleveland Open Cup) 370°F (187.8°C)
Viscosity .... 301 SUS @ 100°F (37.8°C)
Residue .... Liquid
pH .... 9.2 at 2% conc. 9.4 at 5% conc.
Refractometer Factor 1.0

MASTER CHEMICAL CORPORATION
501 West Boundary Street • Perrysburg, Ohio 43551 • 419-874-7902

Master Chemical Corporation
501 West Boundary Street
Perrysburg, OH 43551
(419) 874-7902

TRIM® SOL LC

DATA AND INFORMATION

GENERAL DESCRIPTION:

TRIM® SOL LC is a CHEMICAL EMULSION CONCENTRATE designed for light and moderate machining and grinding. This product is designed for a wide variety of metal removal operations on most FERROUS METALS and many non-ferrous metals such as BRASS, COPPER and ALUMINUM.

ADVANTAGES:

- WIDE RANGE JOB APPLICATION from tough assignments such as gear hobbing and broaching to lighter duties like turning, TRIM® SOL LC has proven EQUALLY EFFECTIVE.
- EASILY ADAPTABLE to NON-FERROUS and FERROUS METALS.
- STABILITY: TRIM® SOL LC forms an extremely tight emulsion of fine particle size. This tight, stable emulsion allows TRIM® SOL LC to be run for extended periods without pump-outs.
- HOUSEKEEPING IS EASY with TRIM® SOL LC's built-in CLEANING ACTION—metal chips and dirt will not build-up, therefore machines stay CLEAN.
- FLUID RESIDUE eliminates sticky ways and slides—a most important consideration in the operation of AUTOMATIC and NUMERICALLY CONTROLLED MACHINES.
- CONTAINS NO NITRITES.

Available in 5 gallon and 54 gallon Drums and Tank Wagon Lots.

PHYSICAL PROPERTIES: (TYPICAL DATA)

Form	Fluid
Color	Dark Green
Specific Gravity	0.95
Odor	Mild
Pour Point	−20 °F
Flash Point	(Cleveland Open Cup) 305 °F (151.7 °C)
Fire Point	370 °F (187.8 °C)
Viscosity	294 SUS @100 °F (37.8 °C)
Viscosity Index	33
Residue	Liquid
pH	9.2 at 2% conc. 9.4 at 5% conc.
Refractometer Factor	1.0

MASTER CHEMICAL CORPORATION
501 West Boundary Street • Perrysburg, Ohio 43551 • 419-874-7902

TRIM SOL S

DATA AND INFORMATION

GENERAL DESCRIPTION:

TRIM® SOL S is a CHEMICAL EMULSION CONCENTRATE containing STABLE SULFUR and CHLORINE additives. This product is designed for EXTREME PRESSURE applications (tough alloys and severe operations) and is especially recommended for tough GRINDING operations. TRIM® SOL S is very efficient due to its combination of strong cooling and high lubricating characteristics.

ADVANTAGES:

- HIGH LUBRICITY—due to SULFUR and CHLORINE ADDITIVES acting as ANTI-WELD AGENTS
- WILL NOT cause CORROSION.
- WILL NOT cause STAINING.
- STABILITY. TRIM® SOL S forms an extremely tight emulsion of fine particle size. This tight, stable emulsion allows TRIM® SOL S to be run for extended periods without pump-outs.
- FLUID RESIDUE prohibits sticky ways and slides—a most important consideration in the operation of AUTOMATIC and NUMERICALLY CONTROLLED MACHINES.
- HOUSEKEEPING IS EASY with the built-in CLEANING ACTION of TRIM® SOL S—metal chips and dirt will not build-up, therefore machines stay CLEAN.

Available in 5 gallon and 54 gallon Drums and Tank Wagon Lots.

PHYSICAL PROPERTIES: (TYPICAL DATA)

Form	Fluid
Color	Dark Green
Specific Gravity	1.0
Odor	Mild
Viscosity	518 SSU @ 100°F (37.8°C)
Residue	Liquid
Flash Point	305°F (COC)
pH	9.2 at 10% concentration

MASTER CHEMICAL CORPORATION
501 West Boundary Street • Perrysburg, Ohio 43551 • 419-874-7902

Master Chemical Corporation
501 West Boundary Street
Perrysburg, OH 43551
(419) 874-7902

TRIM® VX

DATA AND INFORMATION

GENERAL DESCRIPTION:

TRIM® VX is a CHEMICAL EMULSION CONCENTRATE containing EXTREME PRESSURE lubricant additives. This product is designed for EXTREME PRESSURE applications (tough alloys and severe operations) and is especially recommended for tough broaching operations. TRIM® VX is very efficient due to its combination of strong cooling and high lubricating characteristics.

ADVANTAGES:

- HIGH LUBRICITY—due to EXTREME PRESSURE ADDITIVES acting as ANTI-WELD AGENTS.
- WILL NOT cause CORROSION.
- WILL NOT cause STAINING.
- STABILITY. TRIM® VX is a stable chemical emulsion.
- FLUID RESIDUE prohibits sticky ways and slides—a most important consideration in the operation of AUTOMATIC and NUMERICALLY CONTROLLED MACHINES.

Available in 5 gallon and 54 gallon drums and tank wagon lots.

PHYSICAL PROPERTIES: (TYPICAL DATA)

Form	Fluid
Color	Dark Brown
Specific Gravity	.99 @ 75°F
Odor	Mild
Pour Point	−36°C
Flash Point	330°F
Fire Point	355°F
Viscosity	430 SUS @ 100°F
Residue	Fluid
pH	8.9 @ 10%
Refractometer Factor	1.67

TRIM® tap LIGHT

DATA AND INFORMATION

GENERAL DESCRIPTION:

TRIM® TAP LIGHT is a STRAIGHT OIL possessing strong FRICTION REDUCING properties. It is designed for use with METALS OF LOW MACHINABILITY and is specifically recommended at FULL STRENGTH for FLOODING applications involving BROACHING, SAWING, MILLING, TAPPING, THREADING, GEAR CUTTING and SHAVING, GUN DRILLING and REAMING.

ADVANTAGES:

- Promotes EXCELLENT FINISH.
- ASSISTS in parts SIZE CONTROL.
- Delivers EXCEPTIONALLY LONG TOOL LIFE.
- REDUCES RUBBING of the TOOL FLANK.
- Recommended especially for use on TAPS SMALLER THAN ¼ INCH DIAMETER.

Available in 5 gallon, 54 gallon Drums and Tank Wagon Lots.

PHYSICAL PROPERTIES: (TYPICAL DATA)

Form . . . . . Fluid
Color . . . . . Amber
Specific Gravity 1.08
Odor . . . . . Mild

Flash Point . . . . . (Cleveland Open Cup) 277°F (136.1°C)
Viscosity . . . . 64.6 SSU @ 210°F (98.9°C)
Fire Point . . . . . 320°F (160°C)

MASTER CHEMICAL CORPORATION
501 West Boundary Street • Perrysburg, Ohio 43551 • 419-874-7902

Master Chemical Corporation
501 West Boundary Street
Perrysburg, OH 43551
(419) 874-7902

TRIM® tap HEAVY

DATA AND INFORMATION

GENERAL DESCRIPTION:

TRIM® TAP HEAVY is a STRAIGHT OIL possessing strong FRICTION REDUCING properties. It is designed as a TOUGH OPERATION product for use with METALS OF LOW MACHINABILITY. It is recommended for use at FULL STRENGTH for BRUSH ON APPLICATIONS required in REAMING, TAPPING AND THREADING.

ADVANTAGES:

- Promotes EXCELLENT FINISH.
- ASSISTS in parts SIZE CONTROL.
- Delivers EXCEPTIONALLY LONG TOOL LIFE.
- REDUCES RUBBING of the TOOL FLANK.
- Recommended especially for use on TAPS ¼ INCH DIAMETER AND LARGER.

Available in 5 gallon and 54 gallon Drums and Tank Wagon Lots.

PHYSICAL PROPERTIES: (TYPICAL DATA)

Form	Fluid	**Flash Point**	None
Color	Light amber	**Viscosity**	186.4 SSU @ 210°F (98.9°C)
Specific Gravity	1.16	**Fire Point**	None
Odor	Mild		

MASTER CHEMICAL CORPORATION
501 West Boundary Street • Perrysburg, Ohio 43551 • 419-874-7902

MJM Laboratories Incorporated
4524 Parkway Commerce Boulevard
Orlando, FL 32808
(305) 295-4038

MJM Laboratories INC

4524 PARKWAY COMMERCE BOULEVARD • ORLANDO, FLORIDA 32808
(305) 295-4038

PERFORMANCE PLUS LUBRICATING CUTTING OIL

Performance Plus Lubricating Cutting Oil is unique as a wide-application petroleum-based cutting oil, unusually resistant to high pressures and temperatures, and relatively free of smoke or odor in the most difficult applications.

This superior lubricating product is suitable for use in manual and automatic machining operations: lathes, drill presses, grinders, honing, die stamping, mold milling; it is effective with both ferrous and non-ferrous metals as well as with plastics. With this versatility, it reduces the need to stock other oils, simplifying inventory and supply problems in the shop.

Documented usage at major plants have indicated the following results:

- Life of high speed and carbon bits was dramatically lengthened.
- The product worked effectively with stainless steel, cold roll, copper, nickel, brass, aluminum and plastics.
- Build-up was minimized with both hard and soft metals.
- Machine bearing failures were greatly reduced.
- Burring of stainless steel was eliminated.
- Enabled increased RPM rates and feed rates and drastically reduced the required number of tool changes.
- Scrap due to finish or size variations was eliminated.
- Machines were kept cleaner and unpleasant odors were eliminated. The product does not become rancid with repeated use as do some oils.

Performance Plus Lubricating Cutting Oil increases profits with superior results and enhanced productivity. It is available in five gallon containers, 55 gallon drums, and by bulk carrier.

Mobil Oil Corporation
3225 Gallows Road
Fairfax, VA 22037
(703) 849-3000

Mobil Product Data Sheet

Mobilmet® Upsilon, Omicron and Nu

Mineral Oil Type Nonstaining Cutting Oils

In many operations, staining of the machined parts by the cutting oil cannot be allowed. This is particularly true when copper or its alloys (brass, bronze) are being machined. Many of these alloys are difficult to machine, however, and cutting oils that provide high lubricity and good control of the builtup edge are necessary if machining efficiency is to be maintained. To meet the requirements of such operations, Mobil has developed three new nonstaining cutting oils: Mobilmet Upsilon, Mobilmet Omicron and Mobilmet Nu. Incorporating newly developed synthesized additive technology, these mineral oil type nonstaining cutting oils contain a balanced selection of additives that provide a synergistic effect that maximizes performance. They meet the antiweld and lubricity requirements of the most severe machining operations on the most difficult to machine of these alloys, while providing complete freedom from staining. Mobilmet Upsilon, Mobilmet Omicron and Mobilmet Nu oils do not stain either ferrous or nonferrous metals during cutting operations and likewise will not stain after machining due to hydrolysis of additive residues. The nonstaining nature of these oils also permits their use where complementary oils are required for the splash, hydraulic or spindle lubrication systems of metalworking machines, or where bronze or white metal is used in machine components such as gibs or ways that are exposed to the cutting fluid. They will provide good wear protection for such components, will not corrode the bronze or white metal parts included in the system, and will protect ferrous metals against rust. Leakage of these oils into the more active cutting oils will not seriously affect their cutting efficiency. In machines where leakage of the cutting oil into the lubricant reservoir or hydraulic system occurs, Mobilmet Upsilon, Mobilmet Omicron and Mobilmet Nu oils are ideally suited for use as cutting oils because they perform satisfactorily as lubricants.

Mobilmet Upsilon oil is a light colored, low odor cutting oil. It is intended primarily for high speed deep hole drilling and boring of all nonferrous metals, and is also recommended for the general machining of free machining alloys, magnesium and free machining brass, bronze and phosphor bronze as well as high speed, light feed machining of screw stock steels. In addition, it can be used as a lubricating oil for certain high speed spindles.

Mobilmet Omicron oil is formulated primarily as a dual-or tri-purpose cutting oil for automatic screw machine shops where it may be used as a cutting oil, as a lubricating oil, and in moderate duty hydraulic systems and airline oilers. It offers high lubricity and

Characteristic	Mobilmet Upsilon	Mobilmet Omicron	Mobilmet Nu
Gravity, API	35.1	31.8	28.5
Color	4.5	5.0	8.0
Flash, °F	350	405	405
Pour, °F	20	0	10
Viscosity, SUS at 100°F	60	135	195
Total Sulfur, %	0.35	0.5	2.0
Active Sulfur, %	Nil	Nil	Nil
Chlorine, %	Nil	Nil	0.9
Cu Strip Corrosion (3 hrs. at 212°F)	1A	1B	1B
Phosphorous, %	Nil	Nil	Nil
Lubricity (% Fat)	2.0	2.0	8.0
Antimist	No	Yes	Yes

Mobil Oil Corporation
3225 Gallows Road
Fairfax, VA 22037
(703) 849-3000

[redacted]

maximum cooling on hard, clean cutting metals, making it especially suited to high surface speed machining of screw stocks, aluminum and brass. It can also be used for the general machining of nickel-tin-bronze alloys and the free machining steels, as well as for the severe cutting operations of difficult to machine nonferrous alloys including silicon-copper, silicon-bronze and copper-nickel.

The multipurpose uses minimize the problem of contamination caused by leakage from the lubricating system into cutting fluid reservoirs. Its outstanding lubricating qualities provide cool running spindles and long clutch life to maintain productivity.

Mobilmet Omicron provides superior machining and lubricating performance plus the added benefit of the virtual elimination of oil mist formation in high speed machine tools. The oil resists the dispersing effect of the mist generated at the tool face so that particles coalesce into large drops which drop back into the reservoir or catch pans.

This relatively light colored, transparent, low odor oil has excellent oxidation stability at higher temperature operations and outstanding lubricating properties under heavy loads and high speeds. It has excellent resistance to foaming, even with excessive splashing, so that superior performance is provided in the machine tool lubrication system. With a relatively low pour point and a high VI, Mobilmet Omicron oil is not difficult to dispense in cold temperatures, and it will provide adequate film strength in hot running machine tool bearings under load.

Mobilmet Nu oil is a heavy duty, medium viscosity, dual- or tri-purpose cutting oil. It offers superior cutting performance on free machining stocks as well as tough draggy metals even in severe machining operations such as threading, tapping and broaching with the additional advantages of antimist and nonstaining of nonferrous metals. When nonstaining cutting oils are desired, Mobilmet Nu is recommended for all types of grinding operations including heavy duty grinding of stainless steels and high alloy steels. The use of Mobilmet Nu in grinding applications provides the benefits of reduced wheel loading, extended wheel life, improved finishes and lower operating temperatures.

Mobilmet Nu provides excellent lubrication protection of machine tool system components under conditions of heavy loads and high speeds. The multipurpose capability of Mobilmet Nu eliminates contamination problems caused by leakage between the lubricating oil and cutting fluid reservoirs.

TYPICAL CHARACTERISTICS

The physical and chemical characteristics of Mobilmet Upsilon, Mobilmet Omicron and Mobilmet Nu oils are shown in the data sheet table. The values shown are typical characteristics which may vary slightly.

HEALTH AND SAFETY

Based on available toxicology information, when properly handled and used, these products have little or no adverse health effect. No special precautions are suggested beyond attention to good personal hygiene, including avoiding prolonged, repeated skin contact. A detailed Material Safety Data Bulletin discussing these products is available upon request from your local Commercial Division office.

ADVANTAGES

When used as recommended, Mobilmet Upsilon, Mobilmet Omicron and Mobilmet Nu mineral oil type nonstaining cutting oils provide the following outstanding advantages and benefits:

- Better finish
- Closer tolerances
- Increased tool life between regrinds
- Reduced wheel loading and extended wheel life (Mobilmet Nu)
- Fewer cutting oils in the shop
- Will not stain copper, brass or other nonferrous metals
- Less smoke
- Reduced mist (Mobilmet Omicron, Mobilmet Nu)
- Faster machining at higher speeds and/or feeds
- Serves as cutting oil, lubricating oil and hydraulic oil (Mobilmet Omicron and Mobilmet Nu)

Mobil Oil Corporation
3225 Gallows Road
Fairfax, VA 22037
(703) 849-3000

Mobil Product Data Sheet

Mobilmet® Alpha, Sigma, Gamma, Omega

Mineral Oil Type Active Cutting Oils

Mobil offers a series of four mineral oil type active cutting oils: Mobilmet Alpha, Mobilmet Sigma, Mobilmet Gamma, and Mobilmet Omega which incorporate newly developed synthesized additive technology. Oils of this type contain the active antiweld properties which are required for cutting difficult to machine metals. Each is recommended for use with a wide range of metals and operations beyond the capacity of conventional products. Backing up the versatility of each is a balanced selection of additives designed to provide a synergistic effect that maximizes performance. This improved performance is highly effective over the wide ranges of pressures and temperatures produced when machining difficult metals under difficult conditions. These oils are recommended for use with metals varying from tough, draggy alloys to hard, brittle ones. The improved, chemically-synthesized, active additive combination serves an additional purpose by controlling buildup on the tool tip, the result of which is improved surface finish. These Mobilmet products can cause surface discoloration of copper alloys due to their high activity. They are also more transparent on the workpiece, allowing good visual inspection of the machined parts.

PRODUCT DESCRIPTION

Mobilmet Alpha oil is the lowest viscosity product in the series. It is recommended for deep hole or gun drilling where its low viscosity enables it to flush swarf and chips away from the hole and permits fines to settle quickly. It can be used on a variety of metals including stainless steels, high nickel steels and heat-resistant alloys, and in most machining operations.

Mobilmet Sigma oil is a medium viscosity cutting oil that will provide outstanding performance in general-purpose machine operations. It is recommended for nearly all machining operations on ferrous metals, except heavy duty operations where tolerance and finish are critical or for operations such as deep hole drilling where a low viscosity oil is needed. It is suitable for all types of steels under normal cutting duty, and where staining is not critical, it may be used on brass, bronze or other copper-bearing alloys. It is a transparent, low odor cutting oil that gives outstanding results when compared with other cutting oils in the same viscosity range. Its very high antiweld and lubricity compounding make it useful on tough, draggy metals. By incorporating an effective antimist agent, more pleasant working conditions are assured.

Characteristic	Mobilmet Alpha	Mobilmet Sigma	Mobilmet Gamma	Mobilmet Omega
Gravity, API	31.4	30.4	29.8	26.4
Color	8	8	8	8+
Flash, °F	395	405	435	430
Pour, °F	25	15	25	25
Viscosity, SUS at 100 °F	100	155	175	220
Total Sulfur, %	1.2	1.3	1.3	2.4
Active Sulfur, %	0.6	0.6	0.65	0.85
Chlorine, %	Nil	Nil	0.45	1.6
Cu Strip Corrosion (3 hrs. at 212°F)	4C	4C	4C	4C
Phosphorus, %	Nil	Nil	Nil	Nil
Lubricity (% Fat)	5.0	3.0	4.0	12.0
Antimist	No	Yes	Yes	No

Mobil Oil Corporation
3225 Gallows Road
Fairfax, VA 22037
(703) 849-3000

Mobilmet Gamma oil with high antiweld and lubricity compounding, is recommended for more severe operations, such as heavy duty threading, tapping and broaching on difficult to machine metals. Unlike many conventional heavy duty cutting oils, however, Mobilmet Gamma will also operate efficiently on less difficult materials, and where staining is not critical, may be used on brass, bronze or other copper-bearing alloys. This antimist type oil is a transparent, low odor product whose use provides outstanding performance when compared to other so-called "heavy duty" cutting oils.

Mobilmet Omega oil is the extremely heavy duty cutting oil recommended for use whenever hard to machine metals must be broached, tapped or threaded. By comparison to other products normally used in this type of service, this oil is more pleasing to operators.

TYPICAL CHARACTERISTICS

The typical physical and chemical characteristics of Mobilmet Alpha, Mobilmet Sigma, Mobilmet Gamma and Mobilmet Omega mineral oil type active cutting oils are shown in the data sheet table. The values shown are typical characteristics which may vary slightly.

HEALTH AND SAFETY

Based on available toxicology information,when properly handled and used, these products have little or no adverse health effect. No special precautions are suggested beyond attention to good personal hygiene, including avoiding prolonged, repeated skin contact. A detailed Material Safety Data Bulletin discussing these products is available upon request from your local Commercial Division office.

ADVANTAGES

When used as recommended, the Mobilmet Alpha, Sigma, Gamma and Omega series oils will provide the following outstanding advantages and benefits:

- Better finish
- Closer tolerances
- Increased tool life between regrinds
- Fewer cutting oils in the shop
- Reduced chip welding in heavy duty operations
- Reduced misting (Mobilmet Sigma, Mobilmet Gamma)
- Faster machining at higher speeds and/or feeds

Mobil Oil Corporation
3225 Gallows Road
Fairfax, VA 22037
(703) 849-3000

Mobil Product Data Sheet

Mobilmet® 180
Mobilmet® 235

Mobilmet 180 mineral oil based coolant and Mobilmet 235 semi-synthetic coolant represent state-of-the-art coolant formulation technology. Both products are designed to provide excellent performance in all types of machining and grinding operations, with all types of tooling and metals, except magnesium alloys.

PRODUCT DESCRIPTION

Mobilmet 235 semi-synthetic coolant combines carefully balanced additives with selected solvent refined mineral oils to provide a high level of lubricity for machine components and the machining operation. Mobilmet 180 mineral oil based coolant has many of the same, excellent properties as Mobilmet 235. In the neat form, Mobilmet 180 and Mobilmet 235 have a transparent appearance. When added to water and mixed, they produce a stable milky dispersion in either soft or hard water.

The unique combination of additives in Mobilmet 180 and 235 provides outstanding performance benefits, including extended batch life through effective control of bacteria, fungi and mold. They resist the formation of hard and abrasive deposits on machine components and finished parts, reducing component wear. They do not form the tacky deposits left by many semi-synthetic coolants eliminating the need for subsequent cleaning of parts and equipment. After machining, a thin film of coolant is left on the parts providing good rust protection. Mobilmet 180 and 235 will not stain most metals. The excellent stability of these formulations insures that these properties will be retained over extended periods of batch life.

TYPICAL CHARACTERISTICS

The physical characteristics are shown in the data sheet table. The values shown are typical properties which may vary slightly.

APPLICATION

Mobilmet 180 mineral oil based coolant provides excellent machinability in medium-duty applications. Mobilmet 235 semi-synthetic coolant is a

Characteristic	Mobilmet 180	Mobilmet 235
Type of Product	Mineral Oil Based Coolant	Semi-Synthetic Coolant
Gravity, API	26	16
Pour Point, °F	+20	+25
Color, ASTM	3	8
Viscosity @ 100°F, SUS	200	450
Flash Point, COC, °F	350	350
Sulfur, Wt. %	0.3	0.9
Phosphorus, Wt. %	0.15	0.15
Chlorine, Wt. %	NIL	7.5
pH (6% emulsion)	9.4	9.5
Copper Corrosion (6% emulsion)	1B	1B
Foam Test	Pass	Pass

Mobil Oil Corporation
3225 Gallows Road
Fairfax, VA 22037
(703) 849-3000

[redacted]

heavy-duty product which is recommended for any application where coolants are preferred. The suggested concentration for both products ranges from four percent (1:25) to six percent (1:16) in water, to optimize machining performance and batch life. Lower concentrations can be used, but batch life may be reduced. At these suggested concentrations, Mobilmet 180 and Mobilmet 235 have a mild, characteristic odor, and will remain free of odors caused by bacteria and fungi during operation and periods of shutdown. Mobilmet 180 and Mobilmet 235 are compatible with high speed steel (HSS) and carbide tooling, and grinding wheels designed for water systems. Compared to other semi-synthetic coolants, Mobilmet 235 provides superior tool life with either HSS or carbide tooling.

HEALTH AND SAFETY

When Mobilmet 180 and 235 are diluted with water (the dilute forms) and properly handled and used, there is little potential adverse health effect. No special precautions are suggested beyond attention to good personal hygiene, which includes avoiding prolonged repeated skin contact. However, when handled in the form as manufactured (concentrate, before dilution) they are potential skin and eye irritants. Therefore contact with the concentrated form should be avoided as much as possible. In view of these facts, it is recommended that the following precautions be observed in handling the concentrates:

- Wear impervious gloves and aprons.
- Wear chemical type goggles with face shields.
- In case of contact with skin, immediately wash affected area with soap and warm water.
- In case of contact with eyes, flush with copious amounts of water.
- Launder contaminated clothing before reuse and discard shoes if product has penetrated to the inside surface.

For further details, a Material Safety Data Bulletin on both products is available upon request from your local Commercial Division office.

ADVANTAGES

When used as recommended, Mobilmet 180 and Mobilmet 235 coolants offer the following advantages and benefits:

Long batch life by resisting formation of bacteria, fungi and molds.

Increased machine component life due to reduced abrasive deposits.

Excellent operating cleanliness in equipment and finished parts by eliminating tacky deposits.

Superior tool life with Mobilmet 235, and good tool life with Mobilmet 180.

Excellent lubricity to protect machine slides, ways and other components.

Good antirust protection of ferrous metals.

Mobil Oil Corporation TECHNICAL PUBLICATIONS AND TRAINING
3225 GALLOWS ROAD, FAIRFAX, VIRGINIA 22037

PDS6684055
(10-15-84)

Mullen Circle Brand, Incorporated
3514 West Touhy Avenue
Chicago, IL 60645
(312) 676-1880

PRODUCT INFORMATION

FROM . . . MULLEN CIRCLE BRAND, INC. • 3514 West Touhy Avenue • Chicago, IL 60645 • 312/676-1880

SHURCOOL SYNTHETIC COOLANTS

The SHURCOOL SYNTHETIC COOLANTS were developed to provide a broad product line, meeting a variety of metal working requirements. Our seventy years of experience in the field coupled with extensive research and development have resulted in superior products free from chemcial pollutants.

Before a formulation is as adopted into the SHURCOOL line, it has to meet such tests as:

- Lubricity
- Wetting
- Reserve alkalinity
- Temperature stability
- Anti-rust
- Anti-foam
- Biodegradability

In addition to complying with pollution regulation, the SHURCOOL products are also free from:

- Phenolics
- Free Caustic
- Heavy Metals
- Sodium Nitrite
- Para-tertiary Butal Benzoic Acid (PTBBA)
- PCBs

SHURCOOL products available:

Shurcool #4	General Grinding	40-60:1	Ferrous/ Nonferrous
Shurcool #12	High performance grinding	20-40:1	Ferrous/ Nonferrous
	Light machining	10-20:1	
Shurcool #36	General Machining	30-40:1	Ferrous/ Nonferrous
	Heavy Machining	10-20:1	
Shurcool #39	General Grinding	50:1	Ferrous/ Nonferrous
	High performance grinding	30-40:1	
	Light Machining	20-30:1	
	Heavy Machining	10-20:1	

Mullen Circle Brand, Incorporated
3514 West Touhy Avenue
Chicago, IL 60645
(312) 676-1880

PRODUCT INFORMATION

FROM . . . MULLEN CIRCLE BRAND, INC. • 3514 West Touhy Avenue • Chicago, IL 60645 • 312/676-1880

SHURCOOL #36

SHURCOOL #36 is a heavy duty water-based cutting fluid concentrate. Properly diluted, SHURCOOL #36 is suitable for all machining operations on both ferrous and nonferrous metals. For the first time outstanding lubricity has been combined with the superior cooling power of a chemical coolant. Tool wear is reduced in three ways:

1. A unique combination of extreme pressure additives prevent welding at the tool point. This will immediately reduce heat generation and improve surface finish.
2. Chemical lubricity permits the chips to slide easily over the tool at the chip-tool interface. This reduces friction and tool cratering.
3. Reduced surface tension permits wettability and cooling that is superior to water. This increases the rate of heat removal, allowing the tool and workpiece to remain cooler. Better tool life is easily achieved.

SHURCOOL #36 is completely biodegradable. It contains the latest development in bactericides to assure long coolant life and productivity.

SHURCOOL #36 is inherently low foaming. This precludes the use of silicone defoamers which can cause fish-eyeing and cratering on steel surfaces. Other defoamers which leave a heavy cream on the fluid surface are also unnecessary. SHURCOOL #36 is a technologically advanced product for the modern machine shop.

TYPICAL PROPERTIES

Color	Red
Weight/Gallon	8.5 lbs.
Ph 40:1 Solution	8.0
concentrate	8.5
Flash Point	None
Biodegradability	Complete
Soldium Nitrite	None
Hazardous Ingredients	None

TREAT LEVELS:

General Machining	20-40:1
Severe Machining	10-20:1

The information given herein is believed to be reliable, but no guarantee is made or liability assumed.

Mullen Circle Brand, Incorporated
3514 West Touhy Avenue
Chicago, IL 60645
(312) 676-1880

PRODUCT INFORMATION

FROM . . . MULLEN CIRCLE BRAND, INC. • 3514 West Touhy Avenue • Chicago, IL 60645 • 312/676-1880

MCB #1450-10 CUTTING OIL

This heavy duty oil is compounded for practically all ferrous machining operations. #1450-10 cutting oil contains both sulfur and chlorine, unique in that both are on the same molecule. This means the E.P (extreme pressure) qualities of this molecule is available at all times. The result is the oil will handle a multitude of jobs, whether it be a light skin pass, broaching, or gear hobbing. #1450-10 is so versatile it will cut down on your inventory of cutting oils, one cutting oil will handle a variety of jobs. This product is also available with an anti-mist additive.

Over the years, #1450-10 has been our biggest seller, proving its superiority in the marketplace.

TYPICAL PROPERTIES:

Viscosity @ 100°F SUS*	100
Flash Point°F	330 mins.
Pounds per Gallon	7.45 lbs.
Sulfur	yes
Chlorine	yes

*This product can be obtained in higher viscosities.

Mullen Circle Brand, Incorporated
3514 West Touhy Avenue
Chicago, IL 60645
(312) 676-1880

PRODUCT INFORMATION

FROM . . . MULLEN CIRCLE BRAND, INC. • 3514 West Touhy Avenue • Chicago, IL 60645 • 312/676-1880

#1592 SOLUBLE OIL

Description:

#1592 SOLUBLE OIL eliminates the need for many soluble oils by allowing one product to be used across the entire machining spectrum. Suitable for ferrous and non-ferrous metals, this product can be used from heavy duty machining to high dilution grinding. The oil concentrate mixes readily with hard or soft water to form a milky white opaque emulsion.

Application:

As a heavy duty soluble, #1592 can be used in such applications as broaching, tapping, milling, turning, and drilling. The #1592 SOLUBLE OIL gives excellent part finish and tool life when used as a medium duty soluble. At high dilutions, the #1592 is unsurpassed as a grinding soluble.

Advantages:

-Rust inhibition in excess of 75:1 on cast iron chips
-Excellent rancidity control
-Open and free cutting wheels
-Rapid chip settling
-Non-irritating
-No copper or brass stain
-Good odor control
-Superb cleaning of grinding surface
-Long term emulsion stability

Treat Levels:

Severe Machining	10-20:1
General Machining	20-40:1
Grinding	40-60:1

Optional Additive Packages:

#1592-R	Extra rust inhibitor
#1592-B	A Biocide package

ORB Industries Incorporated
2 Race Street/P.O. Box 1067
Upland, PA 19015
(215) 874-2537

ORB aerosol bulletin

ORB #515 CUTTING AND TAPPING OIL

What It Is

ORB #515 Cutting and Tapping Oil is a premium cutting and tapping compound specially formulated for machine shop use and all types of metal cutting.

What It Does

Gives smoother cuts, cleaner threads, and prolongs tool life. Delivers cutting oil direct to cutting surface and will not run or break down.

Uses

Use for boring, reaming, threading, tapping, milling, broaching, lathe cut, sawing, etc. Use on steel, stainless steel, aluminum, brass and other metals. Apply to tools as required.

Advantages

For pinpoint application use the extension nozzle furnished. No waste - no spillage - means real savings - smoother finish. Contains important ingredients which cannot be added to bulk cutting oils. Saves time because it can be kept at your finger tips for use whenever needed.

Price

See enclosed price list (subject to quantity discounts).

Packing

12 - 14 oz. cans per case --- Shipping weight 15 lbs./case.

Page-Wilson Corporation/Allison-Campbell Division
875 Bridgeport Avenue/P.O. Box 353
Shelton, CT 06484
(203) 929-5301

PAGE–WILSON CORPORATION

ALLISON-CAMPBELL DIVISION

Issued 3-1-83 04-310

Campbellene®

Cool-Blue Coolant Concentrate with Odormask

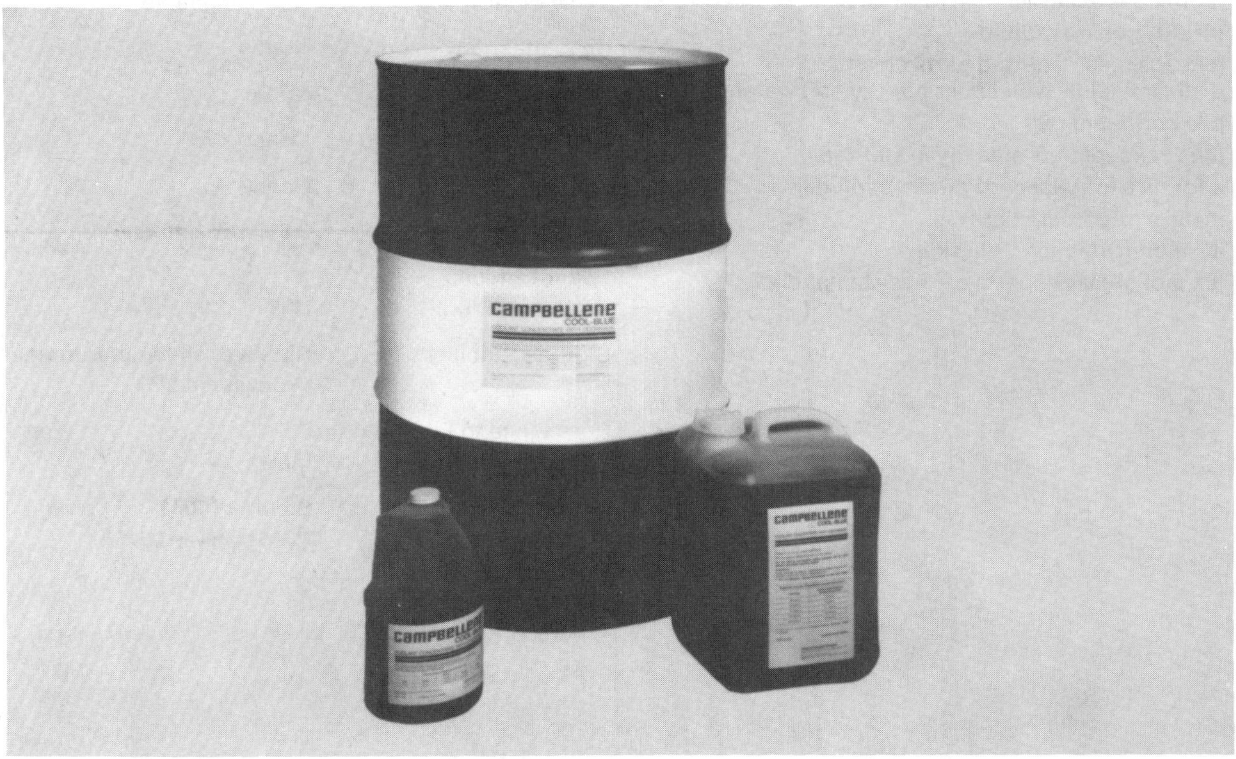

Application

Campbellene Cool-Blue Coolant Concentrate is recommended for use with all wet abrasive cutting and most wet grinding of ferrous metals and titanium. It is not recommended for cutting and grinding operations involving aluminum.

Description

Because *Campbellene* Coolant Concentrate is chemical in nature, it will not support bacterial growth or turn rancid. *Campbellene* Coolant Concentrate contains no oily or greasy substances, providing even more safety to your shop personnel and work areas. Offensive odors are minimized by the inclusion of an exclusive scenting agent — Odormask.

Campbellen Coolant Concenrate retards rust and minimizes the build-up of hard deposits thereby keeping the machine and parts clean and assuring a faster, straighter cut. This quality also promotes longer wheel life and better quality cuts. Because the recirculation of metal chips is reduced to a minimum an extremely rapid chip settling can be maintained.

Non-foaming, *Campbellene* Coolant Concentrate allows close contact of coolant with the cutting wheel and work for maximum cooling. Cooler cutting will add to the life of the wheel.

Campbellene coolant is available in convenient units, including a 4 pack of one-gallon containers, five-gallon containers (with grip handles) and 55-gallon drums. Because it is used in extreme dilutions, this coolant solution is exceptionally economical.

Campbellene Cool-Blue Coolant Concentrate measures up as the finest coolant for the price.

Page-Wilson Corporation/Allison-Campbell Division
875 Bridgeport Avenue/P.O. Box 353
Shelton, CT 06484
(203) 929-5301

Campbellene® Cool-Blue Coolant Concentrate with Odormask

Advantages

Campbellene Cool-Blue Coolant Concentrate is the best choice because —

- it promotes keen cutting and grinding action
- it promotes exceptional accuracy with high production
- it keeps wheels open with less wheel dressing
- it provides excellent rust protection
- it settles chips rapidly
- it keeps machines and parts clean
- it mixes rapidly with either hard or soft waters
- it is non-foaming
- it has exceptional stability in storage
- it forms a long-lasting, "sweet" solution
- it stays free from odors
- it is non-irritating to the skin
- it is not greasy — parts are easy to handle

Technical Data

Campbellene Coolant is a concentrated water solution of inorganic and organic materials selected for their rust inhibiting and cooling properties.

Weight per Gallon	10.7 lbs.
Viscosity	Watery fluid
Flash Point	Non-flammable
Sulphates	None
Sulphur	None
Chlorine	None
Fat	None
pH (1% solution)	9.6 approximately
Foam Tendency (1% solution)	Nil
Storage Stability	Stable at all temperatures down to 20°F
Resistance to Bacterial Decomposition	Inert
Mixing Ratio (Water to Coolant)	Steel — 150/1 Cast Iron — 100/1

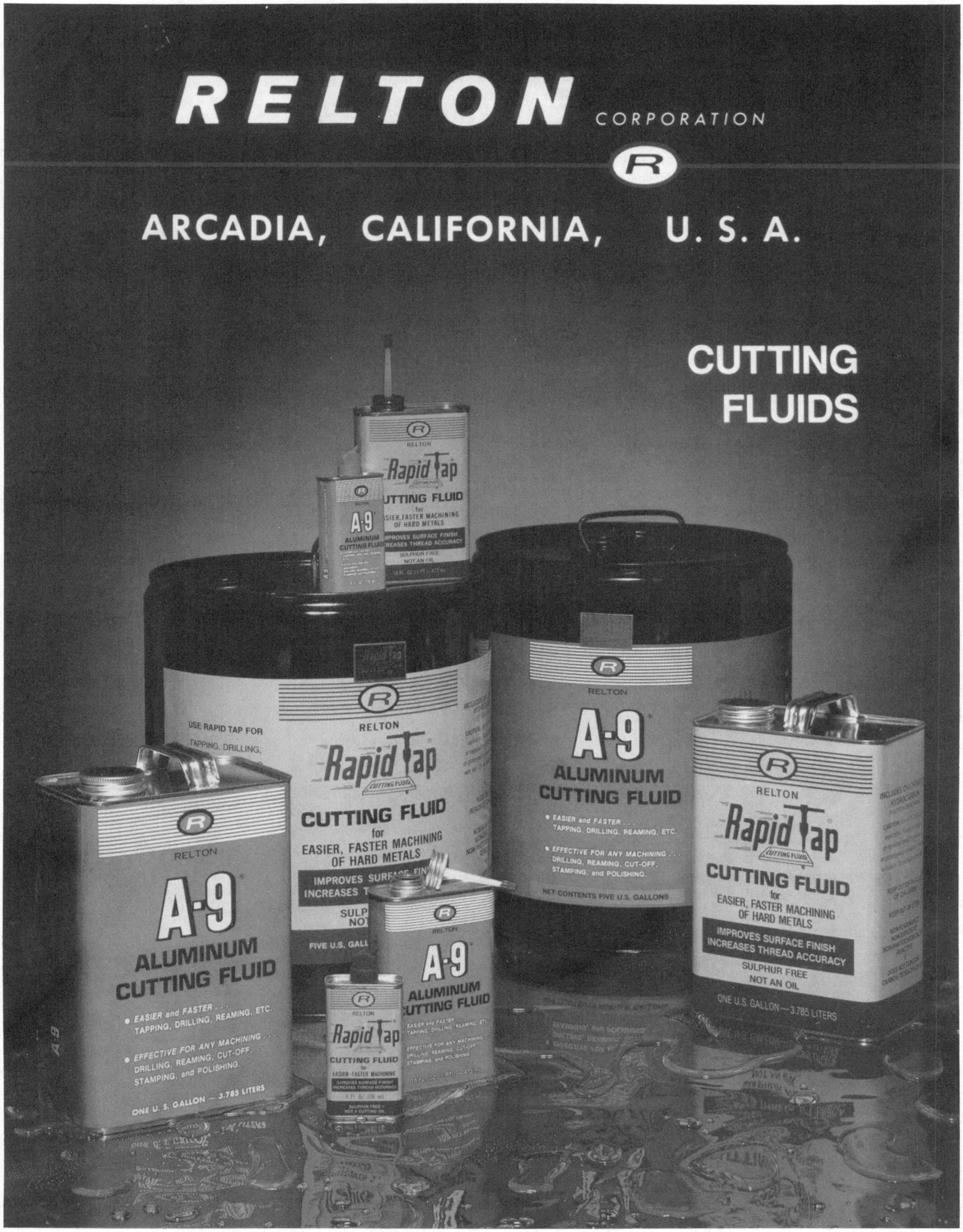
RELTON CORPORATION
R
ARCADIA, CALIFORNIA, U. S. A.
CUTTING FLUIDS
RELTON
Rapid Tap
CUTTING FLUID
for
EASIER, FASTER MACHINING OF HARD METALS
IMPROVES SURFACE FINISH
INCREASES THREAD ACCURACY
SULPHUR FREE
NOT AN OIL
ONE U.S. GALLON—3.785 LITERS
RELTON
A-9
ALUMINUM CUTTING FLUID
EASIER and FASTER . . . TAPPING, DRILLING, REAMING, ETC.
EFFECTIVE FOR ANY MACHINING . . . DRILLING, REAMING, CUT-OFF, STAMPING, and POLISHING.
ONE U. S. GALLON — 3.785 LITERS
NET CONTENTS FIVE U.S. GALLONS
USE RAPID TAP FOR
TAPPING, DRILLING,

Relton Corporation
317 Rolyn Place
Arcadia, CA 91006
(213) 681-2551

Rapid Tap HARD-METAL CUTTING FLUID

Just a drop or two —

- Amazingly effective on tough metals and alloys, including BERYLLIUM COPPER, BRASS, BRONZE, CAST IRON, MAGNESIUM, MOLYBDENUM, STAINLESS STEEL, STELLITE, TITANIUM, and others (not for use on plastics).
- Just a drop or two makes tapping, reaming, drilling, threading, and all other precision machining operations easier, faster, and more accurate. Insures close tolerances and reduces machining costs.
- Breaks down surface tensions and adhesions resulting from tool operations. Frees the tool, reducing tool wear, part rejections, and breakage. Keeps tools sharper longer.
- Improves surface finish by eliminating scuff marks and gouging. Leaves a clean surface.
- Passes Lawrence Electronics Co. Hydrogen Effusion Test.

Non-flammable, non-explosive, contains no carbon tetrachloride or sulphur.

CONTAINER SIZE	NUMBER PER CASE	LIST PRICE			CASE OR DRUM SHIPPING WEIGHT
		CAN	CASE	DRUM	
4 OZ. CAN	24	$ 1.74	$ 41.74		12 LBS.
1 PINT CAN	12	5.38	64.57		20 LBS.
1 GAL. CAN	4	32.28	129.13		48 LBS.
5 GAL. CAN	1	157.83	157.83		58 LBS.
*30 GAL. DRUM				$ 639.13	370 LBS.
*55 GAL. DRUM				1143.48	655 LBS.

*Available on Special Order

A·9 ALUMINUM CUTTING FLUID

Saves Time — Tools — Money

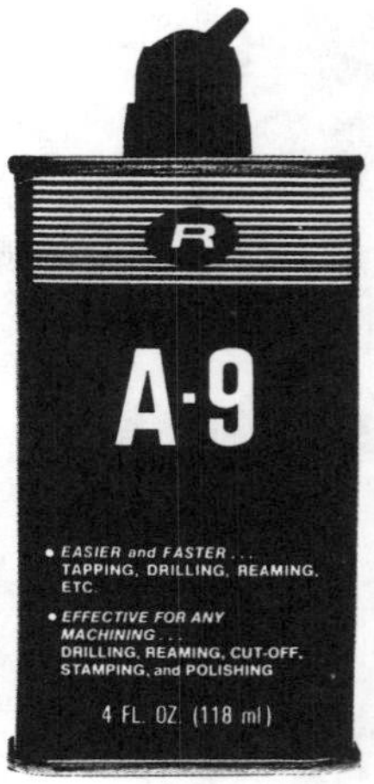

- Does everything for all types of aluminum and non-ferrous metals that RELTON RAPID TAP does for tough metals and alloys.
- Contains a scientific blend of chemicals that attach themselves to aluminum surfaces, forming an effective extreme-pressure boundary lubricant between tool and work surface.
- Extends tool life. Improves surface finish by eliminating scuff marks and gouging.
- Leaves a surface which requires only ordinary cleaning to prepare for subsequent processing.

Non-flammable, non-explosive, contains no carbon tetrachloride or sulphur, or chlorine.

CONTAINER SIZE	NUMBER PER CASE	LIST PRICE			CASE OR DRUM SHIPPING WEIGHT
		CAN	CASE	DRUM	
4 OZ. CAN	24	$ 1.74	$ 41.74		8 LBS.
1 PINT CAN	12	5.38	64.57		15 LBS.
1 GAL. CAN	4	32.28	129.13		35 LBS.
5 GAL. CAN	1	157.83	157.83		43 LBS.
*30 GAL. DRUM				$ 639.13	255 LBS.
*55 GAL. DRUM				1143.48	468 LBS.

*Available on Special Order

RELTON CORPORATION (213) 681-2551

317 ROLYN PLACE, ARCADIA, CALIFORNIA 91006

Manufacturers of . . .
CONCRETE-TERMITE ROTARY MASONRY BITS • CONCRETE-TERMITE HAMMER BITS
TIP-TOP ROTARY MASONRY BITS • RAM-TIP MASONRY BITS
RELTON CARBIDE-TIPPED HOLE SAWS • RELTON CARBIDE-TIPPED PORCELAIN CUTTERS
R4X MASONRY BITS
RAPID TAP HARD-METAL CUTTING FLUID • A-9 ALUMINUM CUTTING FLUID

Renite Company
2500 East Fifth Avenue/P.O. Box 19235
Columbus, OH 43219
(614) 253-5509

RENITE CUTTING AND GRINDING FLUIDS

Renite Company, long known for its lubricants for high temperature metalworking (i.e. forging, extrusion) now offers lubricants for lower temperature metalworking applications, such as cutting and grinding.

The lubricants are of three basic types, described as follows:

RENITE C-SERIES: Oil-based, with viscosity and additives to suit the application. Available with conventional fatty/chlorine/sulfur additives and, particularly where tooling wear is a special problem, with the more recently developed anti-wear/extreme pressure additives, such as organic molybdenum compounds.

RENITE D-SERIES: Soluble oils and oil-in-water emulsions. These are in general water-dilutable variations on the Renite C-Series, some containing suspended solid lubricants.

RENITE Y-SERIES: Synthetic fluid based. These are generally water-soluble, with both the base fluid and the extreme pressure/anti-wear additives having the "inverse solubility" characteristic--tending to separate out of solution at higher temperatures. These may often be diluted to a greater extent than soluble oils, the heat of friction of the cutting or grinding process serving to separate out the active ingredients at the point where needed. In some cases, the metalworking fluid can serve also as a hydraulic fluid.

Consultation is provided on choice of lubricant for the application in question and also on matters such as preservatives, reclaiming and recycling. Renite Company handles special formulation requests, and welcomes small orders as well as large. Your inquiries are invited.

Rust-Lick, Incorporated
72 Morgan Avenue/P.O. Box 1244
Danbury, CT 06810
(203) 792-0052

TAPZOL

Tapping Fluid

Product Code: 20040
Mfgr. No. 698242

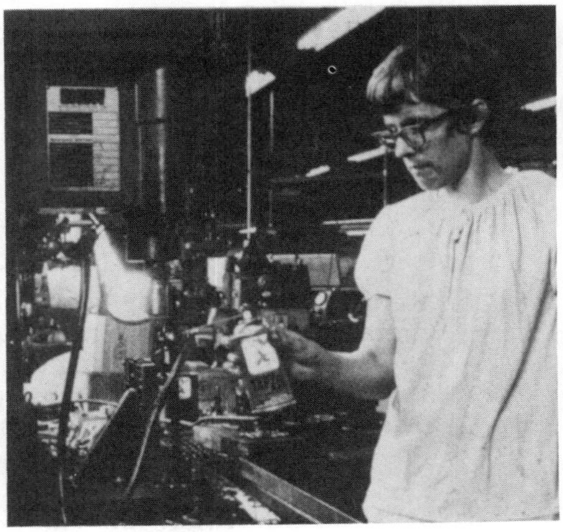

TAPZOL, tapping fluid, is formulated from chemical compounds having the highest anti-weld properties known to the metalworking industry. TAPZOL insures increased tool life and fine finish when working with conventional or hard to machine metal alloys, both ferrous and non-ferrous. In addition, TAPZOL contains specially processed fatty oils that provide the utmost lubrication to relieve tension, and prevent seizing and galling. TAPZOL is also highly recommended for use as an E.P. additive to straight oils to improve finish on tough machining operations. It is transparent and non-staining.

APPLICATIONS:

Tapping, drilling, reaming, threading and milling. Also as an additive to straight oils to improve tool life and provide better finish. Use as penetrant to free frozen threaded parts.

DIRECTIONS:

For tapping, drilling, reaming, threading and milling—use as received. As additive to straight oils—use one gallon of TAPZOL to every 10 gallons of straight oil in use.

AVAILABLE IN:

Pints Cans (12/16 oz. per case) (#21640)
One gallon cans (4 per case) (#20140)
5 gallon containers (#20540)
55 gallon drums (#25540)

ALL RUST-LICK PRODUCTS ARE SOLD THROUGH SELECTED INDUSTRIAL DISTRIBUTORS IN EVERY MAJOR METALWORKING AREA

Distributed by:

Manufactured by:

RUST-LICK INC

P.O. Box 1244
Danbury, CT 06810
Tele. 203-792-0052

303 N. Manchester Ave.
Anaheim, CA 92801
Tele. 714-535-6075

INDUSTRIAL DISTRIBUTORS

Rust-Lick, Incorporated
72 Morgan Avenue/P.O. Box 1244
Danbury, CT 06810
(203) 792-0052

CUTZOL 711

Bulletin No. 20010
Mfgr. No. 698242

Chlorinated Heavy Duty Cutting Oil

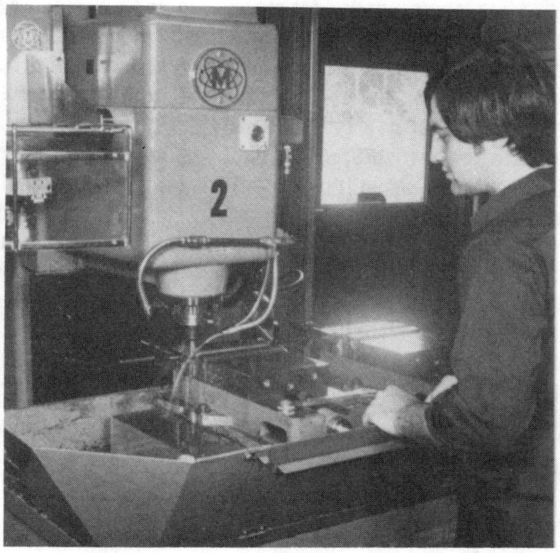

Cutzol 711 is a chlorinated, straight cutting oil recommended for the toughest machining operations on both ferrous and non-ferrous metals. Cutzol 711 does not contain active sulfur, will not stain any metals (including aluminum, brass, copper and beryllium copper) and is inhibited against chlorine stain on steel. Cutzol 711 is also ideal for use on electronic components where sulfur additives are prohibited on exotic metals. A pleasant, peppermint scented material, Cutzol 711 has a light brown, transparent color, a viscosity of 190 SSU @ 100°F and a chlorine content of 10.3%. Cutzol 711 promotes increased tool life, finer finishes and closer tolerances, and will provide excellent rust protection for both machined parts and machines. Cutzol 711 may also be used as a blanking or drawing oil, or as an additive to enhance the E.P. characteristics of the straight oil in use.

BENEFITS

Contains chlorine and proprietary E.P. additives—may be used in the toughest machining operations • Contains no active sulfur—will not stain any metal • Transparent—allows visibility of workpiece • Pleasant smelling—insures operator acceptance • Multipurpose—may be used as a cutting oil, blanking and drawing oil or as E.P. additive.

DIRECTIONS:

For use on all metals, especially high alloys, stainless steels and exotic metals. Use as received for tapping, threading, turning, milling, drilling and broaching. Use as received for blanking and drawing. As an E.P. additive add 1 part Cutzol 711 to 10 parts of straight oil in use.

AVAILABLE IN:

One gallon cans (4 per case) (#20110)
5 gallon pails (#20510)
55 gallon drums (#25510)

ALL RUST-LICK PRODUCTS ARE SOLD THROUGH SELECTED INDUSTRIAL DISTRIBUTORS IN EVERY MAJOR METALWORKING AREA

Distributed by:

Manufactured by:

RUST-LICK INC

P.O. Box 1244
Danbury, CT 06810
Tele. 203-792-0052

303 N. Manchester Ave.
Anaheim, CA 92801
Tele. 714-535-6075

INDUSTRIAL DISTRIBUTORS

CUTTING FLUIDS

Rust-Lick, Incorporated
72 Morgan Avenue/P.O. Box 1244
Danbury, CT 06810
(203) 792-0052

CUTZOL PB-10

Bulletin No. 40010
Mfgr. No. 698242

Heavy Duty Sulfur Chlorinated Soluble Oil

CUTZOL PB-10 is a heavy duty, sulfur-chlorinated soluble oil. PB-10 is recommended for machining and grinding operations on all ferrous metals, high alloys, stainless steels and exotics, especially those considered to have a poor rate of machinibility. Diluted with water, CUTZOL PB-10 forms a stable, light tan emulsion, that can be used to replace straight oils in most applications, resulting in increased tool life and improved finish. Unlike straight oils, PB-10 is both pleasant and safe to work with, eliminating fire hazards, excessive heat, smoke and oil mists. CUTZOL PB-10 may also be diluted with a blending oil to form a heavy duty cutting oil for heavy duty, slow speed and heavy feed work. In addition, PB-10 may be used as a brush-on oil that will emulsify with, and improve the E.P. characteristics of the fluid in use and yet will not cause fluid contamination or deterioration, as do most other brush-on oils and additives.

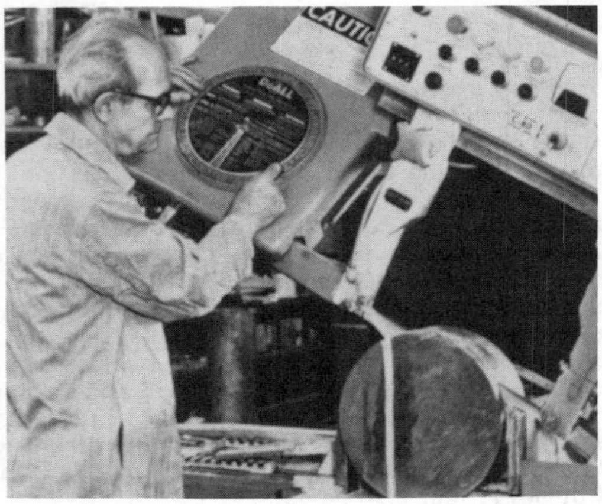

BENEFITS:

Can Replace Straight Oils in Most Applications—Allows Maximum Production Rate—Allows More Accurate Sizing—Has Low Surface Tension Allowing E.P. Oils to Penetrate to Cutting Interface—Pleasant and Safe to Work with, Eliminates Fire Hazards, Smoke, Oily Mists and Excessive Heat.

APPLICATIONS:

Turning, sawing, milling, tapping, drilling, threading, broaching and grinding on ferrous metals, high alloys, stainless steels and exotics.

DILUTIONS:

Turning, milling, drilling	1 part to 10-15 parts water.
Tapping, threading, broaching, sawing, gear hobbing	1 part to 10-20 parts water.
Form and centerless grinding	1 part to 10-20 parts water.
General grinding	1 part to 20-40 parts water.
Heavy duty, slow speed and heavy feed work	1 part to 3-10 parts blending oil.

AVAILABLE IN:

1 gallon cans (4 per case) (#40110)
5 gallon pails (#40510)
55 gallon drums (#45510)

ALL RUST-LICK PRODUCTS ARE AVAILABLE THROUGH SELECTED INDUSTRIAL DISTRIBUTORS IN EVERY MAJOR METALWORKING AREA

Distributed by:

Manufactured by:

RUST-LICK INC

P.O. Box 1244
Danbury, CT 06810
Tele. 203-792-0052

303 N. Manchester Ave.
Anaheim, CA 92801
Tele. 714-535-6075

INDUSTRIAL DISTRIBUTORS

Rust Preventives • Corrosion Inhibitors • Grinding Fluids • Cleaners • Dielectric Oils

3/84

CUTZOL WS-10

Bulletin No. 40020
Mfgr. No. 698242

Multipurpose Chlorinated Soluble Oil

CUTZOL WS-10 is a heavy duty, multi-purpose, chlorinated soluble oil that exhibits excellent lubricity and antiweld properties. When used at recommended dilutions WS-10 will promote longer tool life, higher production rates and better finishes, even on tough to machine metals. In addition, WS-10 dissipates heat effectively, insuring dimensional accuracy because tools and workpieces are cooled efficiently. Exceptionally stable even in hard water, WS-10 is also highly resistant to bacterial decomposition, rancidity and odor.

APPLICATIONS:

Turning, sawing, threading, tapping, milling, drilling and grinding of all metals, especially zinc and aluminum diecast.

DILUTIONS:

Milling, tapping, threading, turning, sawing . . . 1 part to 5-25 parts water. Form and centerless grinding . . . 1 part to 20-30 parts water. General grinding . . . 1 part to 20-40 parts water.

AVAILABLE IN:

One gallon cans (4 per case) (#40120)
Five gallon pails (#40520)
55 gallon drums (#45520)

ALL RUST-LICK PRODUCTS ARE AVAILABLE THROUGH SELECTED INDUSTRIAL DISTRIBUTORS IN EVERY MAJOR METALWORKING AREA

Distributed by:

Manufactured by:

RUST-LICK INC

P.O. Box 1244
Danbury, CT 06810
Tele. 203-792-0052

303 N. Manchester Ave.
Anaheim, CA 92801
Tele. 714-535-6075

INDUSTRIAL DISTRIBUTORS

CUTTING FLUIDS

Rust-Lick, Incorporated
72 Morgan Avenue/P.O. Box 1244
Danbury, CT 06810
(203) 792-0052

CUTZOL WS-11

"No-Frills" Soluble Oil

Bulletin No. 40030
Mfgr. No. 698242

Cutzol WS-11 is a multi-purpose soluble oil formulated for both the machining and grinding of all metals—ferrous and non-ferrous. WS-11 mixes readily in any water to form an extremely rich emulsion allowing for higher dilution ratios than other conventional soluble oils. In addition, WS-11 provides excellent rust protection for both machines and machined parts, and its high lubricity contributes to superior finish and excellent tool life. WS-11 is highly resistant to bacteria and can be used for extended periods of time without rancidity.

BENEFITS:

Very stable-resists oxidation and rancidity • Provides excellent rust protection • Dissipates heat quickly • Extends tool life • Mixes readily in any water • Contains high anti-weld properties • Can be used on all metals • Has twice the dilution ratio of other soluble oils

APPLICATIONS:

General purpose milling, drilling, tapping, turning and grinding.

DILUTIONS:

One part Cutzol WS-11 to 20-30 parts water for milling, drilling, tapping, turning and grinding.
One part Cutzol WS-11 to 40 parts water.

AVAILABLE IN:

One gallon cans (4 per case) (#40130)
5 gallon containers (#40530)
55 gallon drums (#45530)

ALL RUST-LICK PRODUCTS ARE SOLD THROUGH SELECTED INDUSTRIAL DISTRIBUTORS IN EVERY MAJOR METALWORKING AREA

Distributed by:

Manufactured by:

RUST-LICK INC

P.O. Box 1244
Danbury, CT 06810
Tele. 203-792-0052

303 N. Manchester Ave.
Anaheim, CA 92801
Tele. 714-535-6075

INDUSTRIAL DISTRIBUTORS

Rust-Lick, Incorporated
72 Morgan Avenue/P.O. Box 1244
Danbury, CT 06810
(203) 792-0052

CUTZOL WS-15

Bulletin No. 40040
Mfgr. No. 698242

Semi-Synthetic Cutting and Grinding Fluid

CUTZOL WS-15 is a totally unique, water soluble oil that combines the lubricity of traditional soluble oils with the cooling properties and transparency normally only available from chemical type coolants. Its high lubricity results in finer finishes, increased tool life, and is more than adequate for slides and ways. The extremely small particle size of the emulsion allows greater penetration and closer contact with metal surfaces, resulting in a greater rate of cooling. In addition, CUTZOL WS-15 has a high film strength and resists rancidity.

BENEFITS:

Transparent in solution—allows visibility of workpiece • May be used on all metals (except magnesium) • Highly resistant to bacteria • Provides excellent rust protection for tool and workpiece • Does not contain sulfur, chlorine or phosphorous • Nitrite free.

APPLICATIONS:

Turning, milling, broaching, cut-off sawing, form and centerless grinding, tapping, drilling, threading and general grinding of all metals, except magnesium.

DILUTIONS:

Turning and milling	1 part to 10-25 parts water
Broaching and cut-off sawing	1 part to 10-25 parts water
Form and centerless grinding	1 part to 10-25 parts water
Tapping, drilling and threading	1 part to 10-25 parts water
General grinding	1 part to 20-30 parts water

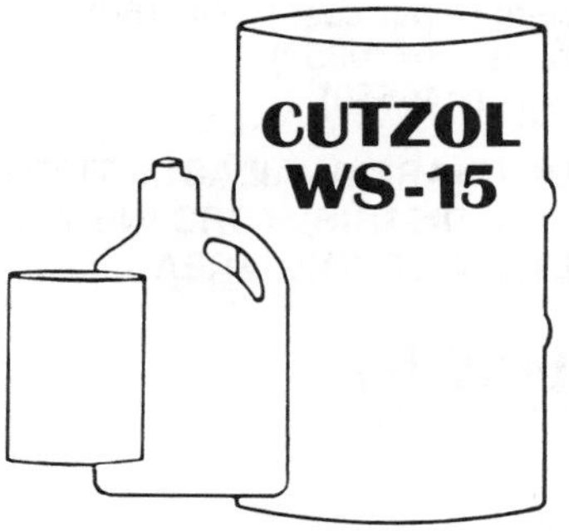

AVAILABLE IN:

One gallon cans (4 per case) (#40140)
5 gallon pails (#40540)
55 gallon drums (#45540)

ALL RUST-LICK PRODUCTS ARE AVAILABLE THROUGH SELECTED INDUSTRIAL DISTRIBUTORS IN EVERY MAJOR METALWORKING AREA.

Distributed by:

Manufactured by:

RUST-LICK INC.

P.O. Box 1244
Danbury, CT 06810
Tele. 203-792-0052

303 N. Manchester Ave.
Anaheim, CA 92801
Tele. 714-535-6075

INDUSTRIAL DISTRIBUTORS

Rust-Lick, Incorporated
72 Morgan Avenue/P.O. Box 1244
Danbury, CT 06810
(203) 792-0052

CUTZOL WS-500-A

Bulletin No. 40050
Mfgr. No. 698242

Heavy Duty Soluble Oil

Cutzol WS-500-A was developed after several years of research involving a total of more than 300 formulas. This research was conducted jointly by an international aircraft company and Rust-Lick's laboratory. The tests were initiated to find a water soluble cutting fluid that would assist in the machining of Titanium without resultant stress corrosion. In addition, this material had to be equally effective in the machining of aluminum, magnesium, brass, carbon and the stainless steels—Vascomax, Inconel X, Rene 41 and AM 355. These are representative of the materials that an aircraft manufacturer would have to machine. Cutzol WS-500-A meets these requirements. It is composed of selected hydrocarbons, emulsifiers, couplers, bacterial inhibiting agents and a proprietory "Extreme Pressure" additive. As a soluble cutting oil it contains additives that have the necessary lubricity and anti-weld properties to permit high stock removal rates, excellent tool life and workpiece finish when machining and grinding all metals.

BENEFITS:

Forms Stable Emulsion In Hard Or Soft Water • Excellent Resistance To Bacterial Growth • Free Of Active Sulfur, Chlorine And Other Halogens That Create Stress Corrosion • For Use On All Metals • Proprietory EP Additive Insures Exceptional Anti-Weld Properties

DILUTIONS:

One part Cutzol WS-500-A to 10-20 parts water for Milling, Drilling, Boring, Turning, Threading and Broaching.
One part Cutzol WS-500-A to 10-15 parts water for Form Grinding.
One part Cutzol WS-500-A to 25-50 parts water for Centerless Grinding.
NOTE: To insure proper emulsion add CUTZOL WS-500-A to water.

AVAILABLE IN:

One gallon cans (4 per case) (#40150)
5 gallon containers (#40550)
55 gallon drums (#45550)

ALL RUST-LICK PRODUCTS ARE AVAILABLE THROUGH SELECTED INDUSTRIAL DISTRIBUTORS IN EVERY MAJOR METALWORKING AREA

Distributed by:

Manufactured by:

RUST-LICK INC

P.O. Box 1244
Danbury, CT 06810
Tele. 203-792-0052

303 N. Manchester Ave.
Anaheim, CA 92801
Tele. 714-535-6075

INDUSTRIAL DISTRIBUTORS

Rust Preventives • Corrosion Inhibitors • Grinding Fluids • Cleaners • Dielectric Oils

CUTZOL WS-5050

Bulletin No. 40065
Mfgr. No. 698242

Multi-Purpose Heavy Duty Soluble Oil

CUTZOL WS-5050 is a multi-purpose, heavy duty, soluble oil formulated for both the machining and grinding of all metals, especially high nickel alloys, zinc and aluminum diecast and stainless steels. In dilution WS-5050 forms a blue tinted, pleasant smelling, long lasting, micro-emulsion that is stable in hard or soft water. Compounded with E.P. additives that increase metal cutting action while minimizing galling, pick-up and welding, WS-5050 also improves tool life and finish even in tough machining operations on metals with low machinability ratings. Extremely fine finishes are also possible in grinding operations. The micro-emulsion of WS-5050 keeps tools and workpieces cool and grinding wheels clean and free cutting. CUTZOL WS-5050 will not cause staining, has excellent lubricating qualities and provides rust protection for both machines and machined parts. CUTZOL WS-5050 contains NO nitrites and NO phenols.

BENEFITS:

Use On All Metals & Operations—Reduces Coolant Inventories • Pleasant Smelling—Insures Operator Acceptance • Long Sump Life—Reduces Coolant Consumption & Downtime • Dissipates Heat Quickly—Increases Tool Life & Dimensional Accuracy • Excellent Rust Inhibiting Agents—Protects Machines & Parts From Rusting • Contains NO Phenols—Eliminates Costly Disposal Problems.

DILUTIONS:

Milling, tapping, threading, turning	1 part WS-5050 to 5-30 parts water
Sawing, drilling, broaching	1 part WS-5050 to 5-30 parts water
Form and centerless grinding	1 part WS-5050 to 20-40 parts water
General grinding	1 part WS-5050 to 20-40 parts water

AVAILABLE IN:

One gallon cans (4 per case) (#40165)
5 gallon pails (#40565)
55 gallon drums (#45565)

ALL RUST-LICK PRODUCTS ARE AVAILABLE THROUGH SELECTED INDUSTRIAL DISTRIBUTORS IN EVERY MAJOR METALWORKING AREA

Distributed by:

Manufactured by:

RUST-LICK INC

P.O. Box 1244
Danbury, CT 06810
Tele. 203-792-0052

303 N. Manchester Ave.
Anaheim, CA 92801
Tele. 714-535-6075

INDUSTRIAL DISTRIBUTORS

Rust-Lick, Incorporated
72 Morgan Avenue/P.O. Box 1244
Danbury, CT 06810
(203) 792-0052

RUST-LICK G-1066D

Bulletin No. 50010
Mfgr. No. 698242

Synthetic Water Soluble Chemical Coolant

Rust-Lick G-1066D is a synthetic, water soluble, chemical coolant recommended for all types of grinding (surface, cylindrical, centerless and internal) and highly recommended for the diamond wheel grinding of Carbide. G-1066D holds Carbide swarf in suspension, allowing it to be carried to the reservoir for settling, and resulting in cleaner operating conditions. G-1066D mixes readily in hard or soft water, forming a transparent, true chemical solution that keeps saws and grinding wheels clean and true cutting, making faster stock removal rates possible. In addition, Rust-Lick G-1066D has proven its superiority as a coolant for the slicing, dicing and grinding of semiconductor wafers. Rust-Lick G-1066D leaves a microscopic, rust preventive film that will provide protection for machine tools, holding fixtures and parts in-process and in storage.

BENEFITS:

Dissipates Heat Quickly - Assures Dimensional Accuracy • Transparent - Allows Visibility of Workpiece • Stable Solution - Not Subject to Bacterial Decomposition • Rust Inhibiting - Provides Protection for Machine Tool, Holding Fixtures and Parts, Not Gummy or Sticky • Mixes Readily in Hard or Soft Water - Forms True Chemical Solution •

APPLICATIONS:

Grinding, drilling, sawing, slicing and dicing of ferrous metals. Highly recommended for the diamond wheel grinding of Carbide. Also for use on exotic material such as Silicon, Germanium, Gallium Arsenide, Sapphire, Quartz and other ceramics.

DILUTIONS:

One part Rust-Lick G-1066D to 15-40 parts water for grinding, slicing and dicing.
One part Rust-Lick G-1066D to 5-20 parts water for drilling and sawing.

AVAILABLE IN:

One gallon cans (4 per case) (#50110)
5 gallon containers (#50510)
55 gallon drums (#55510)

ALL RUST-LICK PRODUCTS ARE AVAILABLE THROUGH SELECTED INDUSTRIAL DISTRIBUTORS IN EVERY MAJOR METALWORKING AREA

Distributed by:

Manufactured by:

RUST-LICK INC.

P.O. Box 1244
Danbury, CT 06810
Tele. 203-792-0052

303 N. Manchester Ave.
Anaheim, CA 92801
Tele. 714-535-6075

INDUSTRIAL DISTRIBUTORS

Rust-Lick, Incorporated
72 Morgan Avenue/P.O. Box 1244
Danbury, CT 06810
(203) 792-0052

Bulletin No. 50020
Mfgr. No. 698242

RUST·LICK G·25·J

Synthetic Water Soluble Chemical Coolant

Rust-Lick G-25J is a synthetic, water soluble chemical coolant which has proven its superiority as a coolant for grinding cast iron, steel and all ferrous metals. Rust-Lick G-25-J forms a transparent, green solution that keeps grinding wheels clean and true cutting. It is guaranteed not to weaken the bond, load, gum or glaze grinding wheels. It will dissipate heat quickly and permit better dimensional accuracy. In addition G-25-J is an excellent rust preventive for parts in-process or during final rinse, and need not be removed prior to painting, plating or assembly.

BENEFITS:

Transparent - Permits Visibility of Work • Odorless - Not Subject to Bacterial Decomposition • Stable Solution - Completely Soluble in Hard or Soft Water • Non-Foaming - Does Not Contain Oil or Soap, Will Not Foam in Single or Multi-Cone Type Filters • Non-Corrosive - Will Not Corrode Table or Magnetic Chuck • Rust Inhibiting - Contains Corrosion Inhibiting Agents, Not Gummy or Sticky •

APPLICATIONS:

All surface grinding, wet abrasive belt grinding and abrasive cut-off wheels. Also as additive to water used conjunction with tumbling operation, and as a rust preventive for parts in-process or during final rinse.

DILUTIONS:

One part of Rust-Lick G-25-J to 100-150 parts water, for steel and other ferrous metals.
One part of Rust-Lick G-25-J to 75-100 parts water, for cast iron.
One part of Rust-Lick G-25-J to 100-200 parts water, for tumbling operation.
One part of Rust-Lick G-25-J to 75-100 parts water, for rust protection.

AVAILABLE IN:

One gallon cans (4 per case) (#50120)
5 gallon containers (#50520)
55 gallon drums (#55520)

ALL RUST-LICK PRODUCTS ARE AVAILABLE THROUGH SELECTED INDUSTRIAL DISTRIBUTORS IN EVERY MAJOR METALWORKING AREA

Distributed by:

Manufactured by:

RUST-LICK INC

P.O. Box 1244
Danbury, CT 06810
Tele. 203-792-0052

303 N. Manchester Ave.
Anaheim, CA 92801
Tele. 714-535-6075

INDUSTRIAL DISTRIBUTORS

CUTTING FLUIDS

Rust-Lick, Incorporated
72 Morgan Avenue/P.O. Box 1244
Danbury, CT 06810
(203) 792-0052

Bulletin No. 50050
Mfgr. No. 698242

RUST•LICK G-25-AH

Synthetic Water Soluble Chemical Coolant

Rust-Lick G-25-AH is a synthetic, water soluble, chemical coolant, specifically developed for use in applications where inorganic salts are prohibited. Rust-Lick G-25-AH contains NO NITRITES. It forms a transparent, stable emulsion in hard or soft water. A multi-purpose fluid, Rust-Lick G-25-AH may be used on all metals. It will provide superior rust protection for ferrous metals, however, non-ferrous metals should be washed soon after exposure to Rust-Lick G-25-AH. Not subject to bacterial decomposition, Rust-Lick G-25-AH is also odorless, non-toxic, non-irritating, non-foaming, non-staining and non-smoking.

BENEFITS:

Keeps Tool and Work Cool, Increasing Tool Life and Assuring Dimensional Accuracy • Transparent, Allows Visibility of Workpiece • Contains NO NITRITES • Contains No Inorganic Salts • Clean to Work With, Keeps Grinding Wheels Clean and True Cutting •

APPLICATIONS:

General machining - drilling, threading, turning, milling; cutting, polishing, slicing, tapping, broaching, cut-off sawing and grinding. For use on all metals: however, non-ferrous metals should be washed soon after exposure to Rust-Lick G-25-AH.

DILUTIONS:

General machining	1 part Rust-Lick G-25-AH to 10-25 parts water
Cutting, polishing & slicing	1 part Rust-Lick G-25-AH to 10-25 parts water
Broaching & cut-off sawing	1 part Rust-Lick G-25-AH to 10-25 parts water
Form grinding	1 part Rust-Lick G-25-AH to 20-30 parts water
General grinding	1 part Rust-Lick G-25-AH to 25-40 parts water

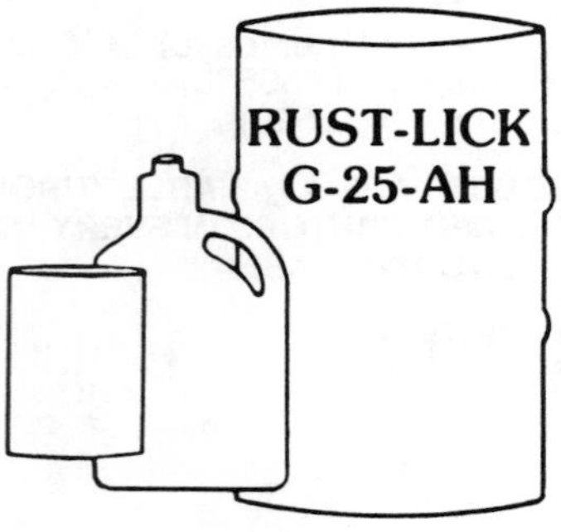

AVAILABLE IN:

1 gallon cans (4 per case) (#50150)
5 gallon containers (#50550)
55 gallon drums (#55550)

ALL RUST-LICK PRODUCTS ARE AVAILABLE THROUGH SELECTED INDUSTRIAL DISTRIBUTORS IN EVERY MAJOR METALWORKING AREA

Distributed by:

Manufactured by:

RUST-LICK INC.

P.O. Box 1244
Danbury, CT 06810
Tele. 203-792-0052

303 N. Manchester Ave.
Anaheim, CA 92801
Tele. 714-535-6075

INDUSTRIAL DISTRIBUTORS

Rust Preventives • Corrosion Inhibitors • Grinding Fluids • Cleaners • Dielectric Oils

Rust-Lick, Incorporated
72 Morgan Avenue/P.O. Box 1244
Danbury, CT 06810
(203) 792-0052

Bulletin No. 50041
Mfgr. No. 698242

VYTRON -N

Synthetic Water Soluble Coolant

VYTRON -N is a synthetic, water soluble chemical coolant. A true chemical solution, it is transparent, contains no petroleum, mineral or vegetable oils and is completely soluble in hard or soft water. Both safe and pleasant to work with, VYTRON -N exhibits an exceptionally long sump life. For use on both ferrous and non-ferrous metals, VYTRON -N is highly recommended for use in spray mist units, and performs equally as well in a wide variety of cutting and grinding operations. In addition, VYTRON -N is absolutely fireproof, is not damaged by freezing temperatures and contains no nitrites.

BENEFITS:

Keeps Tools and Work Cool, Increasing Tool Life and Assuring Dimensional Accuracy • Odorless, Never Turns Rancid • Non-Toxic, Will Not Burn, Cause Irritation or Dermatitis • Provides Excellent Rust Protection • Requires No Degreasing, Washes Off Quickly in Water • Transparent, Allows Visibility of Work • Provides Excellent Chemical Lubricity for Freer Cutting Action and Finer Finishes • Lasts 3 to 4 Times Longer Than Ordinary Coolant. • Contains **no nitrites.**

APPLICATIONS:

Tapping, drilling, threading, broaching, sawing, form and general grinding, turning and milling on both ferrous and non-ferrous metals.

DILUTIONS:

Tapping, drilling, turning & milling .... 1 part VYTRON-N to 10-25 parts water
Cut-off sawing .... 1 part VYTRON-N to 15-25 parts water
Form grinding .... 1 part VYTRON-N to 35-40 parts water
In mist units .... 1 part VYTRON-N to 25 parts water

AVAILABLE IN:

1 gallon cans (4 per case) (#50141)
5 gallon cans (#50541)
55 gallon drums (#55541)

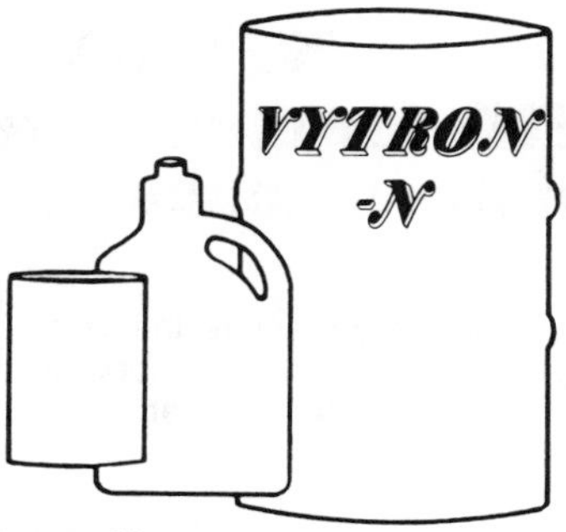

ALL RUST-LICK PRODUCTS ARE SOLD THROUGH SELECTED INDUSTRIAL DISTRIBUTORS IN EVERY MAJOR METALWORKING AREA

Distributed By:

INDUSTRIAL DISTRIBUTORS

Manufactured by:
RUST-LICK INC

P.O. Box 1244
Danbury, CT 06810
Tele. 203-792-0052

303 N. Manchester Ave.
Anaheim, CA 92801
Tele. 714-535-6075

CUTTING FLUIDS

Specialty Products Company
15 Exchange Place/P.O. Box 306
Jersey City, NJ 07303
(201) 434-4700

UNIQUE HEAVY-DUTY OIL CONCENTRATE – TOOLIFE 316 . . . ALLOWS HIGHER FEED RATES – FASTER CUTTING SPEEDS.

EXCEPTIONALLY HIGH COOLING RATE AND SUPERIOR LUBRICITY TOOLIFE 316 . . .

a uniquely different type of heavy duty oil. It is semi-synthetic, water-soluble, and specifically formulated to provide high lubricity and rapid heat transfer properties for machining and grinding ferrous and non-ferrous metals. Toolife 316, with its exceptionally high cooling rate and superior lubricity results in superior work finishes and longer tool life. The resulting higher feed rates and faster cutting speeds offer immediate savings over conventional oils.

EXTRAORDINARILY LONG SUMP LIFE

is to be expected and is further extended by the incorporation of highly effective bacteriacide and fungicide inhibitors. This special anti-fouling quality prevents unpleasant odors from forming. Rancidity will not develop even after prolonged usage and time.

EXCELLENT RUST INHIBITION FOR WORK PIECE AND MACHINERY.

It contains no corrosive sulfur or chlorine elements to stain or attack metals. After evaporation of the water, Toolife 316 leaves an oily protective film residue on work piece, equipment and tools that is neither gummy nor wax-like. This film is readily removed by a water rinse if desired.

LOW-FOAMING . . . NEARLY TRANSPARENT for unimpaired inspection of the machining operation.

In grinding operations the fines settle rapidly, the wheel remains lubricious and will not gum as with many other coolants. This allows faster grinding speeds while maintaining a clean effective grinding wheel surface.

Specialty Products Company
15 Exchange Place/P.O. Box 306
Jersey City, NJ 07303
(201) 434-4700

RECOMMENDATIONS:

Toolife 316 is designed for all types of machine operations and is suggested for ferrous and non ferrous metals, especially for tough machining alloys. Caution should be exercised only when machining certain alloys of aluminum which may darken slightly with prolonged exposure. Prompt rinsing of aluminum machined parts with water will prevent any possible staining.

The suggested dilution range with water for machine operations is offered as follows:

Threading, Tapping, Heavy Turning, Gear Cutting, Broaching	10 to 30 Parts Water
Drilling, Turning, Reaming, Trepanning, Boring, General Cutting, Hack and Circular Sawing	20 to 40 Parts Water
Grinding	30 to 60 Parts Water

Machinability, speed and feed of the operation will further indicate which limit of the dilution range should be initially chosen.

SPECIFICATIONS:

Viscosity, SSU at 100° F.	46
Density, lbs./gals. at 60° F.	8.4
Gravity, API at 60° F.	8.8
Color	Transparent Dark Green
Odor	Mild
Flash Point	None
pH, at 1 to 20 dilution	9.7
1 to 60 dilution	9.3

CAUTION:

At concentrations of 1:15 (Toolife 316 to water) or richer, care should be taken to avoid excessive skin contact.

The concentration of diluted Toolife 316 can increase through evaporative loss. Therefore, provide appropriate make-up to avoid an overrich condition.

As a complimentary service, Specialty Products Company's Laboratory will determine your coolant's dilution ratio to allow dilution adjustment. Please submit a representative sample of at least four ounces for our analysis.

SPECIALTY PRODUCTS COMPANY
15 Exchange Place, Jersey City, New Jersey 07302 (201) 434-4700
Specialty Products Co, Ltd, 227 Norseman St., Toronto Ont. M8Z2R5 (416) 239-6541

Specialty Products Company
15 Exchange Place/P.O. Box 306
Jersey City, NJ 07303
(201) 434-4700

technical data

TOOLIFE 208A

Toolife 208A is a transparent, heavy duty, sulfo-chlorinated fatty cutting oil, formulated especially for demanding Escomatic and Swiss screw machine shops. Although engineered for ferrous hard machining jobs, it performs equally well for free cutting jobs with resultant longer tool life, faster speeds, feeds, and quality output.

Toolife 208A possesses a unique, potent, active sulfur additive which performs as an anti-weld agent, yet is absolutely transparent providing improved operator visual inspection of work piece. Rapid heat dissipation from the work piece and tool for a cool effective cutting operation is achieved by the unusually high tenacious film of Toolife 208A. This film adhesion is further enhanced by fatty compounds which impart extra lubricity and metal wetting properties to assure better surface finishes, longer tool life, reduction of smoke and increased production rates. Extreme pressure resistance is reduced to a minimum by the inclusion of high concentrations of chlorine.

In brief, a high performance, specialized cutting oil with potent fortifiers which translates for machine shops into improved economy, production, and quality.

TYPICAL SPECIFICATIONS

Gravity, $^\circ$API @ 60°F	19.0
Density, lb./gal. @ 60°F	7.83
Viscosity, SUS @ 100°F	250
Color, ASTM	2
Flash Point, $^\circ$F	330
Saponification No.	9.9
Copper Strip Corrosion	4A
Active Chemicals	Sulphur & Chlorine

/mlm

Specialty Products Company
15 Exchange Place/P.O. Box 306
Jersey City, NJ 07303
(201) 434-4700

technical data

specialty products company
15 Exchange Place, P.O. Box 306
Jersey City, N.J. 07303
201-434-4700

TOOLIFE 323

PRODUCT DESCRIPTION:

Toolife 323 is a soluble coolant, dark blue in color, which will produce a milky, micro-emulsion when it is mixed with water. This micro-emulsion will be free of odor, having an exceptionally long service life in the machine sump.

Toolife 323 will have a non-gumming effect when residues are left on machine parts, therefore, it will not interfere with any machining operations.

APPLICATION:

Toolife 323 will serve as a coolant in heavy-duty cutting operations for both ferrous and non-ferrous metals. It will produce superior finishes on the machine parts, giving excellent tool life.

RECOMMENDED DILUTIONS:

The following are the suggested mixes for different machining operations:

Broaching	1:10-20 (Toolife 323 to Water)
Hack and Circular Sawing	1:10-20
Thread Rolling	1:10-20
Gear Hobbing	1:10-20
Tapping	1:10-20
Boring	1:20-30
Turning	1:20-30
Drilling	1:20-30
Threading	1:20-30
Milling	1:20-30
Reaming	1:20-30
Surface Grinding	1:60-80
Centerless Grinding	1:40-60
Cylindrical Grinding	1:40-60
Internal Grinding	1:40-60
Abrasive Cut-Off	1:40-60
Light Gauge Stamping	1:5-10
Roll Forming	1:5-10

PROPERTIES:

Color	Dark Blue
Density, lb/gal	8.3
Viscosity, SSU @ 100°F	390
Flash Point, °F	360

Sprayway, Incorporated
484 Vista Avenue
Addison, IL 60101
(312) 628-0998

Sprayway®

CUTTING OIL

BCO COMPOUND

No. 717

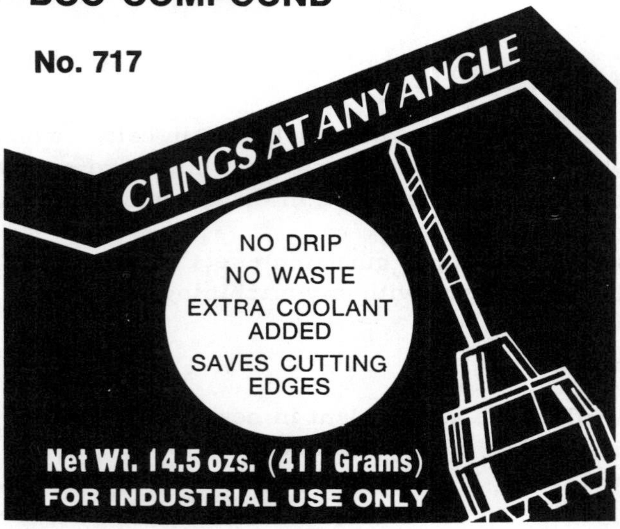

CUTS METAL EASIER — FASTER
FOR HARD OR SOFT METALS

Formulated from reinforced sulphurized mineral oil, chlorinated wax and extreme pressure additives that clings to metal and follows tools down.

Drilling
Spray at the mark to be drilled. It clings to curved and beveled surfaces.

Sawing
Spray along the cutting line using extension tube for continued lubrication all through the cut.

Tapping-Threading
Stays in place for either inside or outside threading. Excellent for tapping.

The Steco Corporation
P.O. Box 2238
Little Rock, AR 72203
(501) 375-5644

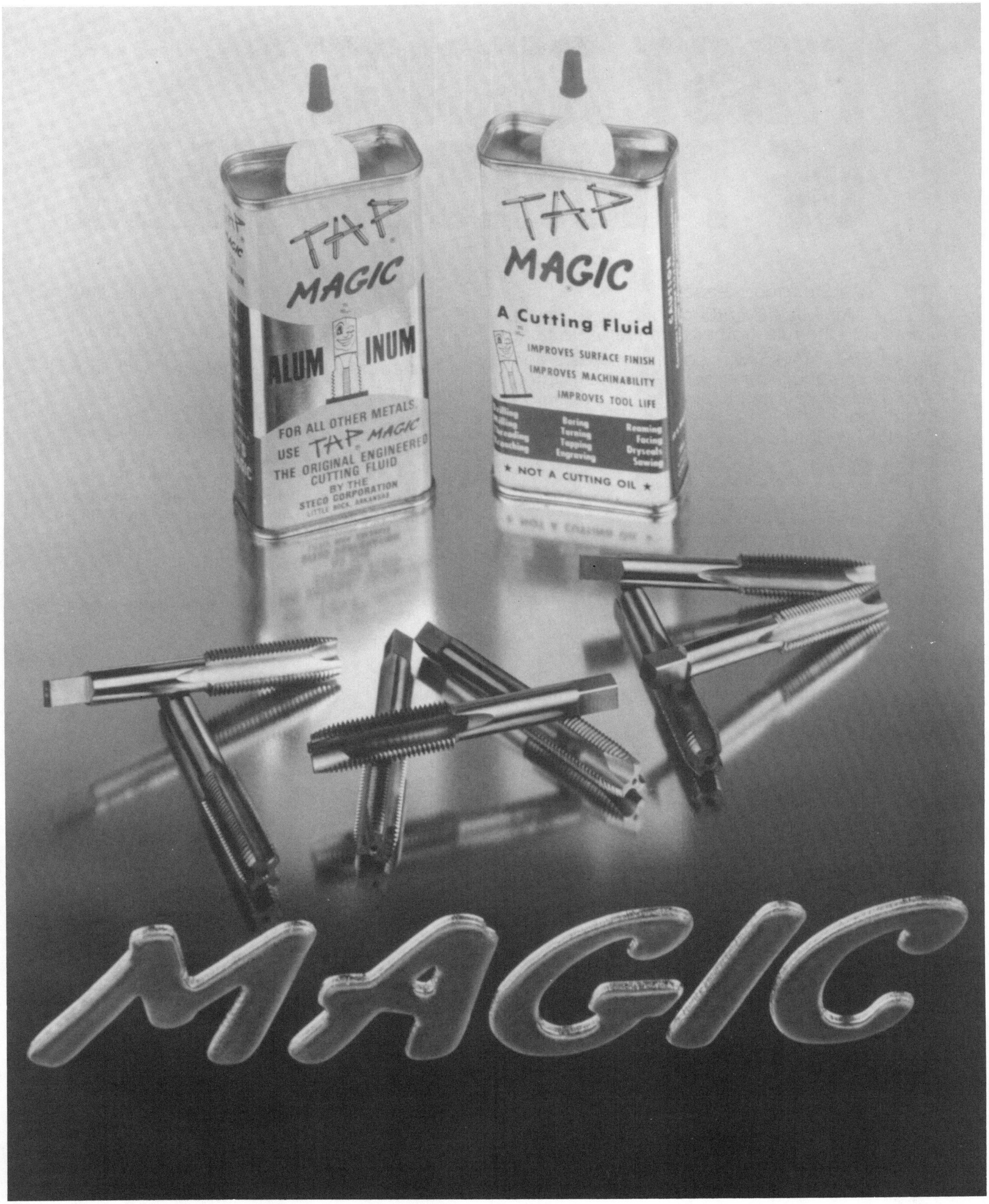

The Steco Corporation
P.O. Box 2238
Little Rock, AR 72203
(501) 375-5644

FIRST CHOICE OF PROFESSIONALS

And for good reason.

TAP MAGIC cutting fluids have made precision the rule–rather than the exception.

TAP MAGIC was developed years ago for a specific user who needed a cutting fluid which allowed far greater precision than conventional cutting oils. Today, TAP MAGIC is America's leading cutting fluid.

TAP MAGIC & TAP MAGIC ALUMINUM

The original TAP MAGIC provides outstanding performance in any machining operation on... stainless... titanium... brass ... monel... nickel... bronze... inconel... beryllium copper... chrome-molly. In fact, it is the recommended cutting fluid for machining operations on all metals except aluminum.

For aluminum, a special TAP MAGIC has been formulated–TAP MAGIC ALUMINUM. Using TAP MAGIC ALUMINUM for aluminum machining operations, you will obtain the same standard of excellence as with the original TAP MAGIC.

LESS COSTLY IN THE LONG RUN

There is a difference between the price of a product and its cost. TAP MAGIC

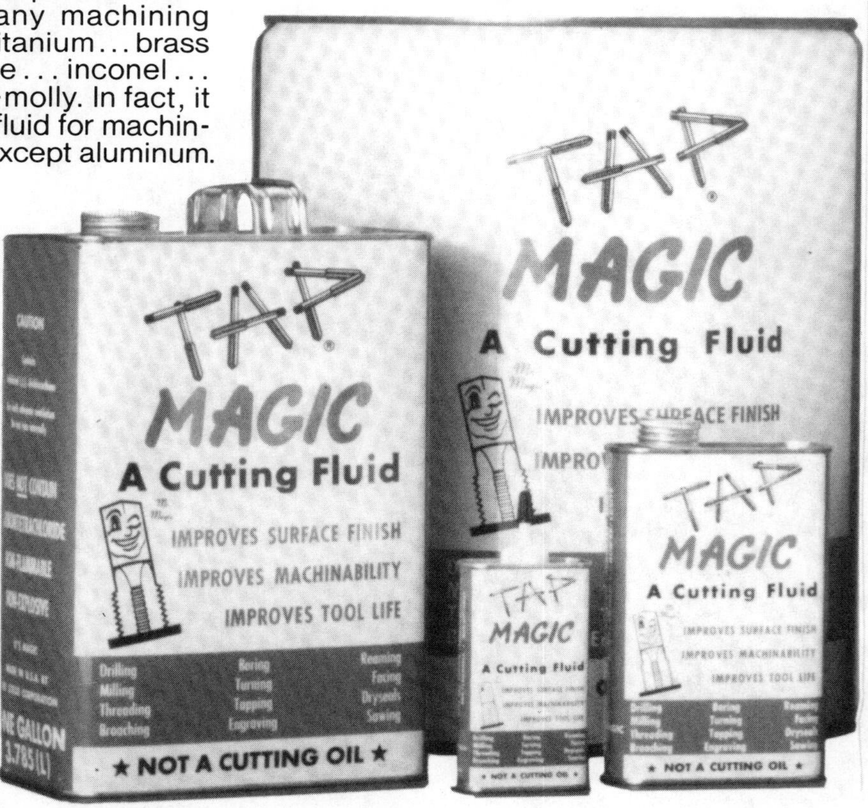

The Steco Corporation
P.O. Box 2238
Little Rock, AR 72203
(501) 375-5644

SINCE 1953.

cutting fluids may be priced higher than conventional cutting oils. But in the long run, you'll find TAP MAGIC will cost far less.

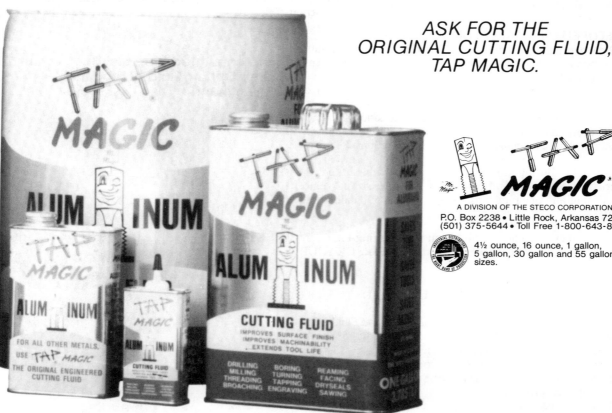

TAP MAGIC encourages longer tool and machine life. Whatever the machining operation, TAP MAGIC does a better job – resulting in a lower rejection rate. And TAP MAGIC is cleaner to use. There's no oily mess left on the hands, machine or workpiece.

TAP MAGIC makes sense. And cents.

ASK FOR THE GENUINE TAP MAGIC

It's available in a variety of sizes from the handy 4½ ounce can all the way up to a 55 gallon drum. We keep it convenient to find too – with an outstanding network of professional industrial distributors, and a "quick-ship" plan that keeps the product available for your immediate use.

Use TAP MAGIC just once and you'll join millions of satisfied users.

ASK FOR THE ORIGINAL CUTTING FLUID, TAP MAGIC.

TAP MAGIC®
A DIVISION OF THE STECO CORPORATION
P.O. Box 2238 • Little Rock, Arkansas 72203
(501) 375-5644 • Toll Free 1-800-643-8026

4½ ounce, 16 ounce, 1 gallon, 5 gallon, 30 gallon and 55 gallon sizes.

Tapmatic Corporation
1851 Kettering Street
Irvine, CA 92714
(714) 979-6080

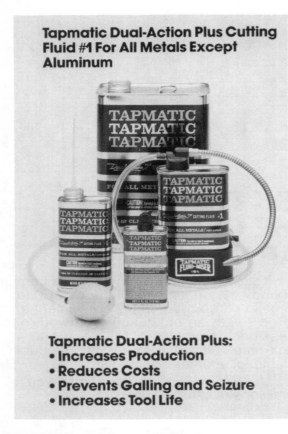

Tapmatic Dual-Action Plus Cutting Fluid #1 For All Metals Except Aluminum

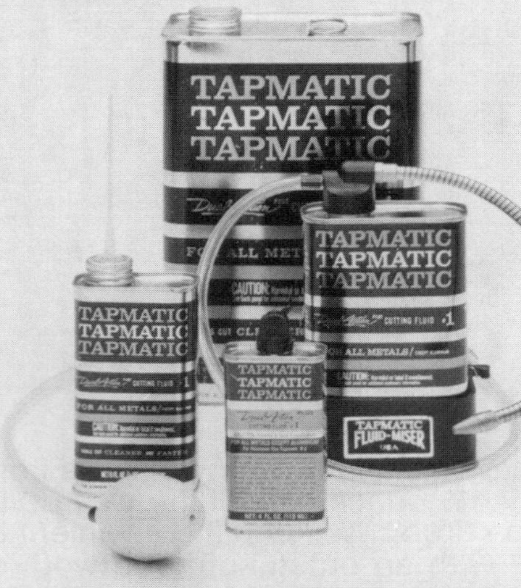

Tapmatic Dual-Action Plus:
- **Increases Production**
- **Reduces Costs**
- **Prevents Galling and Seizure**
- **Increases Tool Life**

DUAL-ACTION PLUS CUTTING FLUIDS

Recognizing the limitations of conventional fluids, which rely on rapid evaporation for cooling or concentrate entirely on a lubricating action, Tapmatic Corporation developed Dual-Action Plus.

These fluids simultaneously REFRIGERATE and LUBRICATE through chemically controlled evaporation, producing unmatched results in finer finishes, closer tolerance and longer tool life.

THE PLUS IN DUAL-ACTION PLUS

The new Tapmatic formula not only refrigerates and lubricates but provides the protection of a space age rust and corrosion inhibitor. It penetrates deep into the metal to displace corrosion causing moisture and maintain a smooth lubricating action for long lasting protection.

ONE DROP CAN DO IT WITH THE NEW FLUID-MISER SPOUT!

A new, four inch long Fluid-Miser spout is packaged with each half-pint and pint can of Dual-Action Plus. Pinpoint application makes for cleaner, more economical machining...drop by drop.

NO UNPLEASANT ODOR

An added bonus in Cutting Fluid #1 is its pleasant scent. Now, more than ever, you will enjoy working with the world's #1 cutting fluid, from Tapmatic.

HIGH PERFORMANCE CUTTING OIL ADDITIVE

Tapmatic Dual-Action Plus #1 may also be employed as a lubricity additive for conventional cutting oils in recirculating systems. The addition of only 3 to 5% of Tapmatic #1 significantly increases extreme pressure capability and results in increased production.

DEVELOPED ESPECIALLY FOR ALUMINUM

New reformulated Dual-Action Plus #2 refrigerates and lubricates. A new blend of fatty methylesters providing superior extreme pressure lubricity and packaged in a refined petroleum distillate carrier offers the absolute maximum performance in the machining of aluminum alloys.

Try a sample run on your most difficult application and you'll notice the difference immediately. No conventional fluid can match the metalworking capabilities of Tapmatic Dual-Action #2.

Tapmatic Cutting Fluids #1 and #2 are available in the **widest variety of convenient and economical container sizes:** 4 ounce, one-half pint and one pint cans. One, 5, 30 and 55, gallon bulk containers.

Remember, Tapmatic Dual-Action Plus Cutting Fluids are GUARANTEED TO OUTPERFORM ALL OTHERS OR YOUR MONEY BACK.

NEW FLUID-MISER APPLICATOR CUTS FLUID COST 75% OR MORE!

The combination of Tapmatic Cutting Fluids and Fluid-Miser applicators can actually reduce your cutting fluid cost 75% while effecting tremendous increases in production.

The Fluid-Miser applies fluid when you want it, where you want it and only in the volume needed. Waste resulting from conventional squirt-on application is eliminated.

Easily attached, the Fluid-Miser is secured to a machine column with a spring attachment. The metering bulb may be hand or foot operated.

See it at your nearest industrial distributor.

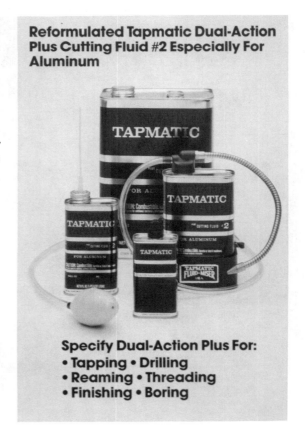

Reformulated Tapmatic Dual-Action Plus Cutting Fluid #2 Especially For Aluminum

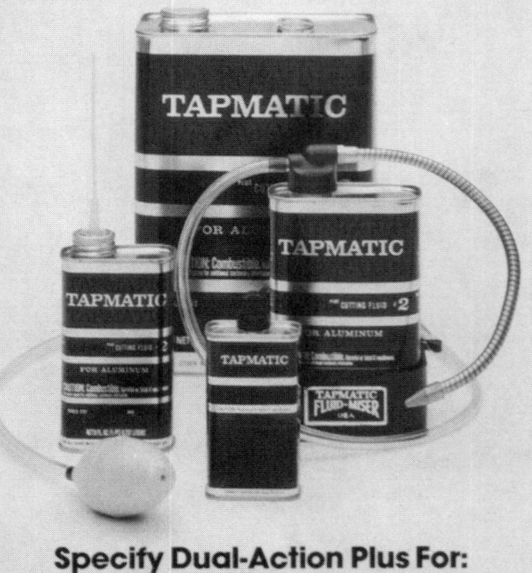

Specify Dual-Action Plus For:
- **Tapping • Drilling**
- **Reaming • Threading**
- **Finishing • Boring**

Tapmatic Corporation
1851 Kettering Street
Irvine, CA 92714
(714) 979-6080

Tapmatic is pleased to introduce a completely new family of versatile machining aids. All are ecological, biodegradable compounds formulated to provide a superior lubricating product in three forms.

EDGE (Liquid)

Edge is a high film strength friction reducing agent for all metal cutting and forming operations. It is recommended for all metals, including the most difficult alloys. There is no known material with which it is not compatible.

For best economy and performance Edge should be used sparingly. The burning of excess fluid will cause smoking, indicating too much is being used. When applied properly, Edge becomes an extremely inexpensive machining aid. Total consumption in an eight hour shift should amount to no more than a few spoonsful.

Available in pint cans, 1 gallon containers, 5 gallon pails and 30 or 55 gallon drums.

EDGE LUBE (Solid)

Edge Lube is a white, waxy, solid bar packaged in a 13 oz. clear plastic dispensing cylinder. It is primarily for hand application to rotating or stationary cutting tools, grinding wheels or belts. A minute amount is all that is required for the most severe cutting or forming operations.

EDGE CREME (Paste)

Offered in a convenient 11 Fl. Oz. squeeze tube with applicator tip, Edge Creme may be applied to the cutting tool or hole prior to tapping or reaming. It is also excellent for forming or bending, precoating fasteners or lubricating rubber seals.

All Edge products are compatible with water or oil systems. They are non-flammable, non-irritating and non-polluting.

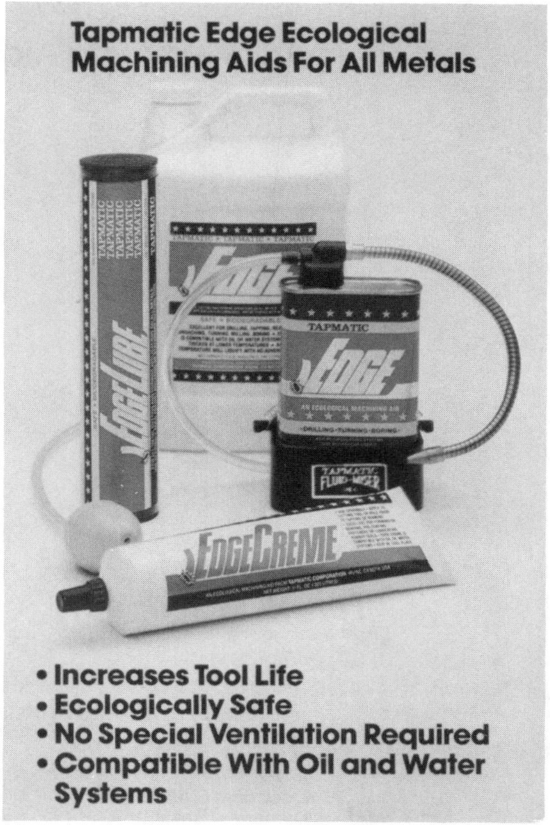

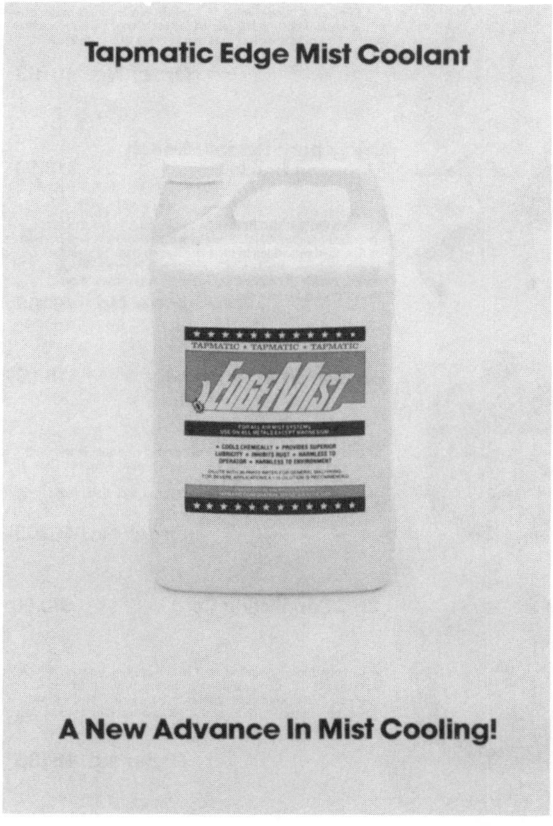

New Tapmatic Edge Mist provides the last word in mist cooling for the most severe metalworking requirements. It will not clog lines, is safe and pleasant to work with and will not fog. It contains no sulphur or chlorine and is not irritating to skin or eyes. Operator safety is assured.

LOWERS TEMPERATURE CHEMICALLY

Liquid vaporization effectively transfers heat as it is generated. Edge Mist goes further and lowers temperature chemically as it delivers its highly effective lubricity package. It is the ultimate synthetic mist coolant offering the proper balance of lubricity and chemical cooling. Superior rust inhibition protects equipment and workpiece.

ECOLOGICAL

Edge Mist is a truly ecological product. It is totally consumed in the machining process thereby eliminating disposal problems. It reduces friction and eliminates chip welding, establishing new standards for finishes and tool life. It may be used in all mist systems.

It is recommended that Edge Mist be diluted with 30 parts of water for general machining. For more severe applications a 1:15 concentration may be required.

Edge Mist is packaged in 1, 5, 15 and 55 gallon containers.

Tapmatic Corporation
1851 Kettering Street
Irvine, CA 92714
(714) 979-6080

EDGE Liquid Machining Aid

For All Materials

EDGE is a high film strength, friction reducing agent for all metal cutting and forming operations. It is recommended for all metals, including the exotic materials and chemically restricted aerospace or nuclear applications. There is no known material with which it is not compatible.
An ecological, biodegradable compound, Edge is compatible with either oil or water based coolant systems. It is guaranteed to increase tool life.

Use Sparingly, More Is Not Better.

For best economy and performance EDGE should be used sparingly. The burning of excess fluid or compound will cause smoking, indicating too much is being used. When applied properly, such as, with the TAPMATIC ELECTRONIC CUTTING FLUID DISPENSER, EDGE becomes an extremely inexpensive machining aid.

CUTTING FLUIDS

ELECTRONIC CUTTING FLUID DISPENSER AND ACCESSORIES

Electronic Cutting Fluid Dispenser Basic Unit **$295.00**

Consists of a housing with electronic control panel.
Comes complete with a 24 volt plug-in transformer and 12 ft. cord, 4 ft. long nylon fluid hose, 6 inch "Lock-Line" with adjustable protective cover at nozzle end and a mounting adapter block for attachment to your magnetic base or one of our accessories.
Two nozzles are furnished. **Order No. 46000**

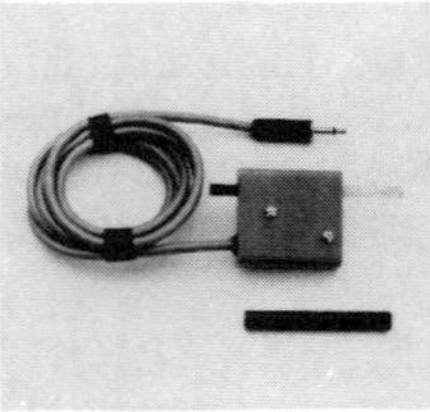

Remote Switch **$25.00**

A housing containing a snap action switch and actuating mechanism with a protruding nylon rod.
Three mounting posts on the back with #10-32 tapped holes attach to a mounting bracket. The nylon rod is spring loaded to provide adequate friction to operate switch when pressed against any movable surface such as a drill press quill, a rotatable shaft or moveable mill table.
Also provided is a 48" electrical cord, with plug for connecting to basic unit and an extension rod for attaching to $\frac{3}{16}$" dia. steel pin for manual operation of switch when attached to Bridgeport type mills. **Order No. 46100**

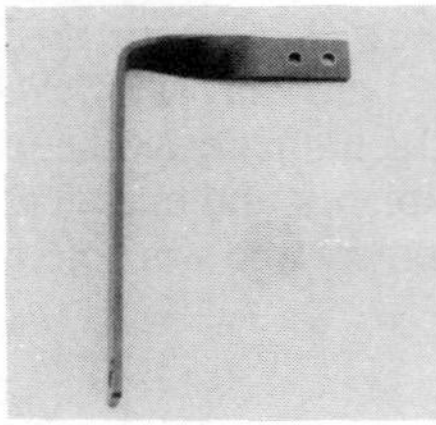

Drill Press: Dispenser Mounting Bracket **$12.50**

$\frac{1}{4}$" thick x 1" wide with one leg twisted 90°. Shorter twisted leg has two $\frac{1}{4}$" holes for attaching to bar on back of Basic Unit. Longer leg has one $\frac{3}{8}$" hole for attaching under one of the motor mounting bolts.

Order No. 46101

Drill Press: Remote Switch Mounting Bracket **$9.00**

Consists Of: Mounting block; $\frac{1}{4}$" dia. x $3\frac{1}{2}$" rod; Adjustable connecting block and a rod with one end flattened and drilled for attaching to remote switch with two #10-32 $\frac{3}{8}$" fillister head screws.

Order No. 46102

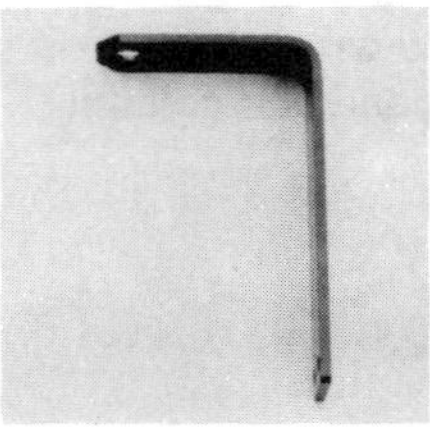

Bridgeport: Dispenser Mounting Bracket **$9.50**

$\frac{1}{4}$" thick x $1\frac{1}{4}$" wide with a $\frac{1}{2}$" hole in one leg to fit under one of the vertical bolts at the rear of the main base pedestal and two $\frac{1}{4}$" holes in the other leg for attaching to bar on back of Basic Unit.

Order No. 46103

Bridgeport: Remote Switch Mounting Bracket **$13.50**

Consists Of: Plate to fit over two bolts on mill head next to quill feed handle; Adjustable rod which accepts round portion of flattened rod which in turn attaches to back of remote switch with two #10-32 x $\frac{3}{8}$" fillister head screws. Nylon rod or remote switch fits behind quill feed handle and rubs on shoulder of quill feed shaft. **Order No. 46104**

Bridgeport Nozzle Holder **$15.00**

A cup which fits over any one of the $\frac{3}{4}$" hex nuts in the quill area with a set screw to lock in place on one of the hex flats. Protruding from the center of the cup is a $\frac{5}{16}$" x $\frac{1}{2}$" long shaft which will accept the fitting on our lock line nozzle locater.

Order No. 46105

CNC Connecting Cord **$10.00**

A two conductor electrical cord with a plug on one end which fits into switch receptacle on side of basic unit. Other end to be connected to a point closing relay to be supplied by customer. Also, the connection of relay to CNC control Unit is to be done by customer.

Order No. 46106

Truly a single, all purpose coolant, Tapmatics M.E. II can be used for ALL MACHINING OPERATIONS and ALL METALS except magnesium

There is no longer any need for 10 or 20 different products. By maintaining proper dilution ratios and adding rarely needed Tapmatic Additives as required, the user can customize M.E. II to the job or machine. This makes M.E. II the ideal coolant for all central systems.

LONG SUMP LIFE

Stringent disposal restrictions and increasingly higher disposal costs make sump life a paramount concern. M.E. II stays in the sump many times longer than conventional coolants. This alone can more than justify M.E. II's higher initial cost.

SAFE FOR OPERATOR AND ENVIRONMENT

M.E. II is completely free of nitrite, chromate, sulphur, phosphate, chlorine, phenol and ptBBA. It is extremely inhospitable to bacteria or fungus, yet it contains NO POISONOUS BIOCIDES.

When kept free of foreign substance M.E. II will not develop odor or rancidity problems. Tramp oils will not emulsify and are readily skimmed off.

HIGHLY CONCENTRATED

M.E. II Super Concentrate contains no fillers, perfumes or dyes. It is comprised of 95% active ingredients.

DILUTION RATIOS

Dilution Ratios are normally 1½ to 2 times greater than competitive products.

General machining: 40 water to 1 M.E. II
General grinding: 80 water to 1 M.E. II
Gear hobbing: 20 water to 1 M.E. II
Broaching: 10 water to 1 M.E. II
Tapping: From 8 to 16 water to 1 M.E. II with Lubadd II added as necessary.

IMPORTANT

M.E. II will soften or peel some paints. It may emulsify certain highly compounded way oils necessitating a change in oil used.

Molecular Edge II

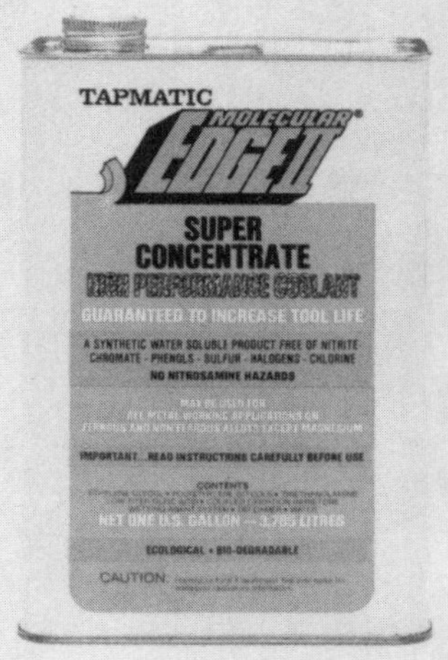

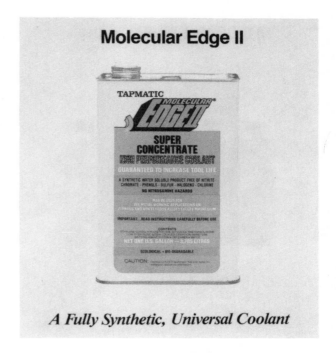

A Fully Synthetic, Universal Coolant

The Coolant Breakthrough That Cuts Cold And Clean!

CONVENTIONAL COOLANTS

Small shear angle causes maximum number of atoms to be disturbed. Stubby, thick chip is produced, maximum heat generated.

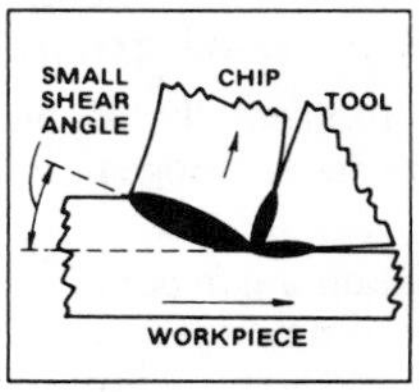

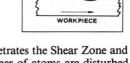

MOLECULAR EDGE II penetrates the Shear Zone and angle is increased. Minimum number of atoms are disturbed, minimum amount of heat produced.

THE DIFFERENCE IS IN THE SHEAR ZONE

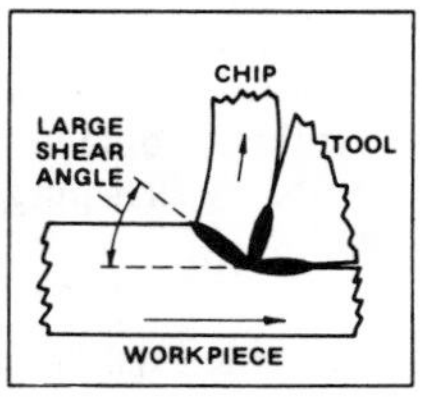

TAPMATIC EDGE COOL

A fully synthetic, water soluble economy coolant. Cools chemically, provides superior lubricity.

Normal dilution ratio: 20:1
Available in 5 gallon packs (2-2½ gallon jugs) and 55 gallon drums

Tapmatic Corporation
1851 Kettering Street
Irvine, CA 92714
(714) 979-6080

ALL TAPMATIC ADDITIVES are available in unbreakable, 17 ounce containers featuring pre-determined dosage markings. A handy 4 pack of HWA, DMA, AFA and LUBADD II is also offered.

Tapmatic Industrial Machine Cleaner is a heavy duty, all purpose, low foaming cleaner for all machine shop applications.

Environmentally safe and fully bio-degradable, it contains no abrasives or halogens. It is not combustible.

While optimum performance will be achieved by using hot water at recommended temperatures, satisfactory results can be obtained with cold water. Stronger solutions may be required in such cases.

MACHINE SUMP AND COOLANT LINE CLEANING

Dual-Action Plus Cleaner may be introduced into old water-based synthetic coolants if desired. However, best results will be obtained by first purging the system of old contaminated coolants.

Extremely fouled systems may require that coolant lines be disassembled for thorough internal cleaning. A bottle or tube brush may be helpful.

CLEANS AND PROTECTS

If protection from rust and corrosion is desired DO NOT RINSE PARTS. A clear, corrosion inhibiting film will be deposited providing effective rust protection in dilutions up to 1:15.

Rinsing is not necessary unless parts are to be painted or plated.

Tapmatic Industrial Machine Cleaner is available in 1 gallon (4 gallons per case), 5 gallon and 55 gallon containers.

In unusual circumstances the use of additives may be indicated. Tapmatic offers four "in-sump" additives and one "pre-mix" additive to help in such situations. It is important that instructions be followed carefully.

LAC Lubricity Additive: Pre-Mix with M.E. II Concentrate when added lubricity is required. This may occur when machining high alloy metals or for extreme high pressure operations, such as tapping, broaching or gear hobbing. (LAC must be added to M.E. II Concentrate ONLY. It should never go directly into mixed water solution).

FOR IN-SUMP APPLICATION:

HWA Hard Water Additive: Hard water may rob a coolant of its effectiveness. Tapmatic's HWA eliminates hardness problems in the sump and maintains M.E. II's unique machining qualities. 20 Hard Water Test Strips are included for on-the-job testing.

AFA Anti-Foaming Additive: Tapmatic's AFA may be added in the sump without interrupting production to eliminate foaming which occurs when exceptionally soft water is encountered, or as a result of poor sump design.

DMA Dissimilar Metals Additive: DMA corrects tarnishing or corrosion caused when dissimilar metals are machined together or on consecutive production runs. This problem may also arise when copper or its alloys (brass, bronze) are present in the coolant system.

LUBADD II: A new, improved formula for in-sump application, Lubadd II provides additional lubricity for smoother, cleaner cutting action and longer tool life.

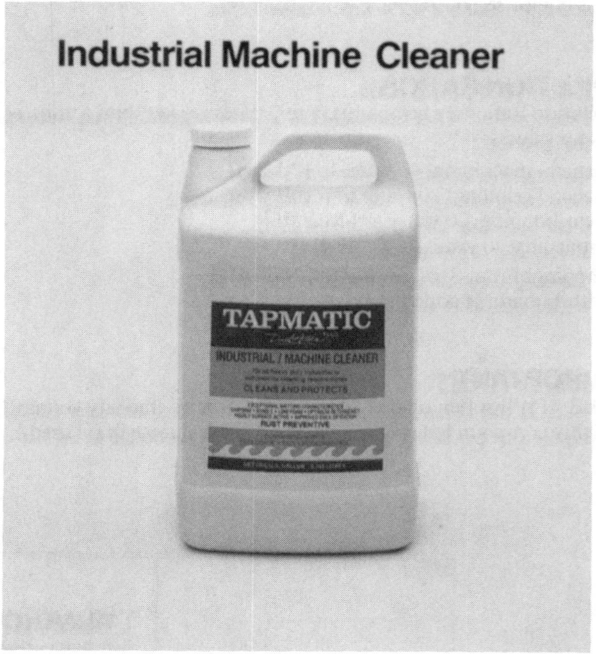

Uni-Mist Incorporated
4134 36th Street S.E./P.O. Box 8344
Grand Rapids, MI 49508
(616) 949-0853

SOME IMPORTANT FACTS CONCERNING COOLUBE 2100

CooLube's cost effectiveness is often overlooked when the price per gallon is first realized. When compared to more conventional flood and mist type lubricants/coolants, CooLube may seem expensive, but this is not the case.

CooLube is a polar lubricant and must be used very sparingly for full lubricating effectiveness. If it is used correctly and the proper amounts are dispensed, CooLube can be the most inexpensive material available to lubricate metalworking operations. Other savings will be realized through longer tool life and lower shop maintenance costs.

Typical usage for a light to medium machining operation:

 1 drop every ten seconds*
=360 drops per hour
=4.8 ounces used in an 8 hr. continuous operation
 (or about .036 gallons)

At the current price for a 55 gal. drum of CooLube this works out to a cost of $2.64 per 8 hr. shift, or 33 to 35 cents per hour.

Typical usage on a heavy machining operation:

 1 drop every eight seconds*
=450 drops per hour
=7.2 ounces per 8 hr. continuous operation

This works out to $3.28 per 8 hr. continuous shift, or 41 to 44 cents per hour.

In comparison the average cost in a typical flood operation includes:

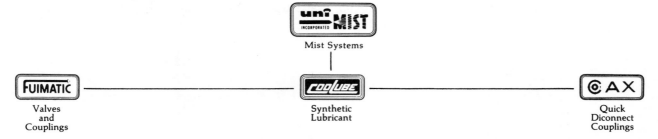

 2.5 to 5 gal. per minute for each nozzle
+5 cents per gal. maintenance costs (recycling and filtering)
+shop maintenance and clean-up costs

Uni-Mist Incorporated
4134 36th Street S.E./P.O. Box 8344
Grand Rapids, MI 49508
(616) 949-0853

*The applications discussed on the previous page are based on the standard drop measurement of Uni-MIST dispensers (1 drop = 1/30th cc) and should be considered an approximation.

CooLube produces dry chips, requires no machine clean-up and little or no part clean-up. It does not stain and is compatible with all known materials. This eliminates the possibility of chemical attack to any base metal or material being used.

CooLube should not be stored below 55 degrees Fahrenheit. Should any separation of the contents become apparent, CooLube can be dissolved by heating to approximately 80°F without harming the material.

ANALYSIS OF COOLUBE FOR DETRIMENTAL MATERIALS

A U.S. Navy Analytical Chemistry Branch has determined the following percentages of detrimental materials in a sample of CooLube 2100.

Test results are as follows:

MATERIAL	FOUND (PPM)
Lead	<50
Antimony	<50
Arsenic	<50
Bismuth	<50
Cadmium	<50
Magnesium	<50
Tin	<50
Zinc	<50
Sulfer	<50
Phosphorus	<50
Chlorine	<50
Florine	<50
Iodine	<50
Bromine	<50
Mercury	< 5

CooLube 2100 meets the requirements of MIL-STD-767B.
CooLube 2100 is insoluble in water.

Price List

1 Gallon..................$75.00
5 Gallons.................$365.00 (Save $2.00 a gal. off the 1 gal. price)
55 Gallon Drum...........$3850.00 (Save $5.00 a gal. off the 1 gal. price)

Union Twist Drill/Litton
134 Chestnut Hill Avenue
Athol, MA 01331-1903
(617) 249-3221

Open up to a new idea in tapping fluid.

Now there's another choice in tapping fluids. And it's from the name industry knows for the finest taps and accessories, Union Twist Drill.

Put this new tapping fluid up against brands of conventional formulations, and you'll discover that it's very *un*conventional.

Safety, Performance, Productivity all in one.

As soon as you open the can you'll notice the difference. But this amazing new tapping fluid has much more to offer besides pleasant fragrance.

- It contains no methyl chloroform or trichloroethane.
- It relies on superior lubrication to prevent high heat generation. Requires less tapping torque.
- It lubricates and prevents corrosion.
- It forms metallic chlorides for cutting purposes.
- It is effective in steel and a wide range of aerospace materials. (Greatly extended tap life is its main attribute.)

Union tapping fluid is a clear, heavy oil which is chlorinated but does not contain chlorinated free solvents. It is a very pure, low acidity fluid. The following are some typical specifications:

Viscosity, Gardner – Holdt	U-V
Free Fatty Acids	1% Max.
Saponification Value	176-184
Moisture & Volatiles	.02% Max.
Hazardous Ingredients	N/A
Reactivity	Stable
Special Protection or Precautions	None

The proof is in the testing.

After extensive testing in the field and laboratory in a variety of applications, Union tapping fluid was found to offer these user benefits:

- Reduced tapping torque
- Fewer scrapped parts
- Improved surface finish on threads
- Accuracy and size control
- Fewer tool changes (longer tool life)
- Less regrinding and resharpening time

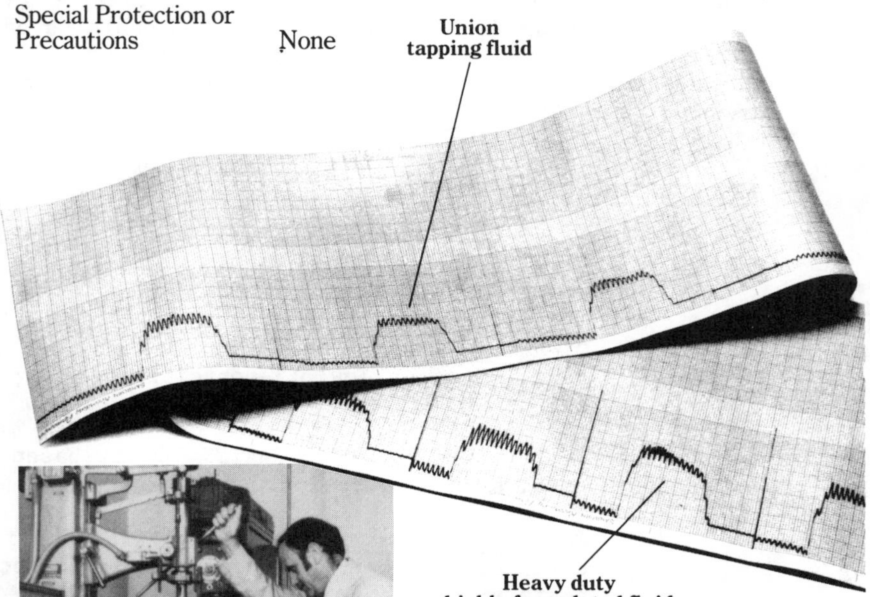

A 33% reduction in tapping torque. Union fluid outperformed a heavy duty tapping fluid normally used by field service engineers. It produced a 33% reduction in tapping torque and noticeable heat reduction while tapping #3 drill holes in 6AL4V Titanium at 150 RPM with a 1/4-28NF-GH4 spiral point tap.

- Excellent boundary lubrication
- Corrosion retardation
- Application to application versatility
- Increased production

Add it all up. Then consider the irritating problems associated with conventional fluids, and the choice is clear. The next time you buy tapping fluid, ask for the best by name. Union Twist Drill. Available through your local distributor now.

UNION TWIST DRILL
Litton Athol, Massachusetts 01331

Van Straaten Chemical Company
630 West Washington Boulevard
Chicago, IL 60606
(312) 454-1000

VAN STRAATEN
Chemical Company

Semi-Synthetic Cutting and Grinding Fluid 560

PRODUCT INFORMATION

HIGH PERFORMANCE SEMI-SYNTHETIC FLUID

VAN STRAATEN 560 is a non-phenolic, non-nitrited high performance semi-synthetic cutting and grinding fluid. VAN STRAATEN 560 contains both chlorine and active sulfur as extreme pressure additives for improved tool/wheel life. VAN STRAATEN 560 has been formulated to offer excellent tank life and rust protection for both piece parts and machine tool surfaces.

KEY PERFORMANCE BENEFITS

Tool Wheel Performance - VAN STRAATEN 560 has been formulated with chlorine and sulfur extreme pressure additives for improved tool/wheel performance.

Cleanliness - VAN STRAATEN 560 provides a clean work environment. It is a low oil-containing semi-synthetic which leaves a slight oil film.

Rust Control - VAN STRAATEN 560 is based on combined polar and passivating protection.

Rancidity Resistance - VAN STRAATEN 560 is formulated using raw materials that have excellent resistance to microbiological degradation.

APPLICATIONS

VAN STRAATEN 560 can be used in both heavy and general machining and grinding applications.

TYPICAL USE CONCENTRATIONS

Milling, Turning	10:1 to 20:1
Drilling, Tapping	10:1 to 20:1
Grinding	20:1 to 40:1

METALS

VAN STRAATEN 560 is recommended for machining and grinding all ferrous metals, aluminum and aluminum alloys, stainless steels and alloy steels which do not contain copper.

Van Straaten Chemical Company
630 West Washington Boulevard
Chicago, IL 60606
(312) 454-1000

VAN STRAATEN
Chemical Company

Semi-Synthetic Cutting and Grinding Fluid 585

PRODUCT INFORMATION

HIGH PERFORMANCE, NITRITE & PHENOL FREE, SEMI-SYNTHETIC

VAN STRAATEN 585 is a high performance semi-synthetic cutting and grinding fluid which has been formulated with chlorine as an extreme pressure additive for tool/wheel life assist on difficult applications. It is a low oil-containing product which leaves a slight oil film resulting in a cleaner work area. This is one of the reasons it has received such outstanding operator acceptance. VAN STRAATEN 585 has been developed to offer excellent tank life and rust protection for both piece parts and machine tool surfaces.

VAN STRAATEN 585 forms a blue emulsion when mixed with most natural waters. If a clear emulsion is preferred, VAN STRAATEN 585-C should be specified.

KEY PERFORMANCE BENEFITS

Tool Wheel Performance - VAN STRAATEN 585 has been formulated with an extreme pressure package for improved tool/wheel life.

Emulsion Stability - VAN STRAATEN 585 forms a stable emulsion in both hard and soft water.

Cleanliness - VAN STRAATEN 585 is a low oil-containing semi-synthetic which leaves a slight oil film resulting in a cleaner work environment.

APPLICATIONS

VAN STRAATEN 585 can be used in severe operations as it contains an extreme pressure lubricant. It is recommended for both ferrous and non-ferrous metals. Use on copper alloys is not suggested.

TYPICAL USE CONCENTRATIONS

Milling, Turning	10:1 to 20:1
Drilling, Tapping	10:1 to 20:1
Grinding	20:1 to 40:1

Van Straaten Chemical Company
630 West Washington Boulevard
Chicago, IL 60606
(312) 454-1000

VAN STRAATEN
Chemical Company

Soluble Cutting and Grinding Fluid 612

PRODUCT INFORMATION

WATER SOLUBLE CUTTING & GRINDING FLUID

VAN STRAATEN 612 is a water soluble cutting and grinding fluid for general machining and grinding. It has been formulated to be easily split in conventional waste treatment systems.

KEY PERFORMANCE BENEFITS

- Emulsion Stability - VAN STRAATEN 612 forms a stable emulsion in either hard or soft water.
- Extended System Life - VAN STRAATEN 612 provides long life in both individual sumps and central systems. It is based on raw materials which have excellent resistance to microbiological degradation.
- Soft Fluid Residue - VAN STRAATEN 612 leaves a soft, non-tacky residue providing lubrication to fixturing and gauges.
- Metal Compatability - VAN STRAATEN 612 can be safely used with both ferrous and non-ferrous metals. It is nonstaining to aluminum and copper.

RECOMMENDED DILUTIONS

Milling, Drilling and Turning	10:1 to 20:1
Boring, Tapping and Threading	10:1
Grinding	20:1 to 40:1

Van Straaten Chemical Company
630 West Washington Boulevard
Chicago, IL 60606
(312) 454-1000

VAN STRAATEN
Chemical Company

Soluble Cutting and Grinding Fluid

653

PRODUCT INFORMATION

WATER SOLUBLE CUTTING AND GRINDING FLUID

VAN STRAATEN 653 is an emulsifiable cutting and grinding fluid for general machining and grinding of ferrous and non-ferrous alloys. The concentrate mixes easily with hard or soft water to form a stable, milky white emulsion that is extremely rancid resistant, and it is formulated to keep machines and parts clean and free of sticky residues. Strong polar rust inhibitors protect machine tools from rust and prevent galvanic corrosion that sometimes occurs around adjusting screws or between machine tables and fixtures. VAN STRAATEN 653 is safe and pleasant to use. It is mild and non-irritating to the skin and has a pleasant, neutral odor. VAN STRAATEN 653 is an excellent choice for general use in shops that handle many different alloys in a variety of machining and grinding operations.

KEY PERFORMANCE BENEFITS

- Stability and Longevity - The concentrate mixes easily into water of various hardnesses and the emulsion provides long life in both individual equipment sumps and central systems, due to the selection of raw materials that have excellent resistance to microbiological degradation.

- Operator Safety and Acceptance - VAN STRAATEN 653 is field proven in terms of operator acceptance and safety.

- Soft Fluid Residue - Offers lubrication to fixturing and gauges making equipment maintenance easy.

- Non-foaming - Even under severe agitation conditions such as a Blanchard Grinder with soft water conditions.

- Compatibility with Work Materials - Use VAN STRAATEN 653 for all machining and grinding of ferrous metals, aluminum and copper alloys. Nonstaining to aluminum and copper.

RECOMMENDED DILUTIONS

Milling, drilling, turning	10:1 to 20:1
Boring, tapping, threading	10:1
General grinding	20:1 to 40:1

Van Straaten Chemical Company
630 West Washington Boulevard
Chicago, IL 60606
(312) 454-1000

VAN STRAATEN
Chemical Company®

Soluble Cutting and Grinding Fluid 709

PRODUCT INFORMATION

EXTREME PRESSURE BELT GRINDING FLUID

VAN STRAATEN 709 is an extremely high performance soluble oil containing a unique additive package specifically formulated for belt grinding and other difficult abrasive machining and grinding operations. VAN STRAATEN 709 has been formulated to provide lubrication which has proven field value in terms of greater grinding ratios obtainable on difficult to machine metals such as stainless and high nickel alloys.

KEY PERFORMANCE BENEFITS

- Rancidity resistance - formulated with an exceptional biocide package to provide excellent bio and fungal resistance.

- Stability - the emulsion system is stable in both hard and soft water and is non-foaming even under the most severe agitation conditions.

- High performance - comparable to conventional straight oils. Contains both polar and chemical lubricants including a unique combination of sulfur and chlorine which results in exceptional surface finish and excellent belt life.

COMPATABILITY WITH WORK MATERIAL

VAN STRAATEN 709 is recommended for belt and conventional grinding of the difficult to machine alloys, stainless steels and tough aluminum alloys. The product can also be used on heavy-duty machining operations of these same metals.

RECOMMENDED DILUTIONS

General grinding	10:1 to 30:1
Belt grinding	10:1 to 40:1
Machining	10:1 to 20:1

Van Straaten Chemical Company
630 West Washington Boulevard
Chicago, IL 60606
(312) 454-1000

VAN STRAATEN
Chemical Company

Synthetic Cutting and Grinding Fluid 826

PRODUCT INFORMATION

OIL ABSORBING SYNTHETIC CUTTING & GRINDING FLUID

VAN STRAATEN 826 is an oil absorbing, nitrite and phenol free synthetic cutting and grinding fluid having the capability of absorbing 1-2% tramp oil. This product provides excellent rust protection. Its performance brackets a wide range of cutting and grinding operations while maintaining the high operator acceptance of synthetic fluids.

KEY PERFORMANCE BENEFITS

- Unique tool/wheel performance - formulated with a high performance package offering excellent tool/wheel life.
- Rancidity resistance - excellent bio and fungal resistance; built with a unique conditioner system which will not break down when contaminated with tramp oil.
- Residue characteristics - oil-absorbing ability, leaves a soft, fluid residue providing lubrication for the machine tool.
- Filtration properties - designed to carry chips and swarf away from the work area. When used with cast or nodular iron, 826 helps prevent "clinkering" associated with classical synthetics.
- Stable product - suitable for use in a wide range of water conditions.

COMPATABILITY WITH WORK MATERIALS

VAN STRAATEN 826 is recommended for machining and grinding all ferrous metals, stainless steels and alloys.

RECOMMENDED DILUTIONS

Heavy Duty Machining	10:1 to 20:1
Heavy Duty Grinding	20:1 to 40:1

Van Straaten Chemical Company
630 West Washington Boulevard
Chicago, IL 60606
(312) 454-1000

VAN STRAATEN
Chemical Company

Synthetic Cutting and Grinding Fluid 850

PRODUCT INFORMATION

HIGH PERFORMANCE SYNTHETIC CUTTING AND GRINDING FLUID

VAN STRAATEN 850 is a synthetic cutting and grinding fluid formulated for non-ferrous metals, particularly aluminum. The formula for VAN STRAATEN 850 incorporates state of the art synthetic technology that provides performance characteristics previously found only in heavy-duty soluble oils. This fluid is both nitrite and phenol free.

KEY PERFORMANCE BENEFITS

- Metal safety - formulated to be non-staining on aluminum and to provide excellent bi-metallic corrosion resistance.
- Rancidity resistance - incorporates a unique biocide and fungicide for extended system life.
- Multiple performance capabilities - can be used for tapping, milling, drilling, reaming and other general machining operations.
- Stable product - mixes easily in both hard and soft water and forms a stable emulsion.
- Cleanliness - emulsifies about 2-3% tramp oil thus leaving a soft fluid residue.

COMPATABILITY WITH WORK MATERIAL

VAN STRAATEN 850 is recommended for machining and grinding aluminum and aluminum alloys. This product can also be used on ferrous materials.

RECOMMENDED DILUTIONS

Grinding	30: to 50:1
Heavy-duty machining	7.5:1 to 20:1
General machining	10:1 to 25:1

VAN STRAATEN
Chemical Company

Synthetic Cutting and Grinding Fluid 902

PRODUCT INFORMATION

SYNTHETIC CUTTING & GRINDING FLUID

VAN STRAATEN 902 is a nitrite and phenol free synthetic, specially formulated for use with hard water. VAN STRAATEN 902 provides excellent rust protection and bioresistance along with high performance characteristics. This product can be used for all types of machining and grinding operations.

KEY PERFORMANCE BENEFITS

- Excellent Rust Protection - VAN STRAATEN 902 offers rust protection comparable to the best nitrite containing synthetics.
- Hard Water Stability - VAN STRAATEN 902 maintains excellent rancidity protection and rust protection in hard water.
- Bioresistance - VAN STRAATEN 902 is formulated to offer extended life in central systems and individual sumps.
- Metal Safety - VAN STRAATEN 902 provides high protection against electrolytic corrosion caused by iron-copper or iron-aluminum couples.
- Excellent Settling Properties - VAN STRAATEN 902 provides rapid swarf settling to maintain clean systems.

RECOMMENDED USE DILUTIONS

Machining	20:1 to 40:1
Grinding	40:1 to 60:1

Van Straaten Chemical Company
630 West Washington Boulevard
Chicago, IL 60606
(312) 454-1000

VAN STRAATEN
Chemical Company

Synthetic Cutting and Grinding Fluid 975

PRODUCT INFORMATION

A PERFORMANCE STANDARD FOR SYNTHETICS

Nitrite-free VAN STRAATEN 975 is a significant step forward in providing corrosion protection comparable to the best nitrite-containing synthetics. In addition to out-performing conventional sulfo-chlorinated fatty solubles in rigorous belt and form grinding operations, 975 has dramatically improved grinding performance and G ratios on various conventional grinding systems and metals. Similarly, VAN STRAATEN 975 has shown high performance on tough machining jobs proving itself in bar machines and chuckers in severe applications including reaming and tapping.

Operator acceptance of VAN STRAATEN 975 is excellent because of its pleasant odor and mildness due to low alkalinity. It is safe to use because it contains no known skin allergens. This product is free of phenolics and formaldehyde preservatives. It is also free of MEA and mineral oils.

In either hard or soft water, VAN STRAATEN 975 forms a stable solution and will not foam. It settles chips and swarf quicker than most solubles or synthetics and its non-tacky, non-gummy, soft residue provides good lubrication for machinery.

KEY PERFORMANCE BENEFITS

- High Performance - VAN STRAATEN 975 outperforms conventional sulfo-chlorinated fatty solubles in rigorous belt and form grinding operations and improves grinding performance and G ratios.

- Rancidity Resistance - VAN STRAATEN 975 eliminates continual rancidity problems associated with solubles.

- Tramp Oil Rejection - VAN STRAATEN 975 totally rejects tramp oil.

- Excellent Rust Protection - Positive rust protection provided by VAN STRAATEN 975 is comparable to that of the best nitrite-containing synthetics.

RECOMMENDED DILUTIONS

Heavy-duty machining	10:1 to 20:1
Heavy-duty grinding	20:1 to 40:1

Van Straaten Chemical Company
630 West Washington Boulevard
Chicago, IL 60606
(312) 454-1000

VAN STRAATEN
Chemical Company

Cutting and Grinding Oil 5299 Series

PRODUCT INFORMATION

HIGH PERFORMANCE CUTTING AND GRINDING OILS

Van Straaten 5299 and 5299–M are high performance straight oils varying in viscosity and offering superior performance that has withstood the test of time in the marketplace. These straight oils are formulated using both sulfurized and chlorinated extreme pressure additives in combination with polar and hydrodynamic lubrication. This formula was selected empirically to offer a fluid which prevents chip welding or metal pick-up on the tool or grinding wheel. Van Straaten 5299 is used extensively where fine finishes and high grinding performance are essential, such as in honing, form grinding and belt polishing. It is also used extensively for gear shaving and broaching operations where metal pick-up must be prevented. Van Straaten 5299–M has demonstrated superior performance on difficult form and crush grinding applications.

KEY PERFORMANCE BENEFITS

- An extreme pressure and polar lubrication package offering excellence in performance, especially on the difficult to machine ductile alloys.

- The polar lubrication package incorporated can actually be considered an emollient. Therefore, it is extremely safe and pleasant to work with 5299 and 5299–M.

- The choice of viscosities offered gives the user a versatile choice from an extremely fine and fluid oil for honing, to an oil offering the fluid film strength necessary for difficult crush and plunge grinding.

- The balanced combination of the performance package makes Van Straaten 5299 and 5299–M suitable for both machining and grinding, offering the user a single material where he has both operations in-house.

Westmont Products
P.O. Box 224049
Dallas, TX 75222
(214) 438-0588

BIO-COOL® 250

WESTMONT PRODUCTS
P.O. Box 224049
Dallas, TX 75222, 214/438-0588

Heavy-duty synthetic coolant with superior lubricity and rust protection

BIO-COOL 250 is an excellent water-soluble metalworking fluid that incorporates corrosion inhibitors, synthetic lubricants and surface active ingredients. It is designed to meet the demands of today's machining needs with ever-increasing speeds and feeds.

BIO-COOL 250 is ideal for machining and grinding all ferrous metals. High lubricity makes it especially good on non-ferrous materials, cast steel, cast iron and cast alloys. BIO-COOL 250 prevents galling when cutting soft alloys and will not rust, stain or sour when machining castings.

Benefits of BIO-COOL 250

- **SUPERIOR COOLING**—Absorbs and transfers heat with top efficiency.
- **EXCELLENT RUST PROTECTION**—Contains a highly effective system of rust inhibitors.
- **TRANSPARENT SOLUTION**—Stays clean and clear for better viewing of tool and workpiece.
- **RESISTS RANCIDITY**—Synthetic composition resists rancidity for longer life and no foul odors.
- **STABLE IN ANY WATER**—Forms no insoluble matter, even in hard water.
- **NON-HAZARDOUS**—Safe to skin and clothing.
- **NON-FOAMING**—In hard or soft water.
- **BIODEGRADABLE**—Requires no special handling or disposal; contains no oil, phenols, phosphates, PCBs, nitrites or heavy metals.
- **VERSATILE APPLICATION**—Works well in a wide range of applications, including cutting, drilling, milling, turning, sawing and grinding.

APPLICATION

For best results with BIO-COOL 250, first dump old coolant and flush sump and coolant system with BIO-CLEAN™ 18 biodegradable machine cleaner. Then charge sump with BIO-COOL 250 mixed 1 part coolant concentrate to 20 parts water for general machining, 1 to 40 for grinding, 1 to 30 for spray mist and 1 to 10 for sawing and other tough machining operations. Use the Westmont HAND-HELD REFRACTOMETER to determine and maintain proper concentrations.

DISTRIBUTED BY:

Another high technology product from WESTMONT PRODUCTS

Westmont Products
P.O. Box 224049
Dallas, TX 75222
(214) 438-0588

WESTMONT PRODUCTS

BIO-COOL® 250

Specifically developed for high productivity and a cleaner shop environment

PHYSICAL PROPERTIES

- Specific gravity......1.086
- Viscosity...........43 SUS
- Water solubility.....100%
- Odor ....Neutral
- pH (Concentrate)....9.5
- Density (per gal.).....9.0
- Flashpoint (open cup) None
- Appearance.........Transparent liquid; color dye added

CHEMICAL PROPERTIES

- Contains no nitrites
- Contains no nitrosamines
- Contains no nitrates
- Contains no chlorine compounds
- Contains no sulfur compounds
- Contains no PCB's
- Contains no phenols
- Contains no phosphates
- Contains no soaps
- Contains no silicones
- Contains no oils
- 100% biodegradable

PREPARATION

- Easy to mix with minimum stirring or agitation
- Starting dilution rate of 20 to 1 for machining and 40 to 1 for grinding
- Mix is stable
- Flush machine before adding BIO-COOL
- Use refractometer to maintain concentrate
- Periodic skimming of contaminants lengthens time between complete changes and reduces coolant costs
- No additives required

TOOL PERFORMANCE

- In actual tests, increased tool life by as much as 100%
- Super-cooling powers prevent heat build-up and heat distortion
- Metal chips come off tool cold to the touch
- Reduces resistance forces at tool point to lower heat, tool wear and chatter

MACHINE PERFORMANCE

- Superior lubricity and coolant power permits machine feed and speed increases from 50 to 100%
- Cleaner machine operation with no oily residues
- Actually cleans out sumps and pumps allowing improved efficiency
- Prevents sludge build-up, extending life of coolant pumps
- Decreases machine downtime
- Overflows, spills or splashes do not require floor compound for soaking because there is no oil
- Translucency allows operator to see work
- No effect on paint, finishes, leather, fabric, rubber, etc.
- Will not leave gummy residue

DISPOSAL

- 100% biodegradable
- Dispose through any primary sewage treatment facility

FINISH PIECE GOODS

- Excellent for both ferrous and non-ferrous metals, also many non-metallic materials
- Better dimensional stability and accuracy
- Actual SFPM more consistent with programmed calculations
- Fewer rejects
- Will not tarnish aluminum, nickel, copper, brass or zinc
- No staining
- No rinsing of parts before heat-treating
- Corrosion resistant even at high dilution ratios

SHOP ENVIRONMENT

- Non-smoking, which can result in increased lighting efficiency and an appreciable decrease in air conditioning or ventilation costs
- Nonflammable, which can result in reduced insurance costs
- Completely odorless
- Not an emulsion so it cannot go sour or turn rancid, which means less frequent changes
- Less clean-up time and expense
- Less downtime for machine maintenance, tool change-over, etc.
- Will not support bacterial or fungal attack
- Non-foaming

OPERATOR SAFETY

- Non-toxic fumes, even when inhaled for extended time periods or at elevated temperatures
- Skin contact is absolutely safe—no tumors or oil boils—it's actually a natural skin cleanser
- Eye contact may have slight irritating effect; flush with water
- Swallowing may cause slight, temporary stomach irritation

LEGAL COMPLIANCE

- Exceeds requirements of Federal Hazardous Substances Act in regard to toxicity, inhalation, eye and skin irritation
- Exceeds requirements of Occupational Safety and Health Administration (OSHA)
- Exceeds requirements of the Environmental Protection Agency (EPA)
- Exceeds requirements of A.S.M.E. Section III

MANAGEMENT CONSIDERATIONS

- Increases machine operation speeds
- Increases tool life
- Decreases downtime
- Can reduce fire insurance costs
- Can reduce health insurance costs
- Indirect savings in lighting, air conditioning and cleaning supplies
- Fewer changes means less inventory
- Universal application to most machines means less inventory
- Less inventory means less procurement, stocking and paperwork
- Decrease in rejects and reworks
- Better worker morale due to more pleasant work environment
- Increased output and efficiency

Westmont Products
P.O. Box 224049
Dallas, TX 75222
(214) 438-0588

BIO-COOL® 500

Universal synthetic coolant

WESTMONT PRODUCTS
P.O. Box 224049
Dallas, TX 75222, 214/438-0588

A biodegradable product for improved machine performance and a cleaner shop environment

NEW, IMPROVED FORMULA with increased rust and corrosion protection

BIO-COOL 500 is a premium heavy-duty, water-soluble coolant and cutting fluid designed for universal application on many metalworking operations. It is excellent for turning, milling, drilling, reaming, boring, threading, grinding, light stamping and any other high-speed machining operation. BIO-COOL 500 contains no oil, phenols, phosphates, PCBs, nitrites or heavy metals—it is 100 percent biodegradable. It is smokeless, odorless, nonflammable and non-polluting.

Improved machine performance

BIO-COOL 500 provides faster machine speeds and feeds, while extending tool life and improving finishes. BIO-COOL 500 has superior lubricity, wetting ability and heat transfer characteristics. These qualities help lower cutting forces at the tool point, substantially reducing excessive heat buildup, tool wear and chatter.

Benefits of BIO-COOL 500

- **EXCELLENT PERFORMANCE**—On both ferrous and non-ferrous metals, as well as many non-metallic materials.
- **RUST AND CORROSION RESISTANCE**—Even at high dilution ratios.
- **NO STICKING**—Won't form gummy residues.
- **PREVENTS RANCIDITY**—Lasts longer for increased economy; no unpleasant odors.
- **NON-FOAMING**—In hard or soft water areas.
- **SAFE**—Keeps machine and work area clean.
- **NO SMOKE**—Or harmful vapors.
- **SAVES TIME AND LABOR**—Parts can be heat treated or plated without rinse.
- **REDUCES INVENTORY**—One product for many applications.
- **NON-HAZARDOUS**—Safe to skin and clothing.
- **TRANSPARENT**—For better viewing of work pieces.

APPLICATION

For best results with BIO-COOL 500, first flush machine with BIO-CLEAN™ 18 biodegradable machine cleaner. Then charge sump with BIO-COOL 500 diluted 20:1 for general machining, 40:1 for grinding, 30:1 for spray mist and 10:1 for sawing and other tough machining. Use the Westmont HAND-HELD REFRACTOMETER to determine and maintain proper concentrations.

DISTRIBUTED BY:

Westmont Products
P.O. Box 224049
Dallas, TX 75222
(214) 438-0588

WESTMONT PRODUCTS

BIO-COOL® 500

Specifically developed for high productivity and a cleaner shop environment

PHYSICAL PROPERTIES

- Specific gravity 1.05
- Viscosity @ 100 32
- Water solubility 100%
- Odor .... Neutral
- pH .... 8.3 (approx.)
- Density (per gal.) 8.8
- Flash point (open cup) . None
- Appearance .... Transparent liquid; color die added

CHEMICAL PROPERTIES

- Contains no nitrites
- Contains no nitrosamines
- Contains no nitrates
- Contains no chlorine compounds
- Contains no sulfur compounds
- Contains no PCB's
- Contains no phenols
- Contains no phosphates
- Contains no soaps
- Contains no silicones
- Contains no oils
- 100% biodegradable

PREPARATION

- Easy to mix with minimum stirring or agitation
- Starting dilution rate of 20 to 1 for machining and 40 to 1 for grinding
- Mix is stable
- Flush machine before adding BIO-COOL
- Use refractometer to maintain concentrate
- Periodic skimming of contaminants lengthens time between complete changes and reduces coolant costs
- No additives required

TOOL PERFORMANCE

- In actual tests, increased tool life by as much as 100%
- Super-cooling powers prevent heat build-up and heat distortion
- Metal chips come off tool cold to the touch
- Reduces resistance forces at tool point to lower heat, tool wear and chatter

MACHINE PERFORMANCE

- Superior lubricity and coolant power permits machine feed and speed increases from 50 to 100%
- Cleaner machine operation with no oily residues
- Actually cleans out sumps and pumps allowing improved efficiency
- Prevents sludge build-up, extending life of coolant pumps
- Decreases machine downtime
- Overflows, spills or splashes do not require floor compound for soaking because there is no oil
- Translucency allows operator to see work
- No effect on paint, finishes, leather, fabric, rubber, etc.
- Will not leave gummy residue

DISPOSAL

- 100% biodegradable
- Dispose through any primary sewage treatment facility

FINISH PIECE GOODS

- Excellent for both ferrous and non-ferrous metals, also many non-metallic materials
- Better dimensional stability and accuracy
- Actual SFPM more consistent with programmed calculations
- Fewer rejects
- Will not tarnish aluminum, nickel, copper, brass or zinc
- No staining
- No rinsing of parts before heat-treating
- Corrosion resistant even at high dilution ratios

SHOP ENVIRONMENT

- Non-smoking, which can result in increased lighting efficiency and an appreciable decrease in air conditioning or ventilation costs
- Nonflammable, which can result in reduced insurance costs
- Completely odorless
- Not an emulsion so it cannot go sour or turn rancid, which means less frequent changes
- Less clean-up time and expense
- Less downtime for machine maintenance, tool change-over, etc.
- Will not support bacterial or fungal attack
- Non-foaming

OPERATOR SAFETY

- Non-toxic fumes, even when inhaled for extended time periods or at elevated temperatures
- Skin contact is absolutely safe—no tumors or oil boils—it's actually a natural skin cleanser
- Eye contact may have slight irritating effect; flush with water
- Swallowing may cause slight, temporary stomach irritation

LEGAL COMPLIANCE

- Exceeds requirements of Federal Hazardous Substances Act in regard to toxicity, inhalation, eye and skin irritation
- Exceeds requirements of Occupational Safety and Health Administration (OSHA)
- Exceeds requirements of the Environmental Protection Agency (EPA)
- Exceeds requirements of A.S.M.E. Section III

MANAGEMENT CONSIDERATIONS

- Increases machine operation speeds
- Increases tool life
- Decreases downtime
- Can reduce fire insurance costs
- Can reduce health insurance costs
- Indirect savings in lighting, air conditioning and cleaning supplies
- Fewer changes means less inventory
- Universal application to most machines means less inventory
- Less inventory means less nrocurement, stocking and paperwork
- Decrease in rejects and reworks
- Better worker morale due to more pleasant work environment
- Increased output and efficiency

Westmont Products
P.O. Box 224049
Dallas, TX 75222
(214) 438-0588

BIO-COOL® 77

High performance chemical coolant

WESTMONT PRODUCTS
P.O. Box 224049
Dallas, Texas 75222, 214/438-0588

Provides extra lubrication and extreme pressure protection for maximum performance in tough machining operations

BIO-COOL 77 is a heavy duty, water soluble, chemical coolant designed for tough applications where highest lubrication and cooling are required.

BIO-COOL 77 has special friction-reducing additives that permit higher speeds and feeds during intricate machining of stainless steel, tough alloys and space age materials.

The high performance characteristics of BIO-COOL 77 mean increased productivity, longer coolant and tool life, less downtime and improved shop environment.

Benefits of BIO-COOL 77

- **LONGER TOOL LIFE**—Extreme pressure agents and high lubricity make cutting tools last longer.
- **HIGHER SPEEDS AND FEEDS**—High performance characteristics permit maximum operation and production.
- **VERSATILE APPLICATION**—Ideal for all machining operations, including turning, milling, boring, drilling, reaming, tapping, broaching, sawing, grinding, light stamping and drawing.
- **RESISTS RANCIDITY**—Remains fresh, clean, odorless when used according to recommendations.
- **NON-IRRITATING**—Won't harm skin.
- **NONFLAMMABLE**—Creates a safer, cleaner, more pleasant shop environment.
- **ECONOMICAL**—Mixes with up to 50 parts water for grinding, up to 20 parts water for general machining.
- **RUST AND CORROSION PROTECTION**—Special ingredients protect machining tools and piece parts from rust and corrosion.
- **NONFOAMING**—Effective in hard and soft water without foaming.

APPLICATION

For best results with BIO-COOL 77, first dump old coolant and flush sump and coolant system with BIO-CLEAN™ 18 machine cleaner.

Dilution with water: Pre-mix BIO-COOL 77 with water before charging sump. (Always add concentrate to water or use an automatic proportioner.) Mix BIO-COOL 77 at a ratio of 1 part concentrate to 20 parts water for general machining, 1 to 10 for bandsawing, tapping and broaching and 1 to 50 for grinding. Use the Westmont HAND-HELD REFRACTOMETER to determine and maintain proper concentrations.

Dilution with oil: Mix BIO-COOL 77 at a rate of 1 part concentrate with 15 parts pale oil for use in automatic chuckers, screw machines, broaching machines, gun-drilling machines and where water is not acceptable.

DISTRIBUTED BY:

Another high technology product from WESTMONT PRODUCTS

12-82

CUTTING FLUIDS

Westmont Products
P.O. Box 224049
Dallas, TX 75222
(214) 438-0588

BIO-CUT™

WESTMONT PRODUCTS
P.O. Box 224049
Dallas, TX 75222, 214/438-0588

Combines the benefits of a coolant, cutting oil and rust inhibitor for faster, more convenient cutting operations.

BIO-CUT is ideal for jobs requiring smaller amounts of cutting fluid—such as pipe threading, drilling, tapping, precision tool and die making and general shop use. It's excellent for ferrous and nonferrous metals, as well as exotic metals. BIO-CUT can also be used on harder and/or more abrasive materials. BIO-CUT can be used on aluminum, and won't react adversely to alloys. For continuous operations requiring larger amounts of cutting fluid, use Westmont BIO-COOL® 77 high-performance coolant.

Benefits of BIO-CUT

- **VERSATILE APPLICATION**—Excellent for threading, tapping, drilling and reaming operations and manual pipe threading on the job site.
- **EXTENDS TOOL LIFE**—Rapidly dissipates heat for cleaner, smoother cuts and less wear on tools.
- **PREVENTS REJECTS**—Effective for metals that are difficult to work or tear easily.
- **PROTECTS FROM RUST AND CORROSION**
- **REDUCES WASTE**—Offers better grade of cut and finish.
- **TRANSPARENT**—For highest work visibility.

Application

Shake well. Attach extension nozzle to aerosol actuator (nozzles included in each case of BIO-CUT). Spray BIO-CUT directly on contact area of work piece and tool while material is being worked.

Consult label for complete directions before using.

Another high technology product from Westmont Products.

DISTRIBUTED BY:

8M408

Wynn Oil Company
P.O. Box 4370/2600 East Nutwood Avenue
Fullerton, CA 92634
(714) 992-2000

WYNN'S INDUSTRIAL PRODUCTS DIVISION

WYNN'S TAPPING COMPOUND

Blend of extreme pressure agents, anti-weld additives and corrosion inhibitors for tapping, drilling and threading for a wide range of ferrous and non-ferrous metals.

WYNN'S TAPPING COMPOUND is a clear amber fluid containing ample amounts of friction reducing agents.

Effective on all metals, including aluminum and titanium.

Increases feeds and speeds.

Improves finish and quality of work.

Provides maximum tool life.

Clean and easy to use.

Prevents tool seizure and galling.

Excellent for blind hole tapping.

TAPPING COMPOUND

Wynn Oil Company
P.O. Box 4370/2600 East Nutwood Avenue
Fullerton, CA 92634
(714) 992-2000

GENERAL APPLICATION AND CONCENTRATION TABLE

MACHINING METHOD	TOOL MATERIAL	STEELS			CAST IRON		NON-FERROUS		
		FREE MACHINING MEDIUM CARBON MALLEABLE CAST	LOW CARBON HIGH CARBON ALLOY STEELS	STAINLESS STEELS TOUGH CARBON TOUGH ALLOY	ALLOYED CHILLED	SOFT GRAY CLOSE GRAIN	ALUMINUM SOFT BRASS COPPER	ALUMINUM ALLOY HARD BRASS BRONZE	TITANIUM & OTHER SULFUR & CHLORINE SENSITIVE ALLOYS
Broaching	HSS	X	X	X	X	X	X	X	X
	CARBIDE	X	X	X	X	X	X	X	X
Threading, Pipe Reaming	HSS	X	X	X	X	X	X	X	X
	CARBIDE	X	X	X	X	X	X	X	X
Threading, Tapping	HSS	X	X	X	X	X	X	X	X
	CARBIDE	X	X	X	X	X	X	X	X
Gears: Hob, Cut, Shape, Shave	HSS								
Drilling, Deep Hole	HSS								
	CARBIDE								
Milling	HSS								
	CARBIDE								
Drilling	HSS	X	X	X	X	X		X	X
	CARBIDE	X	X	X	X	X		X	X
Turret Lathes, Automatic Lathes, Single and Multiple Spindle Automatics, Drill Forming	HSS								
	CARBIDE								
Turning, Forming Boring Lathes	HSS								
	CARBIDE								
Sawing, Circular & Hack	HSS								
	CARBIDE								
Grinding, Plain & Surface									
Grinding, Centerless Thread									
Form Rolling									
Stamping									
Drawing									

WYNN'S TAPPING COMPOUND to be applied on above operations only by aerosol, squirt can or brush.

CAUTION: This product is not to be used in flood applications.

TECHNICAL DATA

Appearance:	Clear
Color:	Amber
Spec. Gravity @ 60° F:	1.07
Flash Point:	None
Base Oil Viscosity @ 100° F:	105
Lb./gal.:	8.91

Wynn Oil Company
P.O. Box 4370/2600 East Nutwood Avenue
Fullerton, CA 92634
(714) 992-2000

wynn's
INDUSTRIAL PRODUCTS DIVISION

WYNN'S 331-1 COOLANT

331-1 COOLANT

This heavy-duty soluble oil contains a substantial amount of extreme pressure and anti-weld additives, giving this product a broad range of metalworking applications.

WYNN'S 331-1 COOLANT is a blend of oils, extreme pressure agents, emulsifiers, rust inhibitors and preservatives which form a stable, blue-green emulsion in water.

- Longer tool life, superior finish and more accurate cuts than general purpose soluble oils.
- Provides excellent rust and corrosion protection on workpiece and machine tool.
- Excellent cooling feature allows increase in cutting speed and feeds.
- Cleaner operating conditions eliminate oil misting and smoke.
- Anti-weld and extreme pressure additives prevent metal build-up on tooling.
- Ideal for milling, drilling, turning and grinding of all metals.
- Certified formulation — no nitrites, no carcinogenic nitrosamines.

Wynn Oil Company
P.O. Box 4370/2600 East Nutwood Avenue
Fullerton, CA 92634
(714) 992-2000

GENERAL APPLICATION AND CONCENTRATION TABLE

MACHINING METHOD	TOOL MATERIAL	STEELS			CAST IRON		NON-FERROUS	
		FREE MACHINING MEDIUM CARBON MALLEABLE CAST	LOW CARBON HIGH CARBON ALLOY STEELS	STAINLESS STEELS TOUGH CARBON TOUGH ALLOY	ALLOYED CHILLED	SOFT GRAY CLOSE GRAIN	ALUMINUM SOFT BRASS COPPER	ALUMINUM ALLOY HARD BRASS BRONZE
Broaching	HSS	20%	20%				10%	10%
	CARBIDE	20%	20%					
Threading, Pipe Reaming	HSS	20%	20%				10%	10%
	CARBIDE	20%	20%					
Threading, Tapping	HSS	20%	20%				10%	10%
	CARBIDE	20%	20%					
Gears: Hob, Cut, Shape, Shave	HSS							
Drilling, Deep Hole	HSS	20%						
	CARBIDE	20%						
Milling	HSS	10%	10%	15%	15%		5%	5%
	CARBIDE	5%	5%	10%	10%		5%	5%
Drilling	HSS	10%	10%	15%	15%		5%	5%
	CARBIDE	10%	10%	15%	15%		5%	5%
Turret Lathes, Automatic Lathes, Single and Multiple Spindle Automatics, Drill Forming	HSS	5%	5%	10%	15%		5%	5%
	CARBIDE	5%	5%	10%	15%		5%	5%
Turning, Forming Boring Lathes	HSS	5%	5%	10%	15%		5%	5%
	CARBIDE	5%	5%	10%	15%		5%	5%
Sawing, Circular & Hack	HSS	20%	20%	20%	20%		10%	10%
	CARBIDE	20%	20%	20%	20%		10%	10%
Grinding, Plain & Surface		5%	5%	5%	5%		5%	5%
Grinding, Centerless Thread		10%						
Form Rolling		100%						
Stamping		20%						
Drawing								

NOTE: These concentrations are starting points only and ultimate use concentration should be determined by actual in-plant testing.

USE INSTRUCTIONS

Always add WYNN'S 331-1 COOLANT to water agitating the solution while additions are being made. Mixing can also be done automatically through the use of an automatic dispensing device.

See your WYNN'S representative for details.

NOTE: We do not recommend the use of soluble oil emulsions in the lubrication system of any equipment.

TECHNICAL DATA

Form:	Liquid
Color:	Dark Green
Spec. Gravity @ 60° F:	0.950
Flash Point COC °F:	380
Freeze Test:	Stable
Emulsion Stability 5%:	Excellent
pH @ 5% Solution:	9.1
Lbs./gal.:	7.91

Wynn Oil Company
P.O. Box 4370/2600 East Nutwood Avenue
Fullerton, CA 92634
(714) 992-2000

WYNN'S
INDUSTRIAL PRODUCTS DIVISION

WYNN'S 381 COOLANT CONCENTRATE

WYNN'S 381 COOLANT CONCENTRATE

Heavy duty, emulsive type soluble oil with a high content of superior anti-weld and Friction Proofing agents.

DESCRIPTION

WYNN'S 381 COOLANT CONCENTRATE contains a high concentration of extreme pressure and anti-wear agents. It can be used as a heavy duty soluble oil or as an additive to straight cutting oils. The blend of emulsifiers in this product forms an emulsion which remains stable in unusually severe conditions of service.

ADVANTAGES

- Multi-purpose concentrate for metalworking.
- Effectively performs well in most metalworking operations.
- Especially effective in severe chip producing operations.
- Compatible with ferrous and non-ferrous metals and alloys.
- Effectively controls odor and foam.
- Protects machine and workpiece from rust and corrosion.
- Lubricates and cleans exposed working surfaces of machining equipment.
- Certified formulation — no nitrites, no carcinogenic nitrosamines.

Wynn Oil Company
P.O. Box 4370/2600 East Nutwood Avenue
Fullerton, CA 92634
(714) 992-2000

GENERAL APPLICATION AND CONCENTRATION TABLE

MACHINING METHOD	TOOL MATERIAL	STEELS			CAST IRON		NON-FERROUS		
		FREE MACHINING MEDIUM CARBON MALLEABLE CAST	LOW CARBON HIGH CARBON ALLOY STEELS	STAINLESS STEELS TOUGH CARBON TOUGH ALLOY	ALLOYED CHILLED	SOFT GRAY CLOSE GRAIN	ALUMINUM SOFT BRASS COPPER	ALUMINUM ALLOY HARD BRASS BRONZE	TITANIUM & OTHER SULFUR & CHLORINE SENSITIVE ALLOYS
Broaching	HSS	10%	15%	20%	20%		5%	5%	10%
	CARBIDE	10%	15%	20%	20%		5%	5%	10%
Threading, Pipe Reaming	HSS	10%	15%				5%	5%	10%
	CARBIDE								
Threading, Tapping	HSS	15%	20%	20%	20%		5%	5%	10%
	CARBIDE	15%	20%	20%	20%				
Gears: Hob, Cut, Shape, Shave	HSS	10%	10%	10%	10%		5%	5%	10%
Drilling, Deep Hole	HSS								
	CARBIDE								
Milling	HSS	3%	3%	5%	8%		3%	3%	5%
	CARBIDE	3%	3%	5%	5%		3%	3%	5%
Drilling	HSS	3%	3%	5%	8%		3%	3%	5%
	CARBIDE	3%	3%	5%	5%				
Turret Lathes, Automatic Lathes, Single and Multiple Spindle Automatics, Drill Forming	HSS								
	CARBIDE								
Turning, Forming Boring Lathes	HSS	3%	3%	5%	8%		3%	3%	5%
	CARBIDE	3%	3%	5%	5%		3%	3%	5%
Sawing, Circular & Hack	HSS	10%	15%	15%	20%		5%	5%	10%
	CARBIDE								
Grinding, Plain & Surface		4%	4%	4%	4%		4%	4%	
Grinding, Centerless Thread									
Form Rolling									
Stamping		20%	20%				20%	20%	
Drawing									

NOTE: These concentrations are recommended starting points only and ultimate use concentration should be determined by actual in-plant testing.

MIXING: WYNN'S 381 COOLANT CONCENTRATE must be added to the water to insure stable emulsion - *do not pour water into concentrate!*

TECHNICAL DATA:

Form:	Liquid	Emulsion:	White, slightly translucent
Color:	Brown	pH of 5% Emulsion:	9.0
Odor:	Typical	% Sulfur:	None
Spec. Gravity @ 60○ F:	0.998	% Chlorine:	10.0
Flash Point COC ○ F:	400 (Conc)	Lb./gal.:	8.31

WYNN OIL COMPANY
2600 EAST NUTWOOD AVENUE • P.O. BOX 4370, FULLERTON, CALIFORNIA 92634

PRINTED IN U.S.A.

Item 7418 US (3/79)

Wynn Oil Company
P.O. Box 4370/2600 East Nutwood Avenue
Fullerton, CA 92634
(714) 992-2000

wynn's
INDUSTRIAL PRODUCTS DIVISION

WYNN'S SEMICOOL 969

SEMICOOL 969

WYNN'S SEMICOOL 969

Semi-synthetic coolant especially formulated for grinding and machining in large or central coolant systems.

DESCRIPTION

WYNN'S SEMICOOL 969 is a clear, dark green liquid which forms a pale green, very stable emulsion.

ADVANTAGES

- Specially formulated for central systems.
- Provides extended tool life.
- Excellent for all grinding applications.
- Will not rust or corrode ferrous or non-ferrous metals.
- Will not attack sound machinery paint.
- Multi-metal safe.
- Translucent emulsion can be seen through easily.
- Certified formulation - no nitrites, no carcinogenic nitrosamines.

Wynn Oil Company
P.O. Box 4370/2600 East Nutwood Avenue
Fullerton, CA 92634
(714) 992-2000

GENERAL APPLICATION AND CONCENTRATION TABLE

MACHINING METHOD	TOOL MATERIAL	STEELS			CAST IRON		NON-FERROUS		
		FREE MACHINING MEDIUM CARBON MALLEABLE CAST	LOW CARBON HIGH CARBON ALLOY STEELS	STAINLESS STEELS TOUGH CARBON TOUGH ALLOY	ALLOYED CHILLED	SOFT GRAY CLOSE GRAIN	ALUMINUM SOFT BRASS COPPER	ALUMINUM ALLOY HARD BRASS BRONZE	TITANIUM & OTHER SULFUR & CHLORINE SENSITIVE ALLOYS
Broaching	HSS	20%	20%	20%			15%	15%	20%
	CARBIDE	20%	20%	20%			15%	15%	20%
Threading, Pipe Reaming	HSS	20%	20%	20%			15%	15%	20%
	CARBIDE	10%	10%	20%			15%	15%	20%
Threading, Tapping	HSS	20%	20%	20%			10%	10%	20%
	CARBIDE	20%	20%	20%			10%	10%	20%
Gears: Hob, Cut, Shape, Shave	HSS	20%	20%	20%			20%	20%	20%
Drilling, Deep Hole	HSS	10%	10%	20%			10%	10%	10%
	CARBIDE	10%	10%	20%			10%	10%	10%
Milling	HSS	5%	10%	15%	15%		5%	5%	10%
	CARBIDE	5%	10%	15%	15%		5%	5%	10%
Drilling	HSS	5%	10%	15%	15%		5%	5%	10%
	CARBIDE	5%	10%	15%	15%		5%	5%	10%
Turret Lathes, Automatic Lathes, Single and Multiple Spindle Automatics, Drill Forming	HSS	10%	10%	15%	15%		5%	5%	10%
	CARBIDE	10%	10%	15%	15%		5%	5%	10%
Turning, Forming Boring Lathes	HSS	10%	10%	15%	15%		5%	5%	10%
	CARBIDE	10%	10%	15%	15%		5%	5%	5%
Sawing, Circular & Hack	HSS	10%	15%	20%	20%		5%	5%	10%
	CARBIDE	10%	15%	20%	20%		5%	5%	10%
Grinding, Plain & Surface		2%	2%	2%	2%		3%	3%	3%
Grinding, Centerless		2%	2%	2%	2%		3%	3%	3%
Grinding, Thread		20%	20%	20%	20%		10%	10%	15%
Form Rolling		10%	10%	20%			10%	20%	20%
Stamping		10%	10%	20%			10%	20%	20%
Drawing		10%	10%	20%			10%	20%	20%

NOTE: These concentrations are recommended starting points only and ultimate use concentration should be determined by actual in-plant testing.

MIXING: WYNN'S SEMICOOL 969 must be added to water to insure stable emulsion - do not pour water into concentrate!

TECHNICAL DATA:

Appearance:	Clear Liquid	pH of 5% Emulsion:	8.8 - 9.0
Color:	Dark Green	Herbert Corrosion Test:	No Rust
Spec. Gravity @ 60/60° F:	1.014	Lbs./gal.:	8.45
5% Emulsion:	Clear Green		

WYNN OIL COMPANY

2600 EAST NUTWOOD AVENUE • P.O. BOX 4370, FULLERTON, CALIFORNIA 92634

PRINTED IN U.S.A.

Item 7436 (3/79)

Wynn Oil Company
P.O. Box 4370/2600 East Nutwood Avenue
Fullerton, CA 92634
(714) 992-2000

WYNN'S

INDUSTRIAL PRODUCTS DIVISION

WYNN'S® ULTRA-SYNTHET™ 941

ULTRA-SYNTHET™ 941

WYNN'S® ULTRA-SYNTHET™ 941

Synthetic chemical coolant with exceptional cooling, wetting and lubricating properties.

DESCRIPTION

WYNN'S® ULTRA-SYNTHET™ 941 is a clear, green liquid. Ideal for a wide range of metalworking applications, including light to heavy-duty machining, grinding, stamping, blanking, tube rolling and forming operations.

ADVANTAGES & BENEFITS

- Multi-metal safe.
- Outstanding cooling and lubricating properties.
- Extends tool life.
- Biodegradable — Does not contain petroleum base oils.
- Excellent sump life.
- Contains special corrosion inhibitors. Will not rust or corrode ferrous or non-ferrous metals.
- Tramp oils will not emulsify (mix) with the coolant.
- Good work visibility.
- Helps keep machine area clean.
- Certified formulation — does not contain nitrites or carcinogenic nitrosamines.

Wynn Oil Company
P.O. Box 4370/2600 East Nutwood Avenue
Fullerton, CA 92634
(714) 992-2000

GENERAL APPLICATION AND CONCENTRATION TABLE

MACHINING METHOD	TOOL MATERIAL	STEELS			CAST IRON		NON-FERROUS		
		FREE MACHINING MEDIUM CARBON MALLEABLE CAST	LOW CARBON HIGH CARBON ALLOY STEELS	STAINLESS STEELS TOUGH CARBON TOUGH ALLOY	ALLOYED CHILLED	SOFT GRAY CLOSE GRAIN	ALUMINUM SOFT BRASS COPPER	ALUMINUM ALLOY HARD BRASS BRONZE	TITANIUM & OTHER SULFUR & CHLORINE SENSITIVE ALLOYS
Broaching	HSS								
	CARBIDE								
Threading, Pipe Reaming	HSS	15%	15%	15%			12%	12%	15%
	CARBIDE	10%	10%	15%			10%	10%	15%
Threading, Tapping	HSS	15%	15%	15%			8%	8%	15%
	CARBIDE	15%	15%	15%			8%	8%	15%
Gears: Hob, Cut, Shape, Shave	HSS								
Drilling, Deep Hole	HSS	10%	10%	15%			8%	8%	8%
	CARBIDE	10%	10%	15%			8%	8%	8%
Milling	HSS	7%	10%	15%	10%	10%	5%	5%	8%
	CARBIDE	7%	10%	15%	10%	10%	5%	5%	8%
Drilling	HSS	7%	10%	15%	10%	10%	5%		10%
	CARBIDE	7%	10%	15%	10%	10%	5%	5%	10%
Turret Lathes, Automatic Lathes, Single and Multiple Spindle Automatics, Drill Forming	HSS	10%	10%	15%	10%	10%	5%	5%	8%
	CARBIDE	10%	10%	15%	10%	10%	5%		8%
Turning, Forming Boring Lathes	HSS	7%	8%	15%	10%	10%	5%	5%	5%
	CARBIDE	7%	8%	12%	10%	8%	5%	5%	5%
Sawing, Circular & Hack	HSS	10%	10%	15%	12%	10%	5%	5%	10%
	CARBIDE	8%	8%	10%	10%	15%	5%	5%	10%
Grinding, Plain & Surface		5%	5%	5%	5%	5%	5%	5%	5%
Grinding, Centerless		5%	5%	5%	5%	5%	5%	5%	5%
Grinding, Thread		15%	15%	15%	15%	15%	7%	8%	10%
Form Rolling		10%	10%	20%			10%	20%	20%
Stamping		20%	20%	30%			20%	30%	30%
Drawing		20%	20%	20%			20%	30%	30%
Tube Rolling		4-5%	4-5%	5-10%			5-10%	5-10%	

NOTE: Percent concentrations in water are recommended starting points only and ultimate use concentration should be determined by actual in-plant testing.

TECHNICAL DATA

Form:	Green, Clear Liquid
pH @ 5% Solution	8.6
Spec. Gravity @ 60%F:	1.031
Flash Point:	None
Herbert Corrosion:	
1:20	Pass
1:50	Pass
Falex: 1:20	Pass 4500 lbs
lbs/gal:	8.59

Avoid freezing - If frozen, prodict must be thawed out at room temperature and mixed before using.

WYNN OIL COMPANY
2600 EAST NUTWOOD AVENUE • P.O. BOX 4370, FULLERTON, CALIFORNIA 92634

PRINTED IN U.S.A.

Item 7461 US(1/83)

Wynn Oil Company
P.O. Box 4370/2600 East Nutwood Avenue
Fullerton, CA 92634
(714) 992-2000

wynn's®

INDUSTRIAL PRODUCTS DIVISION

WYNN'S® ULTRA-SYNTHET™ 951-1

ULTRA-SYNTHET™ 951-1

CUTTING FLUIDS

WYNN'S® ULTRA-SYNTHET™ 951-1

Synthetic chemical coolant with exceptional cooling, wetting and lubricating properties.

DESCRIPTION

WYNN'S® ULTRA-SYNTHET™ 951-1 is a clear, dark green liquid. Ideal for a wide range of metalworking applications.

ADVANTAGES & BENEFITS

- Multi-metal safe.
- Outstanding cooling and lubricating properties.
- Extends tool life.
- Biodegradable — Does not contain petroleum base oils.
- Excellent sump life.
- Contains special corrosion inhibitors. Will not rust or corrode ferrous or non-ferrous metals.
- Good work visibility.
- Certified formulation — does not contain nitrites or carcinogenic nitrosamines.

Wynn Oil Company
P.O. Box 4370/2600 East Nutwood Avenue
Fullerton, CA 92634
(714) 992-2000

GENERAL APPLICATION AND CONCENTRATION TABLE

MACHINING METHOD	TOOL MATERIAL	STEELS: FREE MACHINING MEDIUM CARBON MALLEABLE CAST	STEELS: LOW CARBON HIGH CARBON ALLOY STEELS	STEELS: STAINLESS STEELS TOUGH CARBON TOUGH ALLOY	CAST IRON: ALLOYED CHILLED	CAST IRON: SOFT GRAY CLOSE GRAIN	NON-FERROUS: ALUMINUM SOFT BRASS COPPER	NON-FERROUS: ALUMINUM ALLOY HARD BRASS BRONZE	TITANIUM & OTHER SULFUR & CHLORINE SENSITIVE ALLOYS
Broaching	HSS	15%	15%	15%			10%	12%	15%
	CARBIDE	15%	15%	15%			10%	12%	15%
Threading, Pipe Reaming	HSS	15%	15%	15%			10%	12%	15%
	CARBIDE	10%	10%	15%			10%	12%	15%
Threading, Tapping	HSS	15%	15%	15%			8%	8%	15%
	CARBIDE	15%	15%	15%			8%	8%	15%
Gears: Hob, Cut, Shape, Shave	HSS	15%	15%	15%			15%	15%	15%
Drilling, Deep Hole	HSS	8%	8%	15%			8%	8%	10%
	CARBIDE	8%	8%	10%			8%	8%	10%
Milling	HSS	5%	8%	15%	10%		5%	5%	10%
	CARBIDE	5%	8%	10%	10%		5%	5%	10%
Drilling	HSS	5%	10%	15%	15%		5%	5%	10%
	CARBIDE	5%	10%	10%	10%		5%	5%	10%
Turret Lathes, Automatic Lathes, Single and Multiple Spindle Automatics, Drill Forming	HSS	8%	8%	10%	10%		5%	5%	8%
	CARBIDE	8%	8%	10%	10%		5%	5%	8%
Turning, Forming Boring Lathes	HSS	8%	8%	10%	10%		5%	5%	8%
	CARBIDE	8%	8%	10%	10%		5%	5%	5%
Sawing, Circular & Hack	HSS	8%	10%	12%	10%		5%	5%	8%
	CARBIDE	8%	10%	10%	10%		5%	5%	8%
Grinding, Plain & Surface		5%	5%	5%	5%		5%	5%	5%
Grinding, Centerless		5%	5%	5%	5%		5%	5%	5%
Form Rolling		10%	10%	20%			10%	20%	20%
Stamping		10%	10%	20%			10%	20%	20%
Drawing		10%	10%	20%			10%	20%	20%

NOTE: Percent coolant concentrations in water are recommended starting points only and ultimate use concentration should be determined by actual in-plant testing.

TECHNICAL DATA

Form:	Dark Green, Clear Liquid
pH @ 5% Solution:	8.7
Spec. Gravity @ 60% F:	1.029
Flash Point:	None
Herbert Corrosion	
1:20	Pass
Falex: 120	Pass 4500 lbs
Lbs/gal:	8.57

Avoid Freezing — If frozen, product must be thawed out at room temperature and mixed before using.

WYNN OIL COMPANY
2600 EAST NUTWOOD AVENUE • P.O. BOX 4370, FULLERTON, CALIFORNIA 92634

PRINTED IN U.S.A.

Item 7414 US (1/83)

Wynn Oil Company
P.O. Box 4370/2600 East Nutwood Avenue
Fullerton, CA 92634
(714) 992-2000

wynn's®

INDUSTRIAL PRODUCTS DIVISION

WYNN'S® ULTRA-SYNTHET™ 983

WYNN'S® ULTRA-SYNTHET™ 983

A chemical cutting fluid formulated to provide high performance on many materials, both ferrous and non-ferrous machining operations.

ULTRA-SYNTHET™ 983 is intended for use with light to moderate machining, using high speed steel or carbide tools.

DESCRIPTION

ULTRA-SYNTHET™ is pleasant, safe, and clean to use. It offers superior lubricity and cooling to the metal/tool point of contact.

ULTRA-SYNTHET™ 983 does not contain any added sulphur, chlorine or nitrite. This allows the product to be used where staining is a problem.

ADVANTAGES & BENEFITS

- Superior cutting tool performance. WYNN'S® ULTRA-SYNTHET™ 983 is formulated to give improved long life to cutting tools.
- Excellent sump life. Resists bacterial attack and contamination.
- Low foaming — under most water types and conditions, the product is free of foam.
- Safe to use. Pleasant odor, excellent cleanliness properties, nitrite free, good rust protection.
- Reduces heat — results in high productivity. Eliminates frequent tool changes.

ULTRA-SYNTHET™ 983

Wynn Oil Company
P.O. Box 4370/2600 East Nutwood Avenue
Fullerton, CA 92634
(714) 992-2000

RECOMMENDED WATER DILUTIONS:

LIGHT DUTY:	1:20 to 1:30
MODERATE DUTY:	1:15 to 1:20

NOTE: THESE ARE STARTING POINT DILUTIONS ONLY. ACTUAL DILUTIONS CAN ONLY BE DETERMINED BY IN-PLANT USE.

PHYSICAL CHARACTERISTICS:

FORM	LIQUID
COLOR	DARK GREEN
SPECIFIC GRAVITY @ 60° F	1.018
pH @ 1:20 SOLUTION	8.9/9.2
SULPHUR, % WT.	NONE
CHLORINE, % WT.	NONE
HERBERT CORROSION: 1:20	NO CORROSION
FALEX TEST: 1:20 SOLUTION	Pass 4,500 LBS
FREEZING POINT	20°F
LBS/GAL @ 60°F	8.48

NOTE: WYNN'S TECHINICAL STAFF WILL ASSIST YOU IN MAKING THE CORRECT SELECTION AND USAGE OF OUR RANGE OF HIGH QUALITY PRODUCT.

WARNING:

DO NOT USE WATER BASED PRODUCTS ON MAGNESIUM AND ITS ALLOYS.

AVOID FREEZING - IF FROZEN, PRODUCT MUST BE THAWED OUT AT ROOM TEMPERATURE AND MIXED BEFORE USING.

Wynn Oil Company
P.O. Box 4370/2600 East Nutwood Avenue
Fullerton, CA 92634
(714) 992-2000

WYNN'S PRODUCTS FOR METALWORKING

PRODUCT	DESCRIPTION	APPLICATION	ADVANTAGES	DILUTION RATIO
SOLUBLE OILS 301 COOLANT	General purpose soluble oil. For use in soft to moderately hard water. Forms a stable opaque emulsion.	For free-machining steel and nonferrous metals. Can be used for grinding operations and excellent for machining aluminum.	Contains lubricity additive for better finishes. Good rust protection.	Mix 5-20% into water.
331 COOLANT	Heavy duty soluble oil, for use in soft to moderately hard water. Forms an opaque emulsion.	Moderate to difficult machining of ferrous metals, including stainless and superalloys.	Heavily fortified with lubricity and E.P. additives. Contains sulfur and chlorine for difficult work.	Moderate (5-10%) into water. Difficult (10-20%) into water.
381 COOLANT	Heavy-duty soluble oil, for use in soft to moderately hard water. Opaque emulsion in water.	Moderate to difficult machining of both ferrous and nonferrous metals. Not recommended for grinding operations. Can also be used undiluted for metal forming or as an additive for cutting oils.	Heavily fortified with lubricity, E.P. and anti-weld additives. Contains chlorine for severe machining conditions.	Moderate (3-10%) into water. Difficult (10-20%) into water.
SEMI-SYNTHETIC 969 COOLANT	Moderate duty, semisynthetic coolant, for use in moderately hard water. Forms a stable, translucent emulsion in water.	Light to moderate machining on both ferrous and nonferrous metals. Can also be used for grinding operations.	Contains lubricity and E.P. additives. Easier to dispose of than soluble oils. Longer sump life. Good rust protection without nitrites.	Light duty (5%) into water. Moderate duty (5-10%) into water.
CHEMICAL COOLANTS 983 ULTRA-SYNTHET	Moderate duty chemical coolant. Forms a clear, stable, microemulsion in water. For use in moderately hard water.	Moderate to difficult machining on both ferrous and nonferrous metals. Can also be used for grinding operations.	Excellent lubricity and E.P. Does not contain sulfur or chlorine. Can be used with aerospace materials. Good finishes, nongumming, extended sump life, and good rust protection. Multi-metal safe with no staining.	Light duty (5%) into water. Moderate duty (5-10%) into water. Difficult duty (10-20%) into water.
SYNTHETICS 941 ULTRA-SYNTHET	Light to moderate duty synthetic coolant. Forms a clear solution in water. For use in moderately hard water.	Light to moderate machining on both ferrous and nonferrous metals. Can also be used for grinding, stamping, and rolling operations.	Clear fluid allows operator to see the work piece. Multi-metal safe. Outstanding cooling properties. Easier to dispose of than soluble oils. Extended sump life and good rust protection.	Light duty (5%) into water. Moderate duty (5-10%) into water.
951 ULTRA-SYNTHET (HW & B)	Heavy duty synthetic coolant. Forms a clear solution in water. For use in moderately hard water.* *Also, available for hard water or with a bactericide formulated into product.	Moderate to difficult machining on both ferrous and nonferrous metals. Excellent on aluminum. Can also be used for grinding and stamping operations.	Good work visibility and multi-metal safe. Excellent cooling in high speed operations. Excellent sump life with good rust protection. Rejects tramp oil for easy removal. Easier to dispose of than soluble oils.	Light duty (5%) into water. Moderate duty (5-10%) into water. Difficult duty (10-20%) into water.
440 CUTTING OIL	Heavy duty cutting oil. Low viscosity for better cooling. Brown in color and clear in appearance.	Moderate to difficult machining on all ferrous metals. Not recommended where staining of metal may occur. Can also be used for grinding and metal forming.	Contains sulfur and chlorine E.P. additives for severe machining operations. Outstanding lubricity for improved finishes. Excellent for stainless and superalloys. Low odor and increased tool life.	Use as is undiluted.

In addition to the above, Wynn's carries a full line of greases, rust penetrant, and hydraulic oil additive. Lubrication problems? Contact your local Wynn's representative.

SECTION 2

LUBRICANTS

Acheson Colloids Company
1600 Washington Avenue
Port Huron, MI 48060
(313) 984-5581

Product Data Sheet

Description:

Deltaforge 31 is a smokeless water based graphite die lubricant designed to meet the forging industry's needs in difficult mechanical press work and highly automated operations. This product can be used on a wide variety of part sizes and configurations. It's particularly effective for warm or hot precision forging operations. Deltaforge 31 is used extensively to reduce costs and increase productivity when hot forging plain carbon steel, alloy steel, and stainless steel.

Deltaforge 31 forms a uniform dry film coating when sprayed on dies at temperatures up to and exceeding 600°F (316°C). This product offers many advantages:

- excellent lubricity
- efficient die filling
- good release properties
- uniform wetting of hot dies
- extends die life
- control of die temperatures through adjustment of dilution ratio
- no heavy metals or phosphates
- recommended for automatic systems
- nonflammable—safe to use, safe to store
- noncorrosive
- smokeless—no fumes or toxic gasses

Deltaforge 31 solves tough die lubrication problems while assuring a safer, cleaner, more profitable shop operation. It has repeatedly increased die life by 10 to 30% (as much as 250% in one case). Cost savings due to die life increase are typically greater than the total lubricant cost.

Typical Applications:

Hot forging steel
Warm forming steel
Precision forging
Forging powder preforms

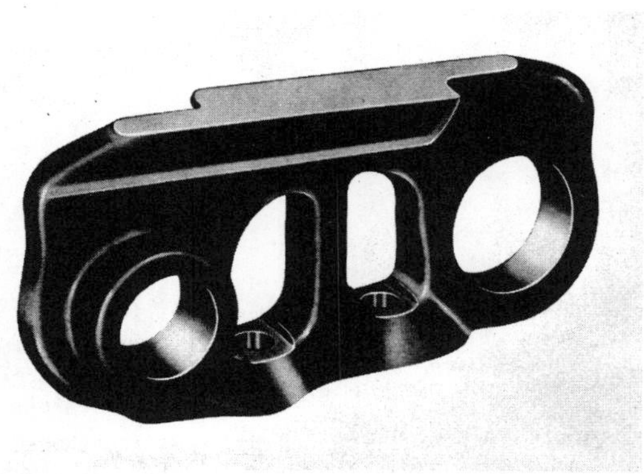

Elimination of smoke and fire hazards and efficient, trouble-free operation of automatic spray systems were the primary reasons for switching to Deltaforge 31 die lubricant, reports a large Midwestern manufacturer of components for earth-moving equipment. Deltaforge 31 is diluted 1:20 with water in production of this track link on a 4000-ton mechanical press.

Acheson Colloids Company
1600 Washington Avenue
Port Huron, MI 48060
(313) 984-5581

Physical Properties:

Lubricant	:	processed micro-graphite
Fluid component	:	water
Consistency	:	liquid
Density	:	9.0 lbs/gal (1.080 kg/liter)
pH	:	10-11
Freezing point	:	32°F (0°C)
Shelf life	:	one year under original seal

Method of Use:

Dilution

Deltaforge 31 is supplied as a concentrate that can normally be diluted with ordinary tap water prior to use. However, if the tap water is high in mineral content (hard), check with your local Acheson representative for assistance.

IMPORTANT:

Add water to the concentrate, not the reverse. Stir the concentrate thoroughly, then add one part water slowly while stirring until the mixture is homogenous. The remaining water may be added rapidly while stirring until the desired dilution and consistency are reached.

Dilution Ratios

The dilution ratio varies according to the difficulty of the job and the degree of cooling required. For initial trials, a ratio of one part Deltaforge 31 to ten parts water by volume is suggested. Ratios of 1:20 and greater are common during production. Use a more concentrated mixture (1:3) to lubricate tooling hotter than 600°F (316°C).

Application

Diluted Deltaforge 31 should be applied by conventional spray techniques for best results. Spray application provides more uniform coating thickness, more complete coverage, and lower friction values. Manual and automatic spray systems specially designed for application of die lubricants are available from Acheson Colloids Company.

Precautions:

Store Deltaforge 31 in a cool place, but do not allow it to freeze. Product should not be used once frozen. Always tightly reseal the container to prevent evaporation or contamination.

Containers:

gallon (3.7 liters)
5-gallon (18.9 liters)
55-gallon (208 liters)

Acheson Colloids Company
1600 Washington Avenue
Port Huron, MI 48060
(313) 984-5581

Product Data Sheet

Description:

Deltaforge 144 is a lubricating and protective coating designed for application to powder metal preforms prior to hot forming or hot densification (up to 1800°F), as well as to wrought slugs prior to warm forming (1000-1500°F). The coating protects P/M preforms from decarburization and provides lubrication during forming. In some operations wrought slugs coated with Deltaforge 144 can be warm formed without the aid of additional die lubrication.

Deltaforge 144 is composed of processed micro-graphite particles and proprietary substances in a water-extendable base. It forms a chemically adherent film which maintains high lubricity in spite of high temperatures. (Actual temperature ranges depend on the length of time and method of heating the workpiece.) Specific advantages offered by Deltaforge 144 include:

- high lubricity
- excellent thermal stability
- protects workpiece from oxidation
- minimizes or eliminates the need for die lubricants
- no pretreatment required
- smokeless
- nonflammable
- no harmful vapors
- improves plant environment

Deltaforge 144 is supplied as a concentrate and is normally diluted with water before use.

Typical Applications:

Warm Forming of Steel
Hot Forming of Powder Preforms

Physical Properties: (as supplied)

Lubricants	: process micro-graphite, proprietary substances
Fluid component	: water
Diluent	: water
Consistency	: fluid
Density	: 10.4 lbs/gal (1.242 kg/liter)
Solids content	: 40%
Freezing point	: 32°F (0°C)
Shelf life	: one year under original seal

Acheson Colloids Company
1600 Washington Avenue
Port Huron, MI 48060
(313) 984-5581

Method of Use:

Surface Preparation
Substrates should be clean (free of oil, scale, etc.) prior to coating. No further pretreatment is required.

Dilution and Mixing
Deltaforge 144 is supplied as a concentrate and is usually diluted with one to two parts water. Always add the water to the well-stirred concentrate. Stir in a small amount of water slowly. When the mixture is homogeneous, add the remaining water more rapidly while stirring until the required dilution is reached.

Application
For best results, parts should be heated to 250-300°F (121-149°C), then dipped briefly in the Deltaforge 144 bath. The water evaporates rapidly, leaving an extremely adherent film. The bath should be continually agitated to assure uniform film thickness from piece to piece.

If cold parts must be used, the immersion time should be longer (about 15 seconds) to allow formation of the chemically adherent film. Blow warm air on the coated parts to facilitate drying.

Precautions:

Store Deltaforge 144 in a cool place, but do not allow it to freeze.

Reseal containers tightly to prevent evaporation and contamination.

Container Sizes:

1 gallon (3.8 liters)
5 gallon (18.9 liters)
55 gallon (108 liters)

Printed in U.S.A. 10-130-R384

Acheson Colloids Company
1600 Washington Avenue
Port Huron, MI 48060
(313) 984-5581

Product Data Sheet

Description:

Deltaforge 950 is a water-based, water dilutable forging lubricant formulated to outperform oil lubricants when forging difficult steel, titanium and super-alloy parts on large hammers. The exceptional lubricating properties found in Deltaforge 950 can significantly reduce the number of blows needed to fill a die cavity. Deltaforge 950 produces better metal movement, good release, and effective scale removal.

Continued use of Deltaforge 950 results in a thin graphoid film on the die surface that lengthens die life by helping to prevent heat checking and metal-to-metal contact.

Deltaforge 950 utilizes conventional spray application techniques or can be swab applied. Specific advantages of Deltaforge 950 water-based hammer lubricant include:

- superior lubrication
- good release
- increased production (fewer blows per piece)
- superior scale blow
- reduced chance of blowouts and die cracking
- cleaner plant environment - nonflammable, smokeless
- longer die life

Water-based Deltaforge 950 provides superior lubrication and release when hammer forging large, difficult steel, titanium or super-alloy parts. Pictured, 119 lb. Propeller Barrel Forging, Courtesy Wyman-Gordon Company, Worcester, Massachusetts.

Typical Applications:

Hammer forging of large and difficult parts

Physical Properties: (as supplied)

Lubricant	:	processed micro-graphite, special polymers
Carrier	:	water
Diluent	:	water
Density	:	9.7 lbs/gal (1.03 kg/l)
Freezing point	:	32°F (0°C)
Flash point	:	none

Acheson Colloids Company
1600 Washington Avenue
Port Huron, MI 48060
(313) 984-5581

Method of Use:

Deltaforge 950 can be applied to forging dies as supplied or it can be diluted with water prior to use. A typical dilution ratio of 1:3 (product:water) is suggested for initial evaluation.

Application

Deltaforge 950 is best applied using conventional spray application techniques; however, swab application can produce acceptable results.

Note

Acheson supplies a complete line of Dag® application equipment for the forging process. Automatic systems and hand-held guns employ a design that features variable spray densities without clogging. For details, contact your local Acheson representative or the Dag Application Equipment Department, Acheson Colloids Company, Port Huron, Michigan.

Containers:

5 gallons (18.9 liters)
55 gallons (208 liters)

Precaution:

Do not freeze. Do not use if product has been frozen. Tightly reseal containers to prevent evaporation.

Printed in U.S.A.

10-149-1079

Acheson Colloids Company
1600 Washington Avenue
Port Huron, MI 48060
(313) 984-5581

Product Data Sheet

Description:

Deltaforge 1001 is a recently developed non-graphited, water-based die lubricant designed especially for warm and hot forging ferrous metals on presses and hammers. It is based on a unique synthetic short chain polymer which forms an excellent lubricating film on the die. Such optimum lubrication allows good metal movement and superior release properties for even the most difficult to forge parts.

Graphite-free Deltaforge 1001 was used to forge this difficult no-draft hub. It provided effective release and lubrication while keeping the shop environment clean and eliminating any possible buildup in the die cavity.

Unparalleled in cleanliness, Deltaforge 1001 allows higher dilution ratios than other current graphite-free products and provides effective lubrication at die temperatures up to and exceeding 800°F (427°C). A number of other advantages are also associated with Deltaforge 1001. It is odorless, non-toxic, smokeless, exceptionally clean and:

- Does not settle after dilution
- Cannot be damaged by freezing
- Eliminates buildup on the tooling and presses
- Does not cause corrosion
- Produces clean finished parts
- Is easily spray applied

Typical Applications:

Hot Forging Steel   Warm Forging Steel

Physical Properties:

Lubricant	:	soluble synthetic
Carrier/Diluent	:	water
Consistency	:	blue creamy liquid
Density	:	9.1 lbs/gal
Freezing point	:	28°F (-2°C)
Shelf life	:	one year under original seal

Acheson Colloids Company
1600 Washington Avenue
Port Huron, MI 48060
(313) 984-5581

Method of Use:

Dilution
To ensure proper mixing, Deltaforge 1001 should be premixed with an equal volume of water, while under high speed mixing. Using gentle agitation, the balance of the water can then be added until the desired dilution ratio is reached. Once properly diluted this product requires no further agitation.

For initial evaluation, a ratio of one part product to five parts water by volume is recommended. During actual production higher dilution is common and will vary according to job severity.

Application
Spray application is recommended, however, product can be successfully swab applied.*

Precautions:

If frozen, allow product to completely thaw and proceed with the described method of use.

Container Sizes:

5 gallon (18.9 liters)
55 gallon (208 liters)

Note:

*Consult your local Acheson representative for additional product and Dag® application equipment information.

Deltaforge 1001 is covered by U.S. Patent No. 4,401,579.

Acheson Colloids Company
1600 Washington Avenue
Port Huron, MI 48060
(313) 984-5581

Product Data Sheet

Description:

Deltaglaze 347 protects and lubricates titanium alloys and superalloys during heating and forging or extrusion processes, assuring superior precision forgings with excellent surface finish. Specially selected glass fuses to form a coating that minimizes oxidation and gas absorption as well as providing effective lubrication and controlled metal flow during forming operations. Deltaglaze 347 coatings air dry with good adhesion and green strength, and are easily removed from finished parts by sandblasting or appropriate salt baths. Specific advantages offered by this product include:

- excellent surface finish
- greater utilization of billet material
- reduced finishing costs
- minimal oxidation and gas absorption
- effective lubricity
- controlled metal flow
- reduced die pressure, container friction
- no heavy metals or phosphates
- easy to apply

Deltaforge 347 coatings should be used in conjunction with a suitable Acheson die lubricant.

Application:

Forging and Extrusion of Titanium Alloys, Superalloys

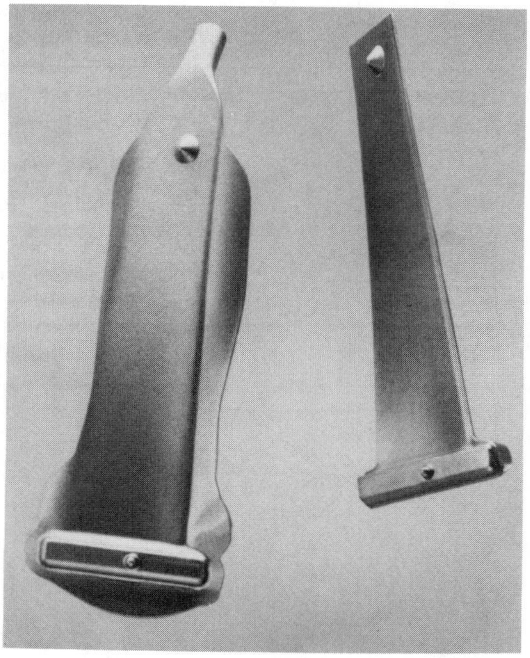

Deltaglaze 347 coatings on titanium alloy stock assure excellent finish on precision forgings such as this compressor blade, shown as forged and trimmed.

Physical Properties: (as supplied)

Fluid component	:	isopropanol
Diluent	:	isopropanol
Color	:	light green
Consistency	:	creamy fluid
Density	:	8.5 lbs./gal. (1.020 KG/liter)
Viscosity	:	33 seconds # 4 Ford cup
Flash point	:	55°F (13°C)
Shelf life	:	six months minimum under original seal

Acheson Colloids Company
1600 Washington Avenue
Port Huron, MI 48060
(313) 984-5581

Method of Use:

Surface Preparation

Substrates must be chemically clean, preferably prepared by chemical etching followed by a water rinse. Sandblasting is acceptable, but metal shot must not be used due to the danger of substrate contamination.

Dilution and Mixing

Deltaglaze 347 is supplied ready-for-use but may be diluted. Add the diluent slowly while stirring gently, and mix thoroughly just prior to use. Excessive agitation (i.e. "vortex" stirring) should be avoided, as air entrapment could seriously impair coating quality. The material may settle slightly on standing; it is easily remixed by stirring.

Application

Standard or electrostatic spraying is recommended for even, consistent coatings. (Deltaforge 347 can be modified for specific electrostatic application techniques.) Dipping is the second choice among application methods; brushing can produce acceptable coatings with sufficient care.

Optimum film thickness is determined primarily by the heating cycle and furnace environment of a particular application.

Curing

Deltaglaze 347 coatings will air dry in 15 to 30 minutes, depending on thickness and ambient temperature. Coatings must be completely dry before entering the heating cycle prior to forging; if solvents are trapped, the coating will blister during the heating cycle.

If a shorter drying time is desired, the billets may be preheated to approximately 120°F (49°C) before coating.

Precautions:

This product contains isopropanol.

Observe the customary safeguards when storing, handling or applying flammable materials of this type.

Reseal containers carefully to prevent evaporation and contamination.

If coated billets are stored, the storage environment should be dry, warm and clean to prevent moisture absorption and/or contamination.

Containers:

Gallon (3.7 liters)
5-gallon (18.9 liters)

Acheson Colloids Company
1600 Washington Avenue
Port Huron, MI 48060
(313) 984-5581

Product Data Sheet

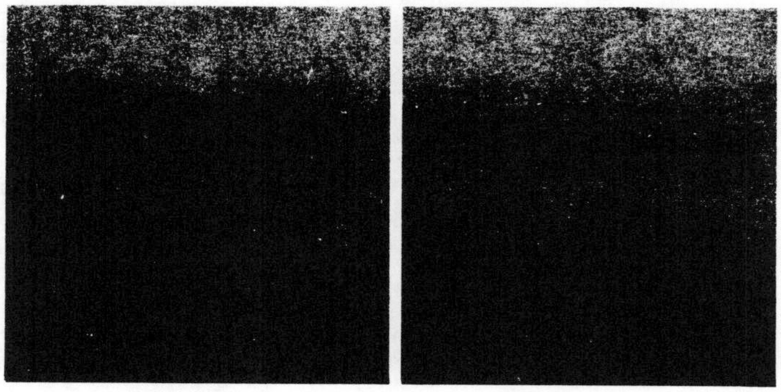

Description:

Molydag® 20 is a unique, water based molybdenum disulfide billet coating designed for use when cold extruding ferrous metals. It forms an excellent lubricating film over zinc and manganese phosphated substrates. Molydag 20 adheres extremely well in mechanical feed mechanisms and moves with the billet surface during deformation, enabling higher production rates and extreme area reductions. It also eliminates the expense and housekeeping problems associated with tumble-applied MoS_2 powder. Using Molydag 20 in cold forming processes offers the following advantages over conventional phosphate soap systems:

- closer tolerances of complex shapes
- precision extrusion of sharp angles and corners
- excellent lubricating film
- permits large reduction in area
- uniform film thickness
- easy control of bath
- no toxic fumes or vapors
- excellent chemical stability in the application bath
- increased tool life

Typical Applications:

Cold extrusion of:

carbon steels
alloy steels
stainless steels

Physical Properties:

Lubricating solid	:	molybdenum disulfide
Carrier	:	water
Consistency	:	viscous fluid
Density	:	10.5 lb/gal (1.26 kg/l)
pH	:	7
pH tolerance	:	4 to 10
Freezing point	:	32°F (0°C)
Flash point	:	none
Shelf life	:	one year under original seal

Acheson Colloids Company
1600 Washington Avenue
Port Huron, MI 48060
(313) 984-5581

Method of Use:	Pretreatment Slugs should be degreased and phosphated before being coated with Molydag 20. Dilution Molydag 20 is supplied as a concentrate and is diluted with ordinary tap water prior to use. However, if tap water in your area is high in mineral content, contact your Acheson representative for assistance. Important: For initial dilution, add water to the concentrate. Stir the concentrate thoroughly before diluting and continue stirring during dilution. Film thickness depends upon the dilution ratio, which varies the solids content of the bath. For extreme extrusions, a dilution ratio of 1 part product to 2 parts water (10% solids) is recommended. For moderate extrusions, dilute Molydag 20 with up to 5 parts water (5% solids). The bath is easily adjusted by adding either water or concentrate. Application Molydag 20 is designed to be dip-applied. Bath temperatures should remain between 160 and 170°F for best coating results. Tumble dry.
Precautions:	Caution! Do not take internally. May cause temporary discomfort on direct contact with eyes. Use with adequate ventilation if product is sprayed. Do not freeze. Tightly reseal containers to maintain shelf life. Store Molydag 20 in a cool place but do not freeze. Do not use if frozen or thawed. Do not store coated slugs outdoors. (See product label for proper first aid instructions.)
Container Sizes:	1 gallon (10 lbs) 5 gallon (52 lbs) 55 gallon (550 lbs)

Printed in U.S.A. 09-102-R1082

Acheson Colloids Company
1600 Washington Avenue
Port Huron, MI 48060
(313) 984-5581

Product Data Sheet

Description:

Aquadag® is an aqueous based colloidal graphite dispersion widely used in the aluminum permanent mold industry. The ultra-fine particle size imparts superior surface finish to the casting together with superb release properties. This product forms a smooth continuous dry film with extra long wear life. Aquadag's unique formulation results in maximum adhesion to most substrates. The fineness and purity of the product produces near perfect suspension properties which substantially reduce stirring or agitation during use. The high covering factor of Aquadag permits for higher dilution ratios than those experienced with commercially available alternatives, thus providing efficient product use. Aquadag is smokeless, nontoxic and nonflammable, which aids in improving plant environment.

Physical Properties:

Pigment	:	processed ultra-micro graphite
Carrier	:	water
Consistency	:	creamy paste
Diluent	:	water
Solids content	:	22%
Density	:	9.3 lbs/gal (1.09 kgs/liter)
pH	:	9.5-11.0
Shelf life	:	one year minimum under original seal

Dilution:

Aquadag is a concentrate and should normally be diluted before use, preferably with distilled, soft or demineralized water. The concentrate should be agitated prior to and during the slow addition of the diluent. A dilution ratio of one part Aquadag to four parts water is recommended for initial trial. Dilution ratios of 1:10 and higher are normal for production.

Application:

Diluted Aquadag should be applied by conventional spray techniques for the most favorable results. Spray application ensures more uniform coating thickness, more complete coverage, and lower friction values. For optimum film formation, the diluted dispersion should be spray applied to surfaces heated to at least 200°F (93°C) to facilitate speedy evaporation of the carrier. The formulation of Aquadag is such that a satisfactory film can be obtained by brush application. Due to the ultra-fine particle size, only minimal stirring is necessary during the standing of the diluted product.

Companion Products:

Dags® 193 and 395, aqueous based refractory dispersions, are frequently used as base coatings for Aquadag. This combination provides a composite film having the characteristics of good thermal insulation, excellent release and imparting a superior surface finish to the casting.

Precautions:

Aquadag ideally should be stored in a cool place, but not allowed to freeze. Containers should be tightly re-sealed after use in order to prevent evaporation and/or contamination.

Containers:

1-gallon (3.8 liters)
5-gallon (18.9 liters)
55-gallon (208 liters)

Note:

Aquadag and Dag are registered trademarks of Acheson Industries, Inc.

Printed in U.S.A.

10-138-R581

Acheson Colloids Company
1600 Washington Avenue
Port Huron, MI 48060
(313) 984-5581

Product Data Sheet

Description:

Dag® 137, a lubricating coating of graphite in water, exhibits excellent release and lubricating qualities. Particularly successful for aluminum metalworking applications, Dag 137 provides a tenacious graphite film and ensures complete wetting of hot metal surfaces. These important characteristics give maximum protection against scoring, galling and soldering.

Typical Uses:

- Non-ferrous extrusion lubricant
- Ferrous forging lubricant
- Non-ferrous forging lubricant
- Aluminum permanent mold coating
- Die casting release agent
- Parting compound

Physical Properties:

Lubricating Solid	Graphite	pH	8
Carrier	Water	pH Tolerance	4-10
Consistency	Soft paste	Particle Size	"B"
Diluent	Water	Freezing Point	32°F
Solids Content	22%	Shelf Life	Indefinite with original seal
Density	9.5 lbs./gal.		

Dilution:

Dag 137 is a concentrate and should be diluted with distilled, demineralized or soft water before application.

The concentrate should be agitated prior to and during the slow addition of water. The dilution ratio will vary between 1:4 and 1:25 (product to water) depending on the particular use and method of application. The higher dilution ratios are employed when Dag 137 is used in aluminum extrusion and die casting. When used as a parting compound or in aluminum permanent molding operations, the lower dilution ratios are normally required.

Application:

To achieve optimum film formation, the diluted dispersion should be spray applied to surfaces heated to at least 200°F. Brush, dip or swab methods may also be employed. If the diluted dispersion stands overnight, or a comparable length of time, it should be mixed prior to use. When using Dag 137 as a pretreat bath for ferrous forging slugs, the material should be diluted 1:4 with water and heated to approximately 150°F. prior to dipping slugs.

Precautions:

The customary methods employed in storing, handling, shipping and applying water-base materials of this type should be employed. If Dag 137 is permitted to freeze, it may take on a granular appearance and settle. Containers should be tightly sealed when not in use to prevent evaporation and drying which is detrimental.

Note:

Dag is a registered trademark of Acheson Industries, Inc.

Acheson Colloids Company
1600 Washington Avenue
Port Huron, MI 48060
(313) 984-5581

Product Data Sheet

Description:

Molydag 225 is an extremely stable dispersion of highly refined, microscopic MoS_2 particles in petroleum oil.

This versatile product is used in a variety of metalworking operations—drawing rod or wire, cold extrusion and cold forming—as both a die lubricant and as a lubricant additive for drawing and extruding oils.

As a DIE LUBRICANT, Molydag 225 provides reliable lubrication under those boundary lubrication conditions associated with high forming loads and high extrusion ratios. Unlike most fluid lubricants, Molydag 225 forms a MoS_2 film that will not squeeze out or break down during metal deformation. Molydag 225 can be applied to dies by swabbing or spraying; or it can be added to a recirculating oil system.

When used as an ADDITIVE for drawing and extruding oils, Molydag 225 improves the oil's anti-wear and anti-friction capabilities. It also prevents premature die failure and die wear.

Specific advantages of Molydag 225 MoS_2 lubricant/additive include:

- reliable lubrication under boundary conditions
- easy to use
- extremely high lubricity
- improved tool life
- improved anti-friction, anti-wear capabilities

Physical Properties: (as supplied)

Lubricating solid	:	molybdenum disulfide
Carrier	:	petroleum oils
Viscosity	:	30 cSt at 93.9°C (typical)
Density	:	8.3 lbs/gal (.996 kg/l)
Flash point	:	385°F (196°C)
Shelf life	:	one year minimum under original seal

Method of Use:

Molydag 225 can be added to most commercially available oils (compatibility should be checked by your Acheson representative).

Dilution

Molydag 225 is supplied as a concentrate and may be diluted prior to use.

Stir Molydag 225 thoroughly to achieve uniform consistency; then premix equal parts of Molydag 225 and the oil before blending with the balance of the diluent. Maintain continuous agitation throughout the dilution operation.

Dilution ratios vary according to the severity of the forming operation. For most applications, a 1:15 ratio (product:oil) is sufficient; however, for initial evaluation, a 1:4 ratio is recommended.

Application

Molydag 225 can be applied by swabbing or spraying the dies. It can also be used in a recirculating oil system. Dipping or flood application onto billets is also acceptable.

Precautions:

Harmful if swallowed. May cause eye irritation. Wash thoroughly after handling. Keep container closed when not in use (see product label for proper first aid instructions).

Containers:

8 pounds (1 gallon)
40 pounds (5 gallons)

Printed in U.S.A.

10-116-R178

Acheson Colloids Company
1600 Washington Avenue
Port Huron, MI 48060
(313) 984-5581

Product Data Sheet

Description:

Oildag® is a highly stable wear and friction-reducing compound of ultrafine graphite particles in a petroleum oil, specially selected for compatibility with most commercially available oils. This material is specifically formulated for applications which approach boundary lubrication conditions.

Consistent use of Oildag produces durable mirror-like graphoid surfaces on friction parts. This graphoid film reduces surface tension between metal and oil and lubricates effectively up to 900°F (480°C). The oil spreads farther and faster over the grahoid film, enabling ruptured oil films to re-establish themselves more rapidly. The solid lubricating film also prevents metal-to-metal contact, greatly facilitating disassembly in many applications. Specific advantages offered by Oildag include:

- excellent adhesion to metal substrates
- eliminates galling, seizing, stick-slip and press-fit distortion
- reduces break-in time for new equipment
- prevents fretting
- protects against corrosion
- lubricates effectively up to 900°F (480°C)
- chemically stable
- extends lubrication service intervals
- reduces maintenance costs through more effective lubrication
- facilitates disassembly

Oildag is used extensively as a superior additive to industrial oils. This highly stable material improves the anti-wear, extreme pressure, and high temperature capabilities of the base oil.

Typical Applications:

High-temperature bearing lubricants
Assembly and break-in lubricants
Extrusion compounds
Glass molding compounds
Oil and grease additives
Forging compounds
Die casting compounds

Physical Properties:

Lubricating solid	:	processed micro-graphite
Fluid component	:	petroleum oil
Diluents	:	petroleum oils, mineral spirits
Solids content	:	10%
Density	:	8.2 lbs/gal (.984 kg/liter)
Flash point	:	385°F (195°C)
Shelf life	:	one year minimum under original seal

Acheson Colloids Company
1600 Washington Avenue
Port Huron, MI 48060
(313) 984-5581

Method of Use:

Dilution and Mixing
Oildag can be blended with most commerically available oils, naphthas and gasolines. For uniform results, heat the oil and/or the Oildag to approximately 150°F (65°C), since blending is most efficient when viscosities are low.

Mix Oildag thoroughly to assure uniform consistency before blending. A pre-mix of equal parts Oildag and diluent should be thoroughly mixed before adding the balance of the diluent. Continuous agitation is desirable throughout the blending operation.

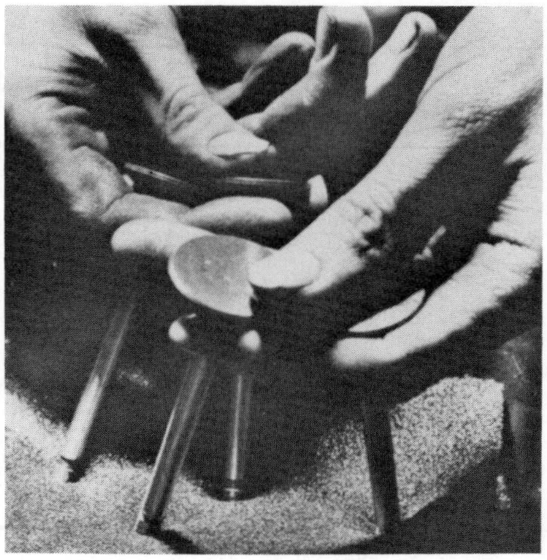

Valve stems for the Vauxhall Viva engine made by the General Motors subsidiary in England are pressed into a sponge impregnated with Oildag. The high-temperature lubricant coating prevents excessive wear of the valve stems during engine break-in.

Application
Oildag and Oildag blends are easily applied by conventional spray, dip, brush and drip methods.

Precautions:

Employ the customary safeguards involved in storing, handling, and applying flammable materials of this type.

Containers:

Pint (.473 liter)
1-Gallon (3.7 liters), 8 lbs (3.6 kg)
5-Gallon (18.9 liters), 40 lbs (18.0 kg)
55-Gallon (208 liters), 440 lbs (198.0 kg)

Note:

Oildag is a registered trademark of Acheson Industries, Inc.

Printed in U.S.A.

13-109-581

American Oil and Supply Company
238 Wilson Avenue
Newark, NJ 07105
(201) 589-0250

TYPE	PRODUCT	NLGI No.	RANGE	BASE	DROPPING POINT	APPLICATION
Hi Temp EP Grease For Heavily Loaded Bearings	PQ C-4005-0 PQ C-4005-1 PQ C-4005-2	0 1 2	0 to 500°F 0 to 500°F 0 to 500°F	Lithium Complex	Over 500°F	EP, water and moisture resistance with outstanding usable temperature range. Where protection is needed for heavy loads and pressures. 40# Timken Heavy Mining Equipment Textile Equipment U.S.D.A. H2
Food Grade "AA" Grease	PQ C-40AA-0 PQ C-40AA-1 PQ C-40AA-2	0 1 2	0 to 300°F 0 to 300°F 0 to 300°F	Aluminum	450°F	For food and beverage plant grease applications meets U.S.D.A. H1 rating requirements.
Open Gear Lubricant	PQ Molygear 92 AOSyn® Open Gear & DSL	2 2	0 to 400°F —30 to 475°F	Complex Inorganic	None Over 500°F	Open gears requiring protection against shock loading — excellent water resistance with tacky-hard film.
Synthetic Grease	AOSyn® 5000-1 AOSyn® 5000-2	1 2	—60 to 500°F —60 to 500°F	Lithium Complex	Over 500°F	Lithium Complex — Synthetic Base Oil suitable for use over a wide temperature range —60°F to 500°F.
Chemical Resistant	AOSyn® Hexane Resistant	2	0 to 450°F	Inorganic	Over 500°F	Resistant to wash out from Hexane. Resistant to most chemicals.
Water Resistant Grease	PQ C-4037 Barium	2	0 to 300°F	Barium	400°F	Bearings operating in wet or submerged conditions. Also excellent for automotive applications and boat trailer bearings.
Wire Rope Lubricant Grease Type	Vitalife No. 2 Plumbago Grease Wire Rope Lubricant	1	—10 to 150°F	Complex 11% Graphite	190°F	Applied cold outstanding protection for slow speed ropes — draw bridges, shovels, cranes - Excellent rail curve grease.

American Oil and Supply Company
238 Wilson Avenue
Newark, NJ 07105
(201) 589-0250

TYPE	PRODUCT	SAE	VISCOSITY (SUS)	APPLICATION
Hydraulics, Bearings, General Purpose	PQ AA10	10	150 at 100°F	Meets USDA rating H1 for incidental contact. Excellent for bearings, chain valves, canning and bottling machinery. 40# Timken load.
	PQ AA20	20	260 at 100°F	
	PQ AA30	30	450 at 100°F	
	PQ AA40	40	700 at 100°F	
Enclosed Gears	PQ AA90	90	88 at 210°F	Meets U.S.D.A. H1 rating 90 & 140 enclosed gear lube with 40#, min. Timken load.
	PQ AA140	140	132 at 210°F	
Synthetic Compressor Fluids	AOSyn® 1510	10	152 at 100°F	U.S.D.A. H1 for screw compressors
	AOSyn® 1520	20	228 at 100°F	U.S.D.A. H1 for sliding vane and screw compressors
	AOSyn® 1530	30	350 at 100°F	U.S.D.A. H1 for reciprocating compressors

TYPE	PRODUCT	SAE	VISCOSITY (SUS)	APPLICATION
Screw Compressor	AOSyn® 992	10	130 at 100°F	Long life with no deposits — major Manufacturers approved. U.S.D.A. H2
Sliding Vane & Screw Compressor	AOSyn® 892	20	309 at 100°F	Extend drain intervals up to 8000 hrs. Diester approved by major Manufacturers. U.S.D.A. H2
Reciprocating Compressor	AOSyn® 882 AOSyn® 750	30 40	491 at 100°F 720 at 100°F	Designed for reciprocating compressors, cylinder lubrication, used successfully in chemical environments. U.S.D.A. H2
Industrial Gear Oils	AOSyn® 1002 (AGMA 2 EP) AOSyn® 1005 (AGMA 5 EP) AOSyn® 1007 (AGMA 7 EP)	80 90 140	50 at 210°F 95 at 210°F 130 at 210°F	Extended drain, long life products designed exclusively for industrial gear box application. For operating temperatures from —40°F to 400°F. 80# Timken OK load. U.S.D.A. H2
Refrigeration Compressor Fluid	AOSyn® RFC	40	70 at 210°F	Designed for use in both ammonia and freon charged compressors. Compatible with neoprene seals. Excellent for use with most industrial gases and chemicals. U.S.D.A. H2.
Chain Lubricants	AOSyn® High Temp Chain	40	856 at 100°F	For chain applications up to 475°F
	AOSyn® 712	50	900 at 100°F	For chain applications up to 500°F
Transformer Oils	AOSyn® Transformer Oil	N/A	50 at 210°F	Provides low flammability, high thermal conductivity and is non-toxic.

American Oil and Supply Company
238 Wilson Avenue
Newark, NJ 07105
(201) 589-0250

TYPE	PRODUCT	SAE	VISCOSITY (SUS)	APPLICATION
Hydraulic Oils	PQ 32 PQ 46 PQ 68 PQ 100	10 20 20 30	150 at 100°F 220 at 100°F 300 at 100°F 500 at 100°F	Hi V.I. Hydraulic Oils — Minimum viscosity index 95 — contains anti wear-anti foam with rust and oxidation inhibitors. U.S.D.A. H2
Enclosed Gear Oils	PQ Gear Oil A.G.M.A. 2EP PQ Gear Oil A.G.M.A. 5EP PQ Gear Oil A.G.M.A. 7EP PQ Gear Oil A.G.M.A. 8EP	80W 90 140 250	50 at 210°F 95 at 210°F 130 at 210°F 200 at 210°F	Oxidation inhibited and fortified with anti wear and anti foam. Heavy load carrying products, 75# Timken. Full line of A.G.M.A. oils available. U.S.D.A. H2
Air Compressor Oils	PQ C-105	20	300 at 100°F	Rotary compressor oil — Outstanding oxidation resistance. U.S.D.A. H2
	PQ C-106	30	500 at 100°F	Paraffinic base for reciprocating machines. U.S.D.A. H2
	PQ C-302	20	300 at 100°F	Naphthenic base for air cooled compressors. U.S.D.A. H2
	PQ C-303	30	500 at 100°F	SAE 30 Naphthenic for air cooled units. U.S.D.A. H2
Refrigeration Oil	PQ C-300	Below 10	110 at 100°F	—40°F pour point for equipment located in low temperature areas. U.S.D.A. H2
Roller Chain Way Oils	PQ L-30 PQ L-50 PQ L-90	20 30 40	300 at 100°F 400 at 100°F 800 at 100°F	Tacky oil — penetrates pin bush joint of roller chain and resists sling off. U.S.D.A. H2
Steam Cylinder Oil	PQ Gear Lube C-116 PQ Gear Lube C-117	A.G.M.A. A.G.M.A.	7 Compounded 8 Compounded	Protect reciprocating engines from super-heated steam — for enclosed heavy pressure worm gears.
Transformer Oil	PQ Transformer Oil	N/A	59 at 100°F	—75° Pour Point Dielectric strength KV (ASTM D877) 35+ Power factor, % at 25°C (ASTM D924) 0.01.
Wire Rope Lubricants	Vitalife 5A	N/A	2200 at 100°F	Field dressing — mine cables for hi film strength — good resistance to acids.
	Vitalife 400	N/A	350 at 100°F	Acid and alkali resistance — water displacing light bodied oil, sprayable — will not sling off.
	Vitalife 201	N/A		Severe cold properties down to —70°F — penetrates inner core.
	Vitalife 92	N/A		Dry film — protects —40 to 160°F.
Spindle	PQ SPINDLE OIL 5 PQ SPINDLE OIL 10	5 10	60 at 100°F 100 at 100°F	Sewing machines. FHP motors. Excellent oxidation resistance.
Needle Oil	PQ 81	10	100 at 100°F	Outstanding scourability — Mist-non fogging — Excellent E.P. protection.
Tenter Lube	PQ 711	40	700 at 100°F	PQ 711 is petroleum product for Tenter Chain operating 0 to 350°F.
Concrete Form Release Agent	PQ FORMOL 100 PQ FORMOL CONCENTRATE 200	10 N/A	40 at 100°F N/A	All concrete forms minimizes bonding for smooth release. Formol 100 is ready to use. Formol 200 can be diluted to make an "on site" release agent.

INDUSTRIAL LUBRICANT VISCOSITY RATINGS

ISO Grade	AGMA[1] Grade No. (Approx.)	S.A.E. Viscosity No. (Approx.)	S.A.E. GEAR Lubricant No. (Approx.)	Viscosity, SUS at 100°F. (Approx.)	Viscosity, Saybolt Universal Seconds 210°F. (Approx.)	ASTM Grade No.
2	—	—	—	29-35	—	32
5	—	—	—	36-44	—	40
10	—	—	—	54-66	—	60
15	—	—	—	68-82	—	75
22	—	—	—	95-115	—	105
32	—	10W	75W	135-165	40	150
46	1	10	—	194-236	43	215
68	2	20	80W	284-346	50	315
100	3	30	—	419-511	60	465
150	4	40	85W	630-770	75	700
220	5	50	90	900-1100	95	1000
320	6	60	—	1350-1650	110	1500
460	7	70	140	1935-2365	130	2150
680	8	—	—	2835-3465	140	3150

[1]American Gear Manufacturers Association.

S.A.E. VISCOSITY GRADES FOR ENGINE OILS

S.A.E. Viscosity Grade	Viscosity Units	Viscosity Range At 0°F. (—18°C.) Min.	Max.	At 210°F. (99°C.) Min.	Max.
5W	Centipoises	—	< 1,200	—	—
	SUS	—	< 6,000	—	—
10W	Centipoises	1,200	< 2,400	—	—
	SUS	6,000	<12,000	—	—
15W	Centipoises	2,400	< 4,800	—	—
	SUS	12,000	<24,000	—	—
20W	Centipoises	4,800	< 9,600	—	—
	SUS	24,000	<48,000	—	—
20	Centistokes	—	—	5.7	< 9.6
	SUS	—	—	45	< 58
30	Centistokes	—	—	9.6	< 12.9
	SUS	—	—	58	< 70
40	Centistokes	—	—	12.9	< 16.8
	SUS	—	—	70	< 85
50	Centistokes	—	—	16.8	< 22.7
	SUS	—	—	85	<110

The "W" designated grades refer to viscosity measured at 0°F. (—18°C.). This is very important when it is necessary to start engines at very low temperatures as found in winter.

S.A.E. VISCOSITY GRADES AXLE AND MANUAL TRANSMISSION LUBRICANTS

S.A.E. Viscosity Grade	Maximum Temperature for Viscosity of 150,000 Centipoise °F.	°C.	Viscosity at 210°F. (99°C.) Min. cSt	SUS	Max. cSt	SUS
75W	—40	—40	4.2	40	—	—
80W	—15	—26	7.0	49	—	—
85W	+10	—12	11.0	63	—	—
90	—	—	14.0	74	<25	120
140	—	—	25	120	<43	200
250	—	—	43	200	—	—

CLASSIFICATION OF GREASES BY NLGI CONSISTENCY NUMBERS

NLGI* Number	ASTM Worked Penetration @ 77°F. (tenths of a millimeter)
000	445 - 475 (Semi-Fluid)
00	440 - 430 (Semi-Fluid)
0	355 - 385 (Soft)
1	310 - 340
2	265 - 295
3	220 - 250
4	175 - 205
5	130 - 160
6	85 - 115 (Hard)

*National Lubricating Grease Institute (NLGI)

Buckeye Lubricants
20801 Salisbury Road
Bedford, OH 44146
(216) 581-3600

BUCKEYE LUBRICANTS
20801 SALISBURY RD. BEDFORD, OHIO 44146
Phone (216) 581-3600

PRODUCT DATA BULLETIN

902H HIGH TEMPERATURE LUBRICANT

902H is a non-toxic, non-flammable water-graphite concentrate. It is a versatile, non-foaming product that has been developed to provide positive lubrication where regular lubricants fail, but at a reasonable cost.

902H is effective on a temperature range of 450°F. It has no flash point and will not smoke.

This product can be diluted up to 7:1 with water and can be applied by the dip, spray or brush methods. # 902H is excellent for Chain Lubrication, Forging, Upsetting, Parting, Die Casting, etc.

PROPERTIES

Non Toxic
pH: 8½ (can be varied easily)
No Flash Point
Non-Flammable
No Bacteriacides Required
Will Not Smoke
Density: 9.2 lbs/gal
Will Not Foam in a Water Solution
30% Anhydrous Solids
Concentrate is a Permanent Dispersion
Contains Proprietory E. P. and Lubricity Agents

Buckeye Lubricants will supply samples of # 902H for evaluation upon request.

Crown Industrial Products Company, Incorporated
100 State Line Road/P.O. Box 326
Hebron, IL 60034
(815) 648-2424

LUBRICANTS

TAP TOOL No. 9106

ONE TAPPING FLUID...
FOR ALL METALS!

Extends tool life. Speeds cutting; reduces wear and breakage. Prevents galling, holds close tolerances. Use on all metals, ferrous and nonferrous including aluminum, stainless steel, titanium, etc., for deep drilling, threading, reaming, tapping, milling, boring, and broaching. Crown Tap Tool contains no sulfur and is not a petroleum base product.

ACTS AS A COOLANT...

It cools the tool and carries heat away from the cutting edge.

ACTS AS A LUBRICANT...

Reduces friction on the surface and acts as an anti-weld on the bottom surface.

Increases tool life as much as 200%
Reduces broading time
Doubles die life
Increases number of pieces per tool grind
Increases number of holes per tap
Increases cutter life

9106 16 oz. can
9101 1 quart
9102 1 gallon
9103 5 gallon can
9104 55 gallon

Obtain better finishes
Obtain closer tolerances
Increased speed
Eliminate hang-up
Reduce breakage
Prevents galling of threads
Prevents excessive tool wear

Pinpoint extension tube included.

MOLY GREASE No. 7041

HEAVY DUTY

A heavy duty petroleum based grease for use where high pressure or temperatures cause breakdown of ordinary grease. For effectively lubricating forklift ways, hoist chains and other cables, small diameter wire rope, screw threads, die set leader pins, flexible couplings, sprockets, linkages . . . any applications requiring periods of lubrication without maintenance.

7041 16 oz. can

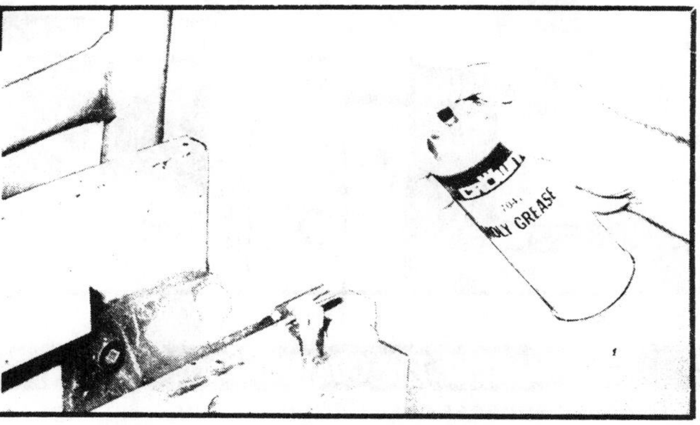

MOLY LIFT TRUCK GREASE No. 6050

LASTING LUBRICATION

A specially formulated Moly Grease, this grease is the finest available. Provides even, lasting lubrication and protection. Use on lift trucks, hoists, chains, screw threads, drive chains, die set leader pins, linkages, sprockets, flexible couplings and any other lubrication where a high grade moly grease can provide longer periods of lubrication without maintenance. For high temperatures and pressures.

6050 16 oz. can

Crown Industrial Products Company, Incorporated
100 State Line Road/P.O. Box 326
Hebron, IL 60034
(815) 648-2424

LUBRICANTS

CROWN

GENERAL PURPOSE SILICONE LUBRICANT No. 8034

LUBRICATES ALL SLIDING SURFACES!

Greaseless lubrication that stops anything from sticking. Will not gum or form unwanted residues. Odorless, colorless, and nonstaining. A universal lubricant that may be used anywhere...except on parts you may wish to paint later...the silicone fluids are dimethyl polysiloxane types. They are clear, water-white, oily fluids; inert, tasteless and odorless. The unique combination of properties associated with Crown silicone fluids makes them suitable for most plastic and rubber lubrication, as dampening or heat transfer fluids, and as an oil defoamer. These fluids are also used in shock absorbers, liquid timers, high-temperature baths, liquid springs, and other similar applications. Acid and alkaline resistance excellent. Moisture and oxidation resistance excellent.

Pinpoint extension tube included.

8034 16 oz. can
8028 1 gallon
8033 5 gallons
8032 55 gallon drum

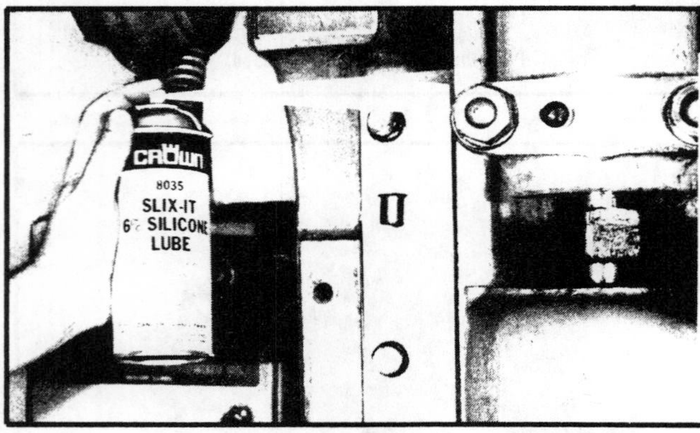

SLIX-IT No. 8035

HIGH SILICONE CONTENT!

6% silicone lubricant with pinpoint spray! This special greaseless lubricant stops anything from sticking...yet leaves no messy residue. Leaves a clear film that is **moisture resistant, long-wearing and odorless.** It won't melt, freeze, gum or become rancid. It's ideal for cabinets, drawers, locks, hinges, ignition wiring, outdoor equipment, canvas tops. Not recommended for surfaces to be painted or plated. High silicone content makes a little go a long way; lasts a long time.

8035 16 oz. can — Pinpoint extension tube included.

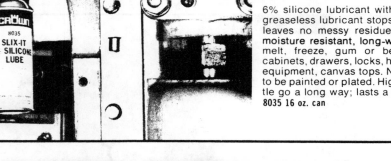

PERMANENT TFE COATING No. 6065

Random Applications: Conveyor parts, copying machines, door locks - hinges, drills, dry bulk handling equipment, fan blades, files - wood and metal, heat sealing bars - plates, lawn mowers, mandrels, molds - ceramic, plastic, metal, etc., saws - electric, hand, shafts, shears, transmission systems.

All new...hard and slick...industrial coating features low friction...high wear resistance and excellent release qualities. It is tougher than any other aerosol TFE fluorocarbon coating. It is easily applied to most hard clean surfaces. It sprays on and air dries or heat cures to a durable, rustproof coating, green in color, that withstands temperatures up to 500°F. It has excellent substrate adhesion, coating hardness, abrasion resistance and load carrying capacity.

6065★ 16 oz. can

★ To Insure Proper Bonding of Crown's No. 6065, Use Crown's SAFETY SOLVENT No. 8060 to Clean and Condition the Substrate.

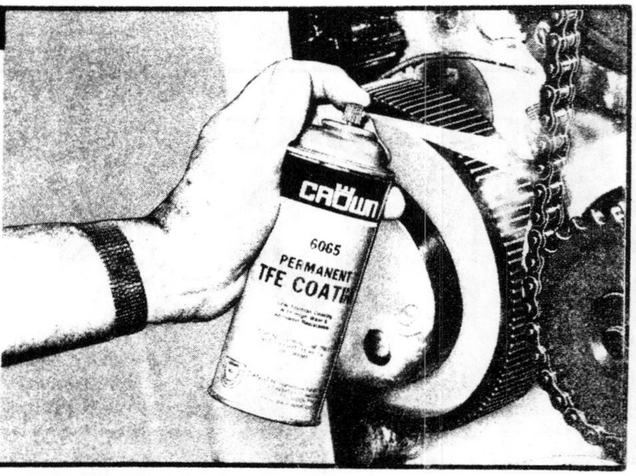

Crown Industrial Products Company, Incorporated
100 State Line Road/P.O. Box 326
Hebron, IL 60034
(815) 648-2424

CROWN

LUBRICANTS

FOOD SAFE LUBE No. 6035

For lubrication near foods. . . does not contain silicones. A food safe lubricant for pans, slides, containers, rollers, belts and other machinery and equipment in close proximity of foods. Spray application is simple, clean and easy. Contains aliphatic and naphthenic hydrocarbons (in the Federal Register of the FDA, paragraph CFR 178.3620, sub-paragraph 172.878) and Propellent.

6035 16 oz. can U.S.D.A. ACCEPTED

FOOD GRADE SILICONE LUBE No. 8036

This colorless, odorless, tasteless aerosol spray is excellent for containers, belts, and other machinery in packaging plants, paper converters, and similar industries related to food processing and preparation. Effective for any materials not to be painted or plated.

8036 16 oz. can

U.S.D.A. ACCEPTED
Authorized By USDA For Use In
Federally Inspected Meat And Poultry Plants

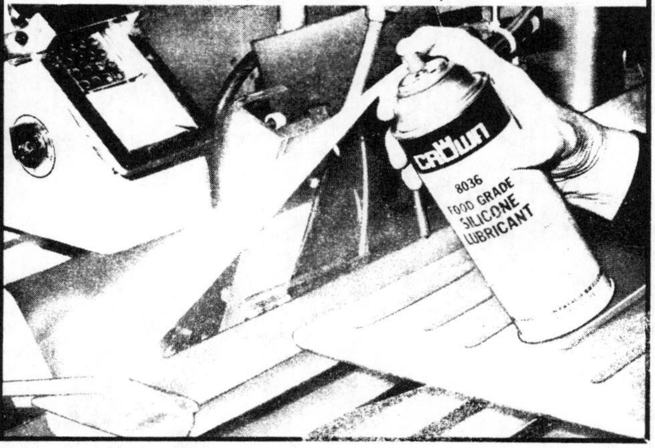

DRY FILM LUBRICANT (TFE) No. 6075

GREASELESS LUBRICATION!

Lubricant and release agent. . . nonstaining; chemically inert. Offers extreme adhesion and durability, provides efficient lubrication well beyond normal break in periods. Contains an active lubricant that gives unique, dry, nonstaining lubrication and release qualities up to 500°F. Forms a slippery, dry, translucent film on rubber, wood and metals. Won't interfere with post finishing operations; reduces clean up and labor time. Ideal for parts that are impractical to relubricate after assembly such as electrical equipment, timing devices, etc. . . . Effective under dusty or abrasive conditions such as mining operations, will not pick up dust.

6075 16 oz. can **6076 1 gallon**

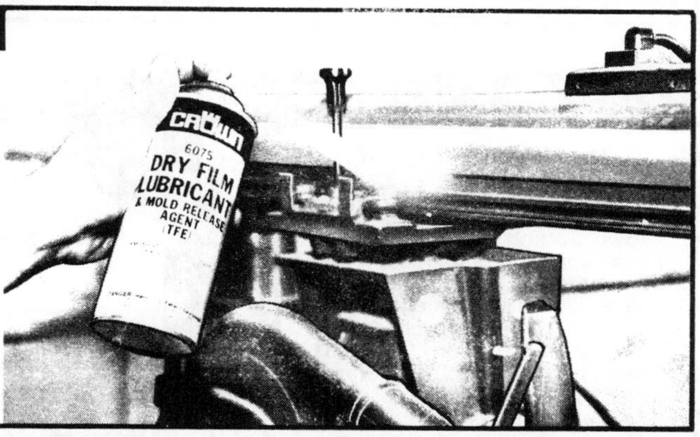

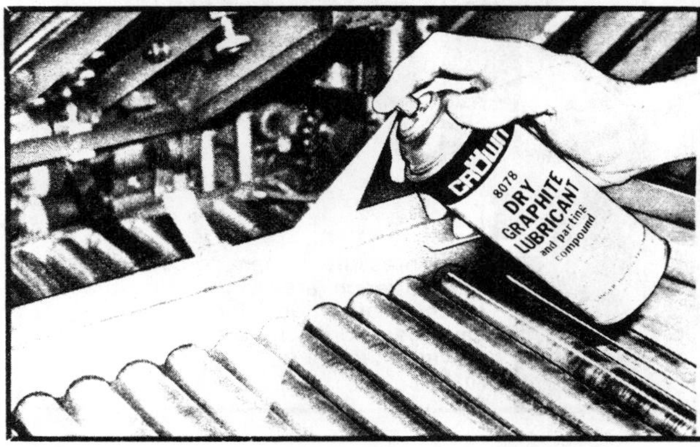

DRY GRAPHITE No. 8078

A DRY LUBRICANT!

Lube and parting compound. Contains no oil or moisture based diluents; is inert to water, oils, alkalies. Effective from -100°F. to +1000°F. . . on hydraulic pumps, internal combustion engines, business machines. . . wherever heat and friction are a problem. Dries rapidly at room temperature. Ideal where a combination of good electrical characteristics and positive lubrication are required. Clings to most surfaces with a minimum amount of pre-treatment. A graphite lubricant which can be sprayed for even dispersion. For use where a "dry" lubricant is required, where petroleum products must be avoided, where temperatures are extreme.

8078 16 oz. can

Crown Industrial Products Company, Incorporated
100 State Line Road/P.O. Box 326
Hebron, IL 60034
(815) 648-2424

LUBRICANTS

CROWN

Super PENETRATING OIL No. 6030

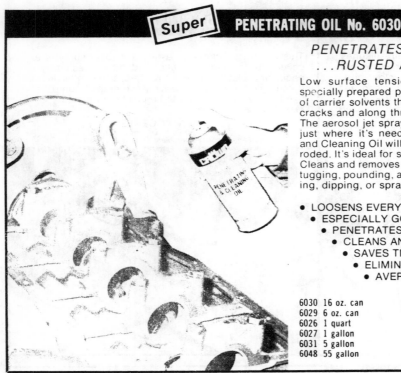

PENETRATES... LOOSENS... CLEANS...
...RUSTED AND CORRODED PARTS...

Low surface tension gives complete penetration... This specially prepared penetrating oil is formulated with two types of carrier solvents that permit the oil to travel easily into small cracks and along threaded fasteners for fast, thorough action. The aerosol jet spray applicator permits accurate coverage... just where it's needed... prevents waste. Crown Penetrating and Cleaning Oil will loosen most anything that is rusted or corroded. It's ideal for salt water corrosion. Penetrates on contact. Cleans and removes light rust. Saves time and labor. Eliminates tugging, pounding, and breakage. Apply bulk product by brushing, dipping, or spraying.

- LOOSENS EVERYTHING RUSTED OR CORRODED
- ESPECIALLY GOOD FOR SALT WATER CORROSION
- PENETRATES AND WORKS ON CONTACT
- CLEANS AND REMOVES LIGHT RUST
- SAVES TIME AND LABOR COSTS
- ELIMINATES TUGGING AND POUNDING
- AVERAGE JOB NEEDS JUST A FEW DROPS

6030	16 oz. can	Pinpoint extension tube included.
6029	6 oz. can	Pinpoint extension tube included.
6026	1 quart	
6027	1 gallon	
6031	5 gallon	
6048	55 gallon	

CUTTING OIL No. 7020

POSITIVE COOLING AND LUBRICATION!

Crown Cutting Oil is an all purpose lubricant and coolant. It gives maximum lubricity for longer tool life and means less tool sharpening, less down time. Crown Cutting Oil is a specially formulated compound that can be used for all ferrous and nonferrous machining operations such as grinding and cutting in addition to deep drawing and forming. Washes off easily with commercial type cleaners. Applications: turret lathes, gear cutters, broaches, grinders, automatic screw machines, all machine tools where a quality cutting oil is needed.

7020 16 oz. can
7021 1 gallon
7022 5 gallons
7023 55 gallon drum

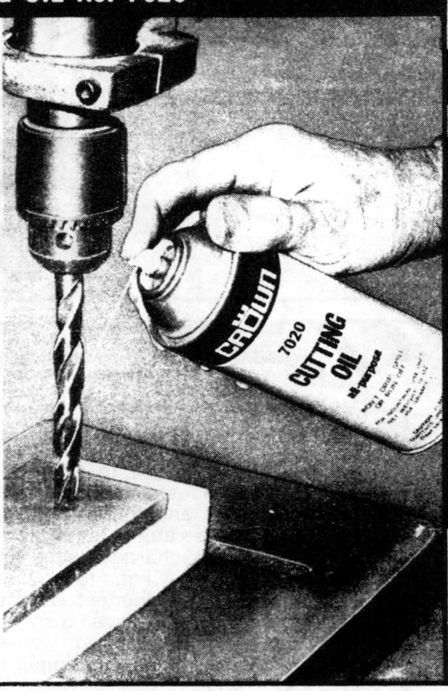

LUBRICATING OIL No. 6028

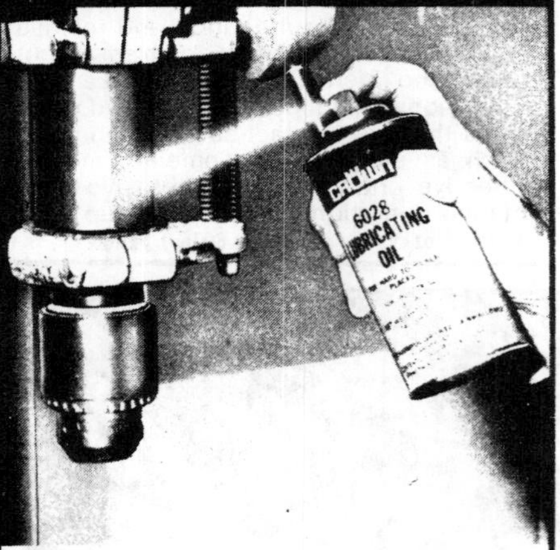

High pressure stream forces oil into hard-to-reach areas. This top quality machine oil is ideal for ball bearings, slide bars, drill chucks, etc. Ideal for moving machinery.

6028 6 oz. can

Crown Industrial Products Company, Incorporated
100 State Line Road/P.O. Box 326
Hebron, IL 60034
(815) 648-2424

CROWN LUBRICANTS

HEAVY-DUTY OPEN GEAR AND WIRE ROPE LUBE No. 7045

CLINGS & PENETRATES

High performance lubricant with excellent mastic and wicking properties. Penetrates wire rope . . . requires no heating. Has outstanding water and pressure resistance. Takes extreme pressure loads. Ideal for outdoor application, will not weather away. Unaffected by temperature extremes . . . protects against rust and corrosion. Noncorrosive . . . easy to apply in hard-to-reach areas.

7045 16 oz. can
7038 1 gallon
7039 5 gallon
7076 55 gallon

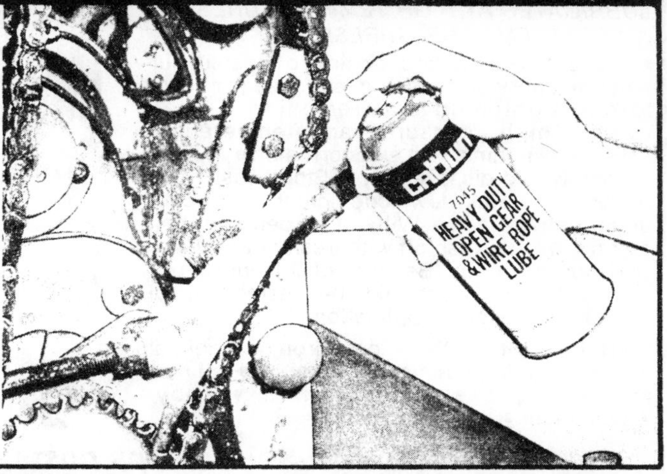

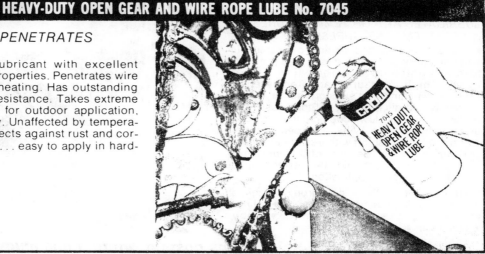

TFE LUBRICANT PERMANENT FILM No. 6078

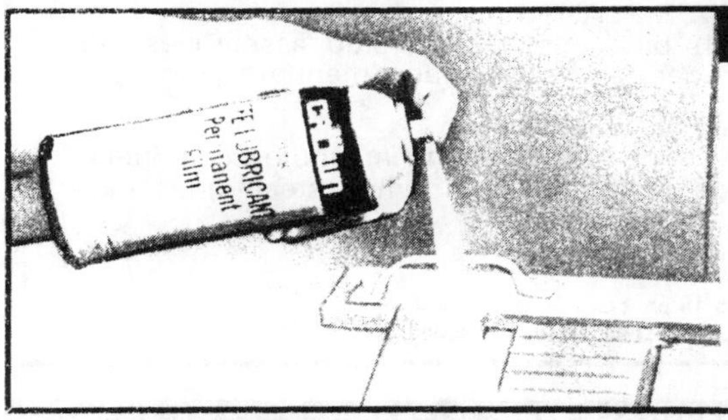

NON-YELLOWING/RESIN BONDED

A chemically inert lubricant for use on wood, metals, any smooth surface. Forms a slippery, transparent film. Contains an active fluorocarbon Telomer dispersion that gives unique nonstaining qualities and permits excellent lubrication up to 500°F. TFE lube forms a continuous film of lubrication providing an extremely low coefficient of friction.

6078 16 oz. can

MOLY OIL/OPEN CHAIN LUBE No. 7043

MOLYBDENUM DISULFIDE ADDED

Heavy duty, low viscosity oil for superior lubrication under high pressures or temperatures. Molybdenum Disulfide added. Use on hoist chains, conveyors, chain drives, cutter bars, clippers, chain saws, link and roller assemblies, sprockets, hinges, any application requiring a high grade light oil.

7043 16 oz. can

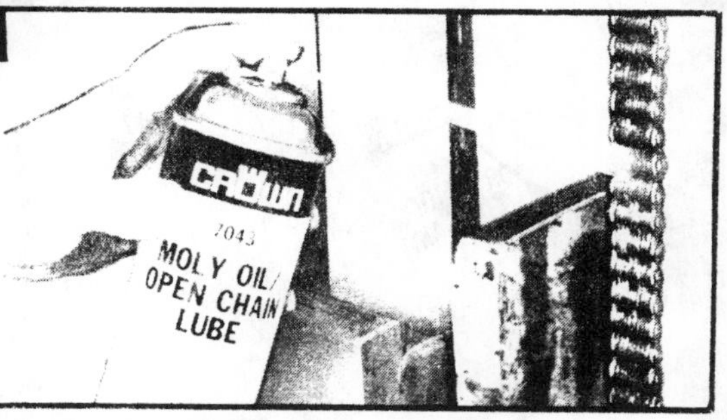

DRIPLESS OIL No. 7047

HAS THE FLUIDITY OF OIL . . .
THE DRIPLESSNESS OF GREASE

A specially treated oil that does not drip readily. Use where conventional oils cannot be retained, where it is desirable to lessen the amount of oil that drips off a surface. Use on bearings, wire gear mechanisms, textile industries, food industry — helps prevent spotting, leakage, blowing. (Do not use as an engine or motor oil.)

7047 16 oz. can

Crown Industrial Products Company, Incorporated
100 State Line Road/P.O. Box 326
Hebron, IL 60034
(815) 648-2424

LUBRICANTS

ANTI-SEIZE COMPOUND No. 9105

LUBRICATES AT HIGH TEMPERATURES...
AND EXTREME PRESSURES!

• Prevent seizing and galling of mated surfaces • Lubricates mating surfaces for lower torque tightening and removing • Seals against high pressures and temperatures • Protects against rust and corrosion • Speeds assembly and disassembly • Stops costly stud breakage • Has low coefficient of friction under heavy loads • Will not harden after heat aging • Won't wash off with fresh or salt water. Use indoors or out. (Because of solvents in product...Anti-Seize should be tested on all unusual or specific applications.)

CAN BE USED ON: • cast iron • steel, all alloys including stainless • titanium • all non-metallic gasketing materials • alloys • aluminum • copper • brass.

Crown Anti-Seize keeps costs down to a minimum in...

CHEMICAL PLANTS AND OIL REFINERIES
For pump housing flange studs, on cat crackers, stainless steel assemblies, and on all high temperature connections.

STEEL MILLS
Wherever threaded connections are subject to high pressures and high temperatures.

AIRCRAFT AND MISSILES
For jet engines, piston engines, rockets, missiles, on magnesium, aluminum, etc.

POWER PLANTS
On turbine inlet header stud assemblies, on cylinder head and exhaust manifolds, etc.

MARINE SERVICE
On boiler handhole and manhole cover plate studs, flange bolts on pumps, steam lines, etc.

9105 16 oz. can
Pinpoint extension tube included.

DRY MOLY LUBE No. 6080

- **WITHSTANDS PRESSURES TO 100,000 P.S.I.**
- **TEMPERATURES FROM -100°F. TO +700°F.**
- **QUICK DRYING "TIGHT" ADHESION**

Lubrication of all moving parts. It is a dry molybdenum disulfide (MoS_2), containing no oils or moisture-based diluents. It is inert to water, oils, alkalies, and some acids. It will withstand operating pressures up to 100,000 lbs. per square inch with no appreciable breakdown or loss of lubricating properties. Suited for use involving extreme ambient temperatures...provides lubrication from —100°F. to +700°F. Forms a dry, thin, black film imparting antifriction and antiseizure properties.

6080 16 oz. can

LITHIUM GREASE No. 7037

ALL PURPOSE...UNIVERSAL LUBE

Provides excellent all weather lubrication for any moving parts. Penetrates and forms a protective film that withstands all loads and speeds. Prevents rust and corrosion on metal parts. Eliminates squeaks and squeals, won't run or wash away. Crown Lithium Grease is completely versatile...combines the best qualities of all single purpose lubricants into one all purpose grease...Temperature range —70°F. to +360°F.

7037 16 oz. can **Pinpoint extension tube included.**

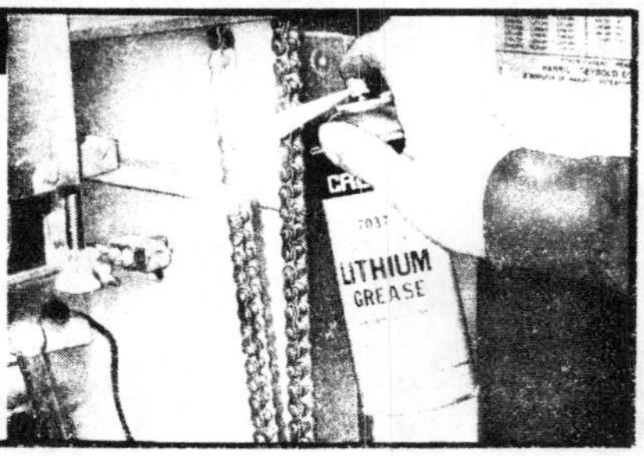

Darmex
71 Jane Street
Roslyn, NY 11577
(516) 621-3000

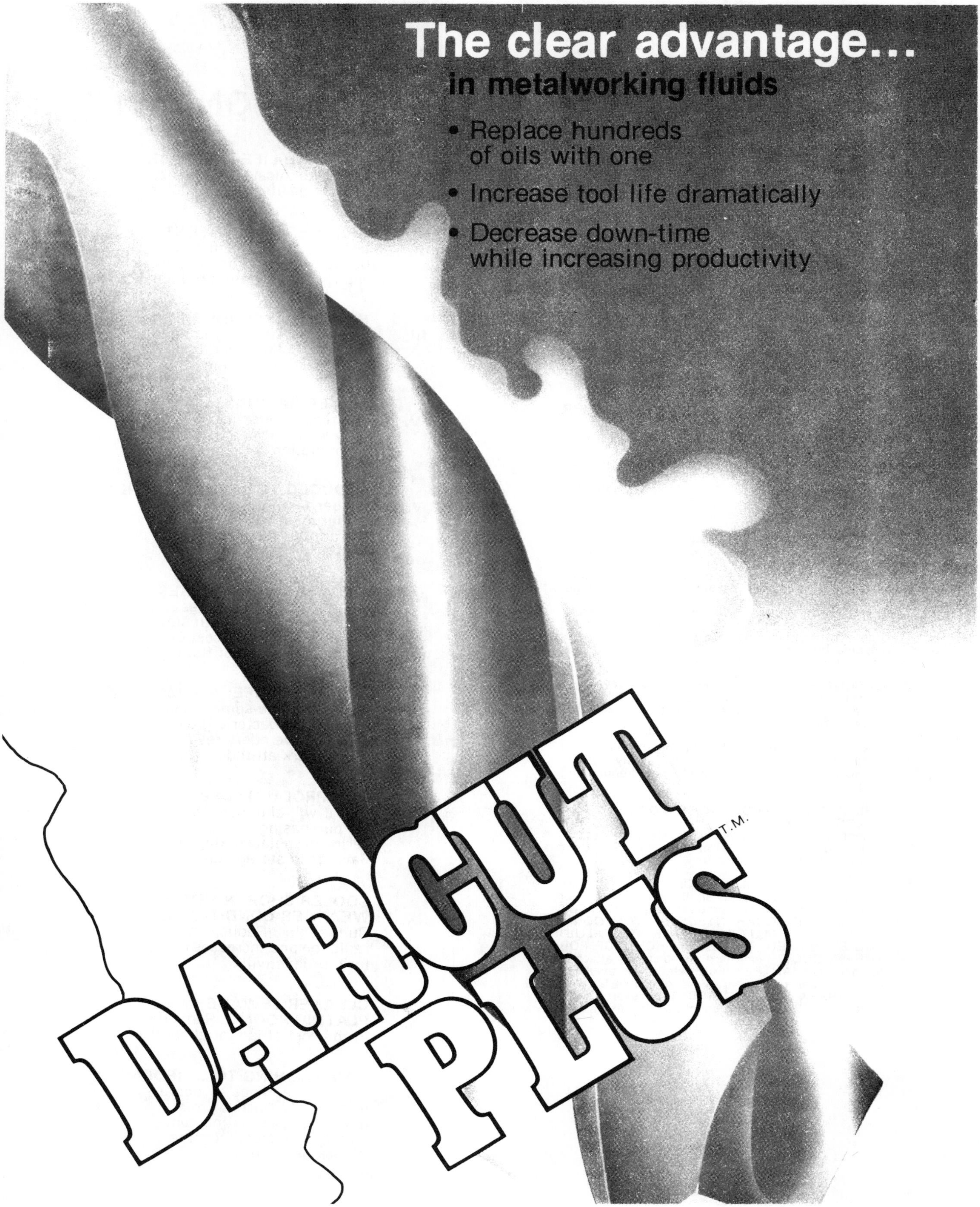

Darmex
71 Jane Street
Roslyn, NY 11577
(516) 621-3000

DARMEX...the name you've looked to for superior lubricants...

DARMEX has established a solid reputation as the leading manufacturer of a complete range of lubricants for industrial, institutional and automotive maintenance. Thousands of customers in all corners of the world, ranging from hospitals and colleges to major manufacturers in all types of industries, are benefiting from these exclusive "problem solving" formulations. DARMEX fluids keep working long after man and machinery have stopped. DARMEX products are sold by technically skilled specialists who are trained to identify customer problems and demonstrate how a change in product or procedure will reduce operating costs, extend equipment life and increase productivity.

...brings you DARCUT PLUS the one Metal-working fluid that replaces hundreds

DARCUT metalworking fluids incorporate the latest developments in modern manufacturing technology. DARCUT PLUS is the ONE METALWORKING FLUID that replaces hundreds of similar formulations, straight or water-soluble!

DARCUT PLUS has been tested in most types of machine-tool operations, including drilling, tapping, grinding, reaming, forming, champhering, honing, lapping, lathing, polishing, drawing, milling, stamping, sawing, hobbing, boring, turning, broaching, cutting, threading, punching and pressing. It has proven successful on all types of ferrous metals, including: iron, steel, stainless, forgings, maleable iron, castings; on non-ferrous metals, such as copper, bronze, aluminum; even on non-metallic materials, such as hard plastics and glass.

DARCUT PLUS excels beyond the recognized parameters of the industry. User's acceptance is rapidly growing as the product surpasses all expectations.

The world's increasing recognition of the critical need to improve productivity is rapidly turning DARCUT PLUS **into a must** to all types of equipment operators.

THIS SUPERIOR PRODUCT OF MODERN TECHNOLOGY...IS HERE TODAY! YOU, THE USER, cannot delay any longer using DARCUT PLUS. Increased tool life, superior finishes and better profitability are the dividends.

While most metalworking fluid manufacturers clearly disclaim all responsibilities on the use of their products, we are proud to say: "DARCUT PLUS IS 100% GUARANTEED. IF YOU DON'T FIND DARCUT PLUS TO BE THE BEST FLUID YOU EVER USED, THERE IS NO CHARGE FOR THE PRODUCT", and there are no strings attached to this GUARANTEE! (Full details are available from your Sales Engineer.) And, of course, as with all DARMEX products, DARCUT is covered by a multi-million dollar full product liability insurance.

THE REASONS WHY

I. **TOOL/DIE LIFE INCREASE:** Reports indicate an increase in tool/die life of up to 3000%.

II. **REDUCED TOOL INVENTORY:** Longer tool life results in reduced tool inventories. Less capital is tied-up unproductively.

III. **INCREASED PRODUCTION:** Fewer tool changes. Longer real production time. Ideal for automated machining.

IV. **POWER SAVINGS:** Latest additives reduce friction. Proven savings in energy consumption.

V. **REDUCED TOOL SHARPENING:** Longer tool life means less resharpenings with longer real life for tools and dies.

VI. **EXTENDED INTERVALS BETWEEN OIL CHANGES:** Longer fluid life reduces cost of purchased oils, with subsequent savings in cost of handling/storage of fluids.

VII. **BACTERIA SAFE:** Assures fewer machine-sump cleanings due to bacteria growth. No offensive odors result in better work attitude.

VIII. **ONE PRODUCT:** So effective that it will eliminate the need for purchasing and handling a series of metalworking fluids. Savings in storage and handling.

IX. **COOLER SHOP IN HOT WEATHER CONDITIONS:** Runs cooler. Reduces heat radiation and increases operator's productivity.

X. **CLEANER SEWAGE IN PLATING ROOMS:** Reduction in the cost of janitorial services.

XI. **NON STAINING TO UNIFORMS:** Better acceptance by operators. Easily washed from uniforms, eliminating common stains usual with the "black oil" types.

Darmex
71 Jane Street
Roslyn, NY 11577
(516) 621-3000

DARCUT PLUS IS BETTER:

XII. **INCREASED MACHINING RATES:** Higher cutting/grinding rates can be used. A plus for carbide and diamond tools.

XIII. **SAVINGS IN USED OIL REMOVAL:** As oil lasts much longer. Saves in the expensive cost of disposing of used fluids.

XIV. **EXTRA BTU OF POWER:** Used oil, if separated from water, can be added to fuel-oil tanks to generate additional heat/power.

XV. **REDUCED NEED FOR PRE-MIXING TANKS:** As a ONE PRODUCT, need for pre-mixing tanks is reduced to a minimum.

XVI. **TOTAL OPERATIONAL RANGE:** No need to inventory several oils, as one product will be effective in most all machine operations and all types of materials, including steels of all specifications, stainless, case hardened alloys, aluminum, brass, copper, even plastics or glass.

XVII. **SAFE FOR THE ENVIRONMENT:** No hazardous materials are found in this formulation: No nitrates, salts, carbon tetrachloride or other banned substances.

XVIII. **SUPERIOR RUST AND OXIDATION PROTECTION:** Tests indicate that even at high water dilutions, rust protection is superior to that found in some straight cutting oils.

XIX. **EXCEPTION CLEANLINESS:** No wax, gum or salty residues will occur. No ''welding'' of metal chips to tools/dies/grinding stones.

XX. **PLEASANTLY SCENTED:** For maximum operators acceptance of the fluid.

XXI. **NON-FOAMING:** Truly non-foaming, even at high production rates.

XXII. **NO FOGGING:** To avoid oil mist that affects operators health and performance.

LUBRICANTS

DARCUT PLUS IS BEST BY TEST

REPORTS:
(Data resulting from actual customer's testing. Full details are available from your Sales Engineer.)

	TAPPING: Cold rolled steel, 54 strokes per minute. Oil mixture: 10:1		**PUNCH PRESSES:** Tapping heads, progressive die with three tapping heads. Oil mixture: Competitive oil: Straight. DARCUT: 10:1		**KAUFMAN AUTOMATIC DIAL MACHINE:** 8 spindle, tapping, reaming and threading. 1¼'' n.p.t. dies, 2'' reamers on malleable iron. Oil mixture: 10:1		**TURRET LATHES:** Turning steel, aluminum and brass. Oil dilution: 15:1		**HORIZONTAL MILLING MACHINES:** Milling of steel, aluminum, brass. Oil dilution: 15:1	
	Before:	WITH DARCUT:	Before:	WITH DARCUT:	Before:	WITH DARCUT:	Before:	WITH DARCUT:	Before:	WITH DARCUT:
Life expectancy of oil:	2 weeks	4 months	4 months	6 months (at this time, still in use)	2 weeks	3 months	2 weeks	3 months	2 weeks	3 months
Tool life:	½ hour	8 hours	14,000 pieces	28,000 pieces	1 to 1½ Hrs. 300 pcs/hr.	24 Hrs. 464 pcs./hr.	4 hrs.	12 hrs.	4 Hrs.	12 Hrs.

Darmex
71 Jane Street
Roslyn, NY 11577
(516) 621-3000

- WATER SOLUBLE. Replaces most straight and water soluble oils.
- EFFECTIVE ON ALL METALS. Including all steels, stainless, hardened alloys, aluminum, brass, copper, even plastics and glass.
- UP TO 3000% INCREASE IN TOOL LIFE
- DOES NOT CONTAIN NITRATES, SALTS, CARBON TETRACHLORIDE.
- ALL COMPONENTS ARE ENVIRONMENTALLY SAFE!
- EXCELLENT RUST AND CORROSION PROTECTION.
- SUPERIOR COOLING PROPERTIES.
- LONG TANK LIFE.
- EXCEPTIONAL CLEANLINESS. No wax, gum or salty residues.
- PLEASANTLY SCENTED.
- NON-FOAMING.
- NO FOGGING.

RECOMMENDED DILUTIONS:
(EXACT MIXTURES TO BE DETERMINED ACCORDING TO OPERATING CONDITIONS)

Grinding:	30:1	to	75:1
Light Machining:	15:1	to	40:1
General Machining: (light alloys)	15:1	to	50:1
General Machining: (medium alloys)	15:1	to	30:1
Heavy Machining: (tough alloys)	5:1	to	20:1
Super Alloy Machining:	5:1	to	15:1

TYPICAL SPECIFICATIONS:

Weight: (Pounds per gallon): 7.96

pH: Straight: 9.05 10:1 8.83 20:1 8.74

Concentrate Color: Dark Amber

Diluted: Opaque white (milky emussion)

Viscosity: 100°F: SUS 650

FROM DARMEX TECHNOLOGY, OTHER AVAILABLE LUBRICANTS INCLUDE:

LONG LASTING BEARING GREASES (for all temperature ranges)
NON-LEAK GEAR BOX OILS
LEAK REDUCING HYDRAULIC FLUIDS
NON CARBONIZING AIR COMPRESSOR OILS
AIR LINE/TOOL LUBRICANTS
COMPLETE RANGE OF INDUSTRIAL LUBRICANTS FOR ALL TYPES OF EQUIPMENT

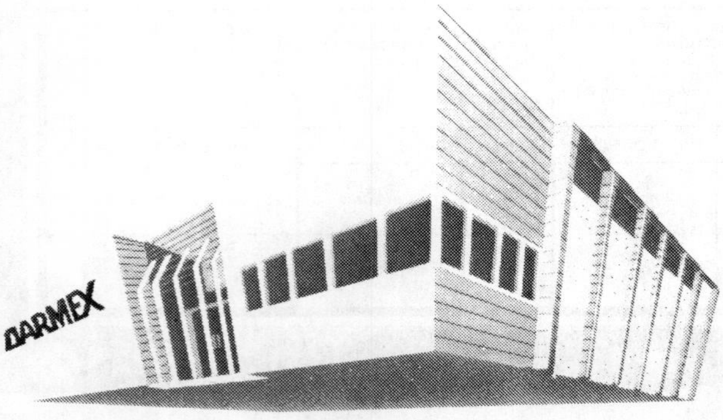

DARCUT, a division of Darmex Corp.

YOUR LOCAL REPRESENTATIVE:

71 Jane Street, Roslyn, L.I., N.Y. 11577 - USA
Telephone: 212-528-5800 • 516-621-3000 Telex: 14-4520
Watts: 800-645-6368—OUTSIDE N.Y. STATE

Darmex
71 Jane Street
Roslyn, NY 11577
(516) 621-3000

DARMEX®

OILS

DARCUT SPF [GRADES L, M, H]

Three spindle oils for high speed equipment. Specially formulated fluids that assure maximum antiwear protection and eliminate rust and corrosion while reducing noise levels. Exact temperature/viscosity slope for maximum precision. A must, to protect expensive high production machinery, where downtime is extremely expensive. Equipment will run cooler, by reducing the frictional heat that contributes to accelerated wear. SPF fluids are non-gumming and will not produce varnish residues.

DARCUT ALF

''Instrument type'' oil designed for all types of airline and air tool equipment.
DARCUT ALF provides complete ''wetting'' of the wall of thin, capilary type, narrow tubings assuring permanent displacement of water and humidity. Thus maintaining airlines and air systems free of rust and corrosion. ALF features exact viscosity for constant oil flow and extreme low pour point for superior results.

DARCUT GEROIL

Single and multiviscosity gear box oils. The most popular type being GEROIL 60/90 formulated for all types of gear boxes, speed reducers, transmissions, and differentials requiring oils within the viscosity range of SAE 60 to SAE 90. The use of submicrometric particle sized Molybdenum Disulfide (MoS2) and a wide temperature range provides over 185,000 P.S.I. of long lasting protection. Additionally, high shear stability and non-drip formulation counteracts centrifugal forces and reduces possible leaks on worn gear boxes.
DARCUT GEROILS will cut operating temperature and show KW/Hour savings averaging up to 20%. Reduced wear and tear, and less expensive downtime are proven in all applications.
DARCUT GEROILS are available with similar features on a ''non-molly'' version (GEROIL CL) and in other viscosities such as SAE 75W/90, 90, 90/140, 140, 250.

DARCUT ALL

The ONE OIL that just DOES IT ALL!!!
Multigrade fluid that will efficiently replace many of the lubricants used in a machine shop.
DARCUT ALL is recommended as a machine oil, hydraulic fluid, way and slide lubricant and high-speed gear oil.
This superior lubricating film, with maximum rust and corrosion resistant properties, assures effective protection in all applications where oils of viscosities from the SAE 10 to the SAE 40 are required (except air compressors and vacuum pumps). It is not an internal combustion engine lubricant.
DARCUT ALL's non-drip formulation reduces oil consumption on systems subject to leaks, but remains pumpable by all automatic lube systems.
DARCUT ALL assures superior shear stability and is WATER SOLUBLE and totally compatible with DARCUT PLUS. If mixed with DARCUT PLUS, the addition of water will result in additional metalworking fluid and NOT in tramp oil.

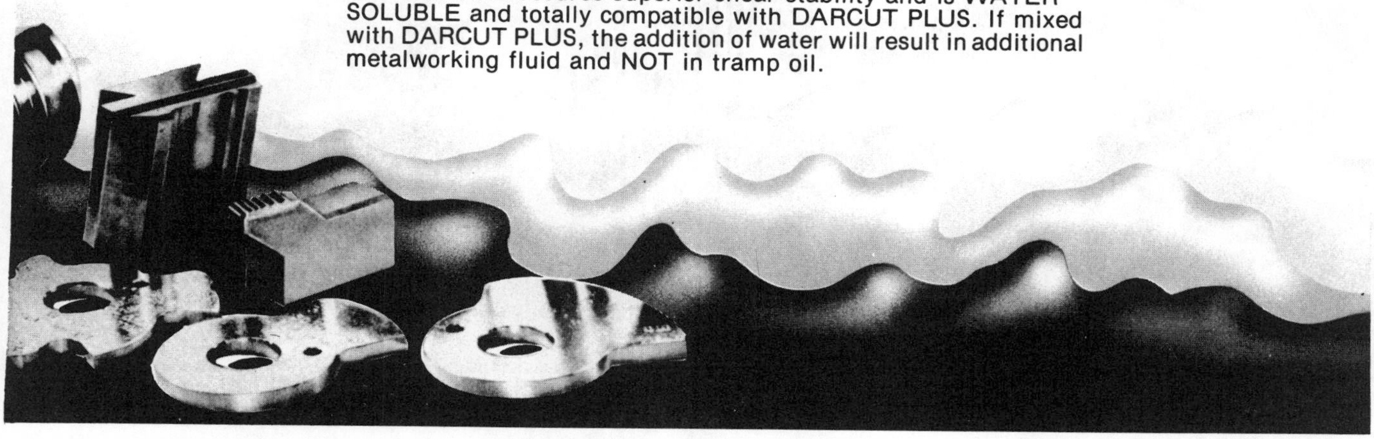

Darmex
71 Jane Street
Roslyn, NY 11577
(516) 621-3000

GREASES

DARCUT GR [#1, #2, #3]

Long lasting, industrial lubricating greases available in three grades, corresponding to NLGI #1, #2 and #3.

DARCUT GR greases contain molybdenum disulfide and are formulated for long intervals between relubrication while providing superior antiwear, rust and corrosion protection. These heavy-duty greases are suitable for all types of roller and ball bearings. They are water and chemical resistant and will provide maximum protection against abrasion, heavy loads and shock impact over a wide temperature range. Selection of grade will depend on the operating speeds.

DARCUT GR/G grades H and L

Open gear fluids for all types of exposed mechanisms such as open gears, sprockets and cams. They are easy to apply as they are self-carrying on gear trains of all sizes. These extreme pressure, antiwear greases are 100% lubricant, and contain no textile fibers or any of the other non-lubricating fillers used in similar compounds. They last longer, protect better, with a tenacious, continuous, monofilament that reheals itself under shock-loads or heavily loaded gears.

DARCUT PK

Antiseize, paste-like compound, used as a ''slide'' lubricant for easy assembly and disassembly of molds, tools, jigs, dies. A wide temperature range formulation containing a synthetic based fluid, that is thickened with an inorganic, non carbonizing gel, and contains submicrometric particle sized metallic powders (MoS_2, graphite, bronze). It can be brushed-on or aplied manually, assuring a continuous, tough and long lasting film. (DARCUT PK is not a bearing lubricant).

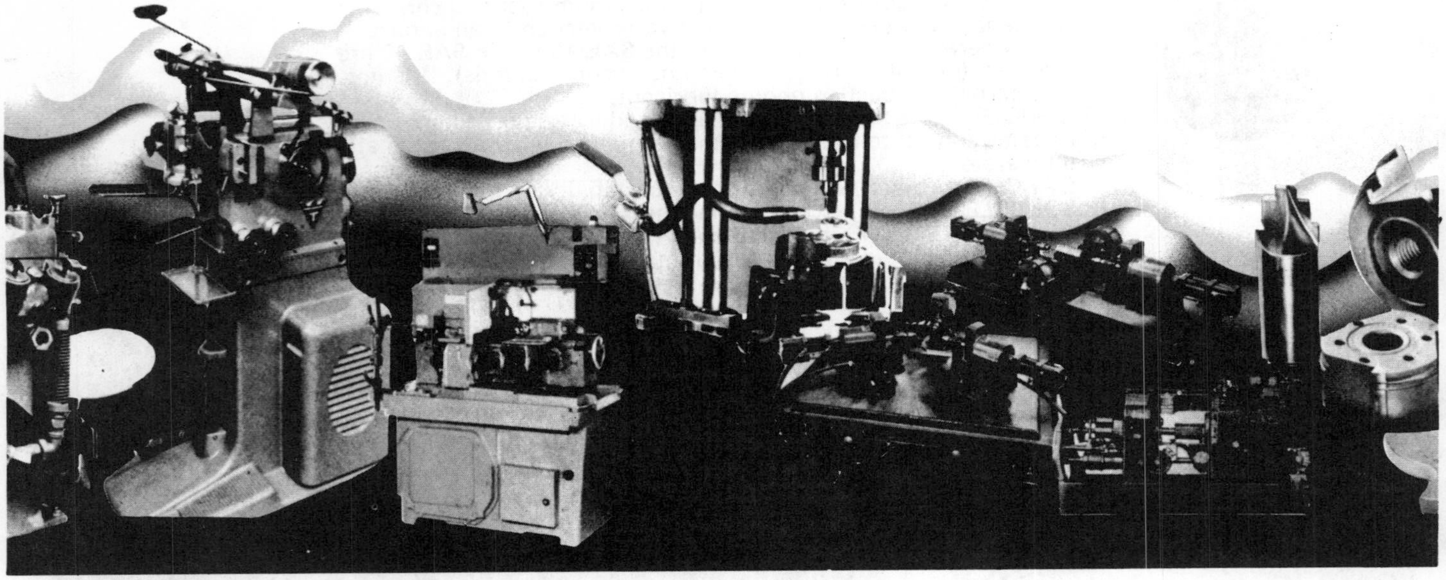

TYPICAL SPECIFICATIONS

GREASES	GR 1	GR 2	GR 3	GRG "H"	GRG "L"	PK
NLGI Grade	1	2	3	3	1½	1
Penetration, unworked at 77°C, average	325	295	250	255	270	325
Worked Penetration, 100,000 strokes, average	330	298	252	260	273	328
Drop Point, °F	550 +	None	None	None	500 +	None
Viscosity Base oil, SUS at 100°F	175	1200	1200	1650	1750	2600
Bearing Corrosion Test, ASTM D-1743	PASS					
Timken Load, LBS/MIN	70	85	75	75	70	80
Max. Bearing Speed (RPM)	7000	2200	1400	Not recommended for bearings		
Operating Temperature °F Constant	-20/300	-10/325	0/300	0/300	0/300	-25/600
Operating Temperature °F Intermittent	-30/325	-10/375	0/325	0/350	0/350	-40/800

OILS	SPF "L"	SPF "M"	SPF "H"	ALF	ALL	60/90	90/140
Viscosity SUS at 100°F	79	105	205	120	300	680	1075
Viscosity SUS at 210°F	35	42	49	49	66	90	125
Viscosity Index	101	100	102	103	146	133	136
Rust Test ASTM D-665, 48 hrs.	PASS NO RUST						
Copper Strip at 210°F, 3 hrs. (ASTM D-130)	NEGATIVE						
Pour Point °F	-40	-35	-35	-50	-20	-20	-20
Flash Point °F	420	440	450	430	440	450	475
Fire Point °F	450	470	480	470	480	495	525
Timken OK Load, LBS/MIN	70	72	75	75	80	90	95
Operating Temperature, °F Constant	-20/300	-20/300	-15/320	-40/350	-20/300	-20/300	-10/325
Operating Temperature °F Intermittent	-25/325	-25/325	-20/350	-45/375	-25/325	-20/325	-15/350

LUBRICANTS

Dri-Slide, Incorporated
411 North Darling
Fremont, MI 49412
(616) 924-3950

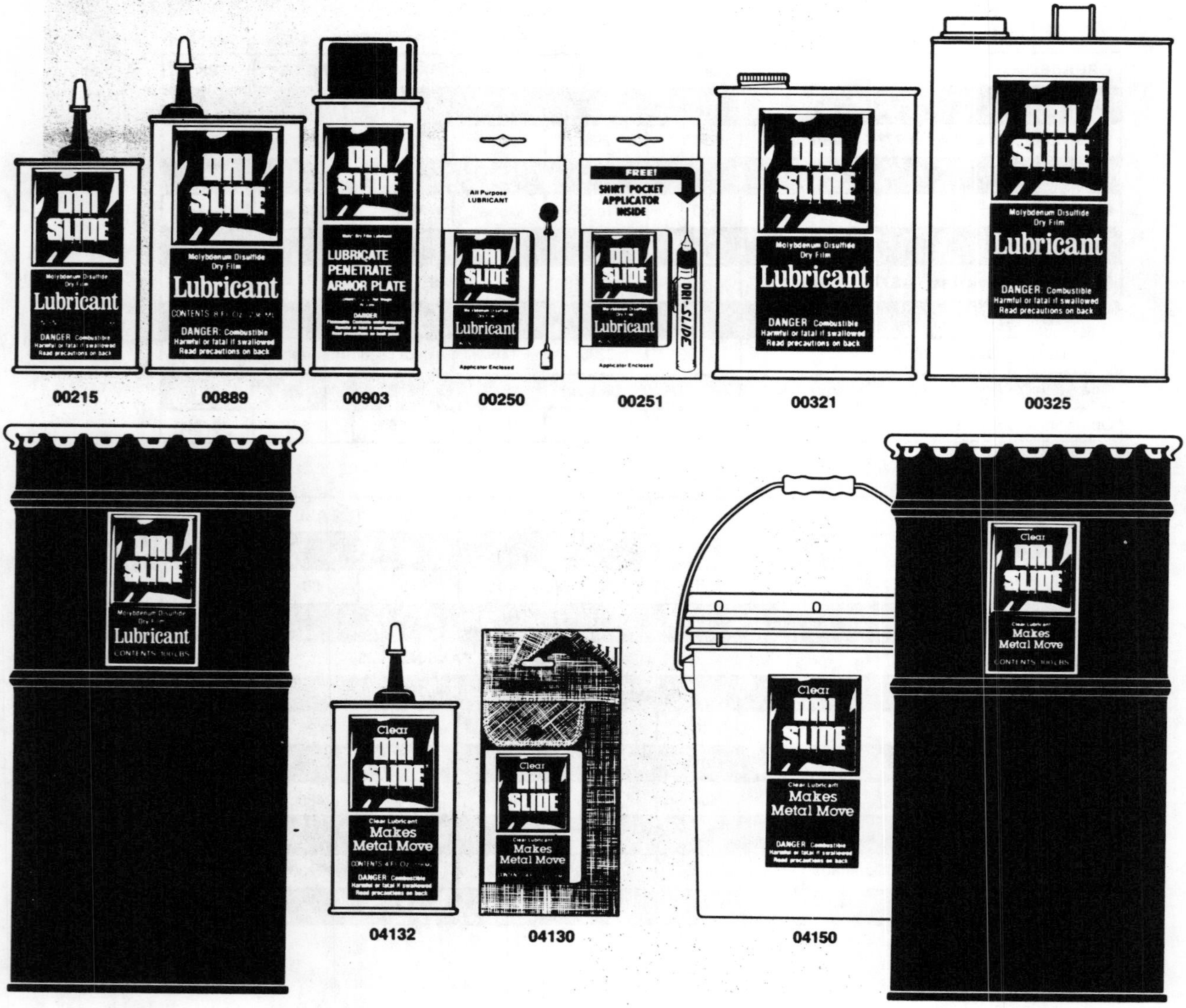

00215 00889 00903 00250 00251 00321 00325

03100 04132 04130 04150 04156

Dri-Slide
00215 4 oz. Spout Can
00889 8 oz. Spout Can
00903 5¼ oz. Aerosol
00250 4 oz. Shelf Pak (needle enclosed)
00251 4 oz. Shelf Pak/Free Applicator
00321 32 oz. Can
00325 1 Gallon Can
03100 100 Lb. Drum

Dri-Slide contains molybdenum disulfide (MoS_2) which withstands pressure of up to 100,000 p.s.i. and can be used in temperatures ranging from −150°F to 750°F. Resists dirt, dust and water. Dry film leaves long lasting protection.

Clear Dri-Slide
04132 4 oz. Spout Can
04130 4 oz. Shelf Pak (needle enclosed)
04150 5 Gallon Pail
04156 16 Gallon Drum

The same fine qualities as regular Dri-Slide, but clearer. Can be used on delicate equipment where the staining characteristics of regular Dri-Slide should be avoided. Synthetic Moly* Base.

*Molybdenum Disulfide.

Dri-Slide, Incorporated
411 North Darling
Fremont, MI 49412
(616) 924-3950

00204 00206 00208

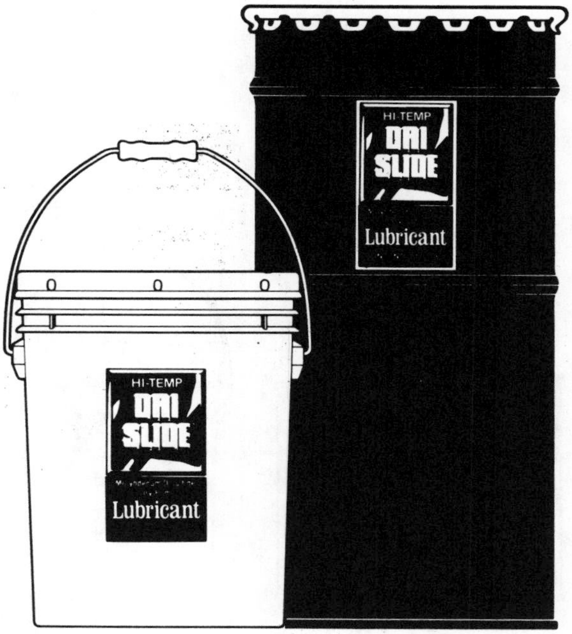

03235 03200

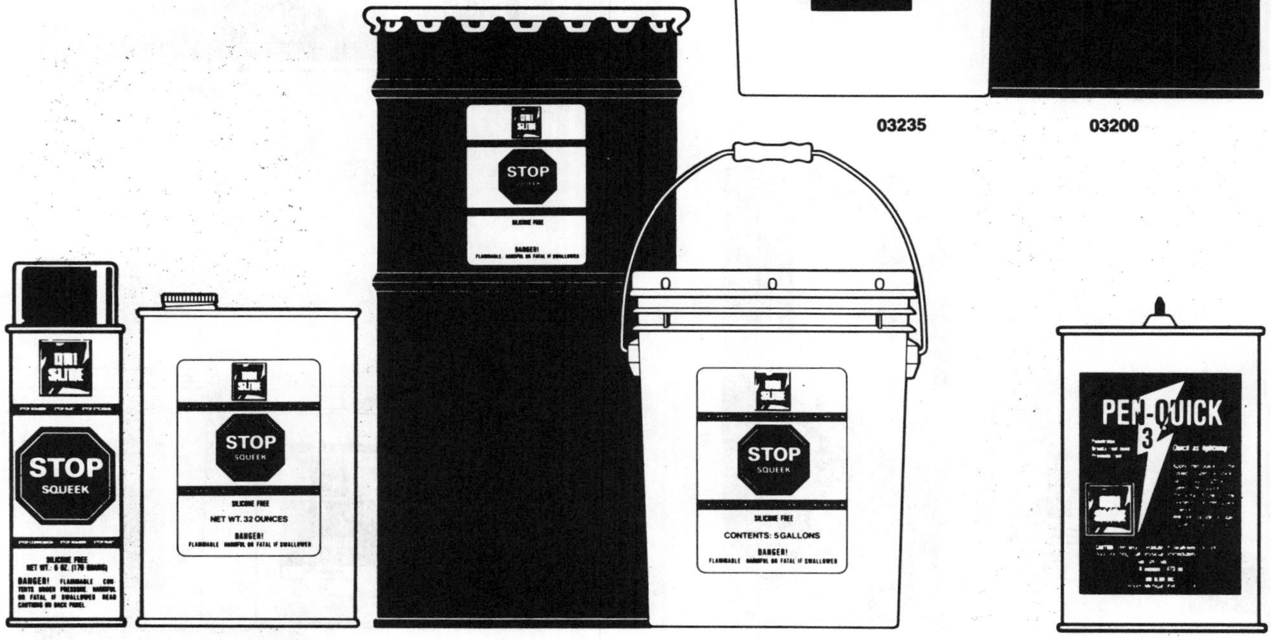

04134 04135 04138 04137 30333

Bike Aid
00204 4 oz. Spout Can
00206 4 oz. Shelf Pak (needle enclosed)
00208 5¼ oz. Aerosol

Made especially for bicycles, motorcycles and mopeds. Resists rust and road dirt. Use on cables, derailleurs, shift controls, chain, pivot points or anyplace that needs lubrication. Contains Moly*. Needle in shelf pak allows you to reach into difficult areas.

Special Hi-Temp Dri-Slide
03235 35 Lb. Pail
03200 100 Lb. Drum

Excellent high temperature characteristics. Will not oxidize to form carbon deposits. Not affected by acids, salts or water. High flash point 450°F (c.o.c.) Contains Moly*. Works up to 750°F.

Stop Squeek
04134 6 oz. Aerosol
04135 32 oz. Can
04137 5 Gallon Pail
04138 16 Gallon Drum

Really stops squeaks, rust and corrosion. Displaces water. Use to stop squeaks and free rusted-shut parts and tools.

Pen-Quick 3
30333 16 oz. Tip-up Spout Can

Made for stubborn rusted parts. Will *not* corrode or stain any metal.

*Molybdenum Disulfide.

Dri-Slide, Incorporated
411 North Darling
Fremont, MI 49412
(616) 924-3950

00770 00774 00775 00776 00777

00709 00704 00706 00702 00695 00694

Dirty Mess, Goo Remover
00770 6 oz. Aerosol
00774 32 oz. Can
00775 5 Gallon Pail
00776 16 Gallon Drum
00777 55 Gallon Drum

The name says it all. An all purpose cleaning solvent which breaks through grime and grease fast. Industrial Strength.

FXR
00709 5¼ oz. Aerosol
00704 32 oz. Can
00706 100 Lb. Drum

Contains extra moly* for added protection. Additional anti-rust and corrosion inhibitors. Can be painted over after drying. Revitalizes frozen parts. Good for corrosive atmosphere and for equipment stored out door.

EP Liquid Grease
00702 8 oz. Aerosol
00695 32 oz. Can
00694 45 Lb. Pail

40% Molybdenum Disulfide for use in *EXTREME PRESSURE* situations. Extra doses of anti-rust and corrosion agents. Excellent on chains and open gears.

*Molybdenum Disulfide.

Dri-Slide, Incorporated
411 North Darling
Fremont, MI 49412
(616) 924-3950

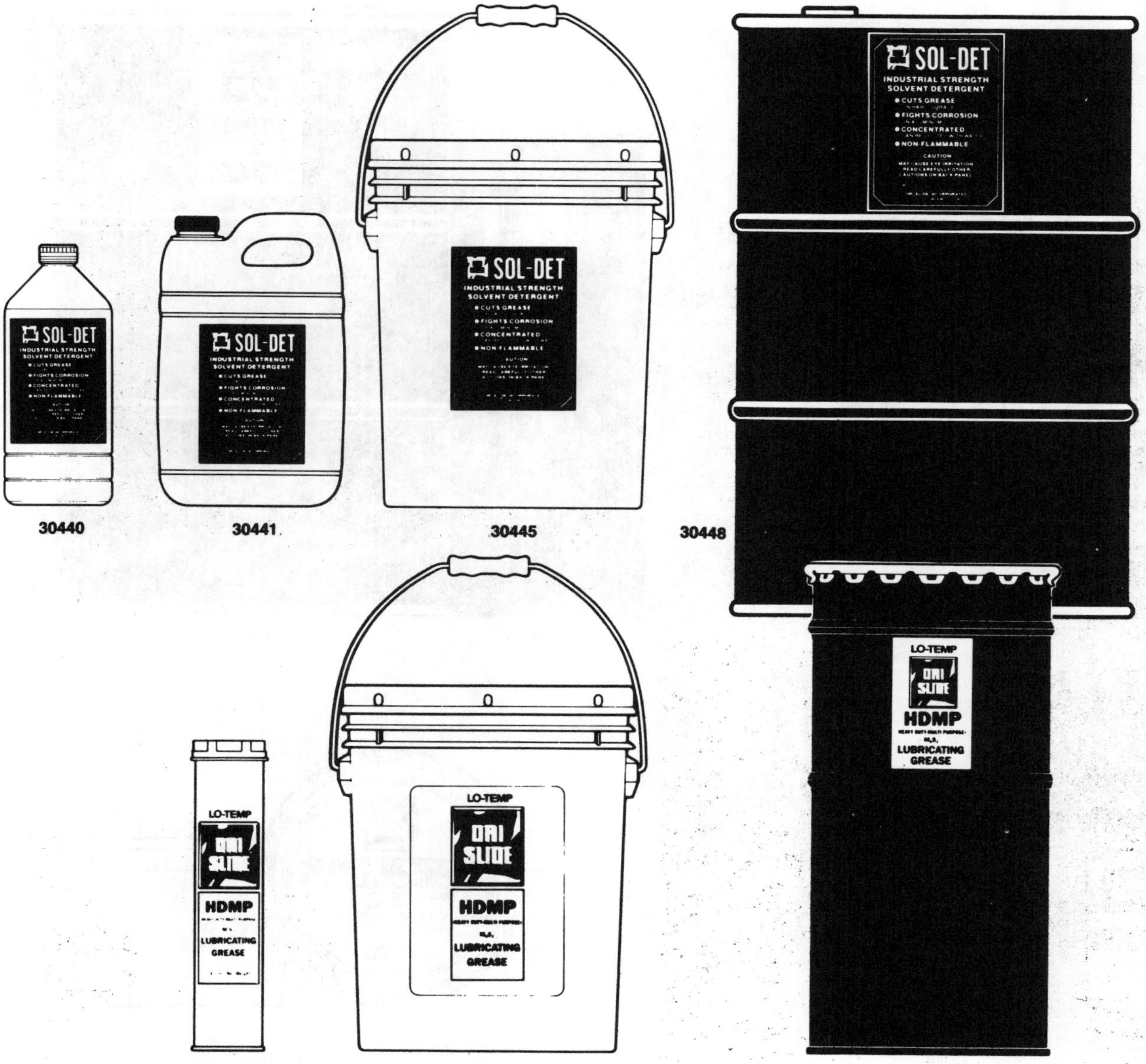

30440 30441 30445 30448

00373 00374 00375

Sol-Det
30440 32 oz. Plastic Bottle
30441 1 Gallon Plastic Bottle
30445 5 Gallon Pail
30448 55 Gallon Drum

A general purpose solvent/detergent *concentrate*. Excellent on walls, floors, counter tops or anything else that needs cleaning. Economical for shop use.

HDMP Grease—Lo-Temp
00373 14½ oz. Cartridge
00374 35 Lb. Pail
00375 120 Lb. Drum

Heavy Duty Multi Purpose Grease designed for heavy duty low temperature applications. Pumpable at temperatures from −40°F to 250°F. Not subject to water wash-out. Contains Moly*.

*Molybdenum Disulfide.

Dri-Slide, Incorporated
411 North Darling
Fremont, MI 49412
(616) 924-3950

00343

00350

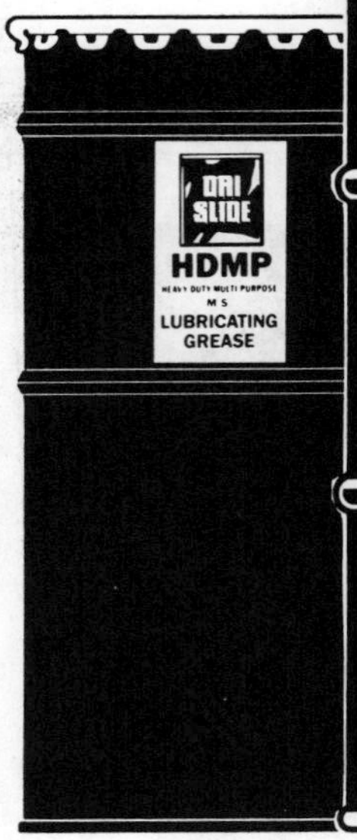

00351

30352

00383

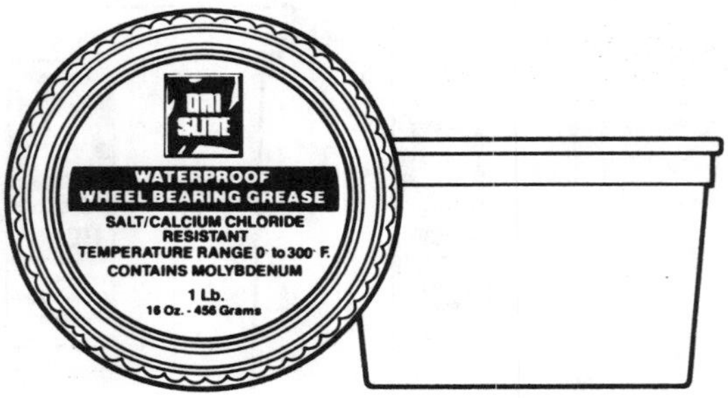

00390

HDMP Grease—Regular
00343 14¼ oz. Cartridge
00350 35 Lb. Pail
00351 120 Lb. Drum
30352 400 Lb. Drum

To be used where temperatures range from 0°F to 300°F. Contains Moly*.

HDMP Grease—Hi-Temp
00383 20 oz. Cartridge

Made to be used where temperatures range from 0°F to 600°F. Greater stability. Can be used in temperatures from 0°F to 400°F to extended lubrication cycles.

Combination of lubricating silicone oil and fluorinated hydrocarbon in a non-melting organic gel system.

Wheel Bearing Grease
00390 16 oz. Plastic Tub

Contains Molybdenum disulfide so you know it's good. Waterproof and salt resistant. Operates in sub freezing temperatures.

*Molybdenum Disulfide.

Dri-Slide, Incorporated
411 North Darling
Fremont, MI 49412
(616) 924-3950

00363

00360

00361

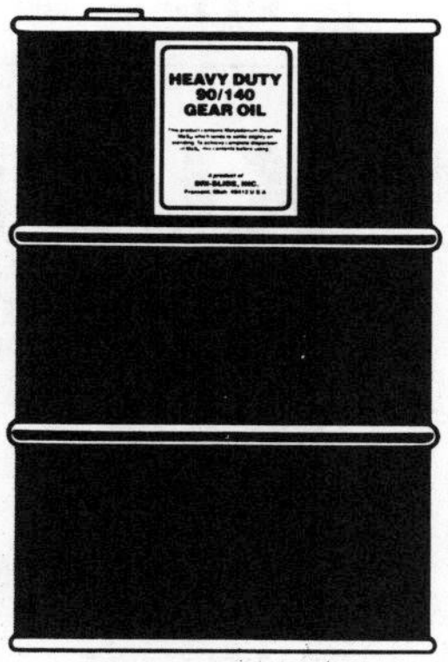

30460

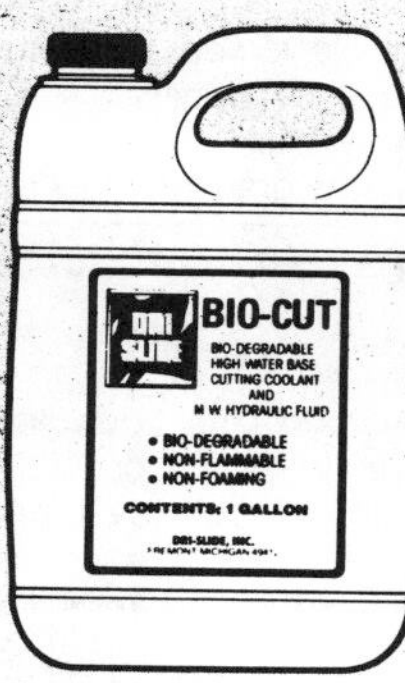

30461

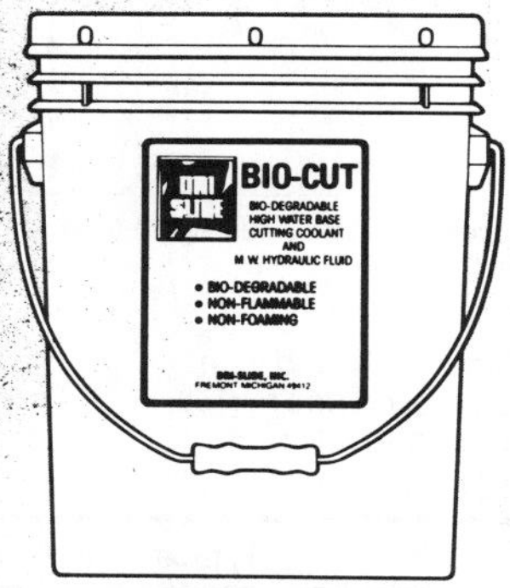

30462

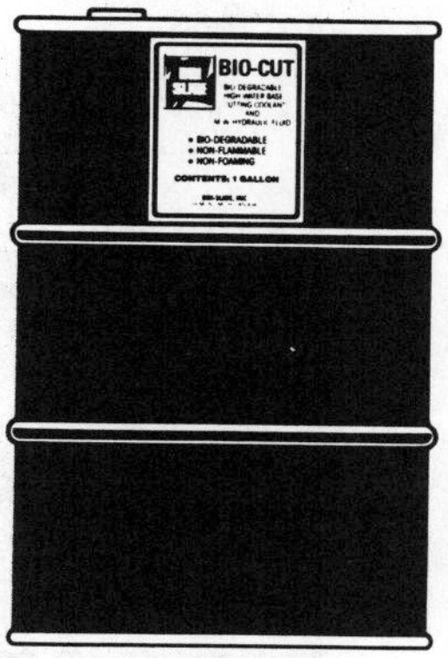

30465

30470

30475

30480

90/140 Gear Oil
00363 32 oz. Plastic Bottle
00360 37 Lb. Pail
00361 120 Lb. Drum
00362 412 Lb. Drum

A special gear oil with Moly* for heavy duty industrial use where maximum gear protection is needed. Improves viscosity and reduces friction better than ordinary oils. Clinging action.

Bio-Cut Cutting Coolant
30460 32 oz. Plastic Bottle
30461 1 Gallon Plastic Bottle
30462 5 Gallon Pail
30465 55 Gallon Drum

Concentrated water base cutting coolant. Cools better than oil alone because of its higher capacity to remove heat. Non-flammable and non-foaming. Economical — can be diluted as much as 50-1.

Bar/Chain Oil
30470 32 oz. Plastic Bottle
30475 1 Gallon Plastic Bottle
30480 55 Gallon Drum

Our know-how in lubricants lead to a Bar & Chain Oil that woodsmen rave about. Stays on the chain and prolongs saw life because of a heavy dose of tacking agents.

*Molybdenum Disulfide.

Keystone/Pennwalt Corporation
21st and Lippincott Streets
Philadelphia, PA 19132
(215) 225-7473

KEYSTONE PRODUCT INFORMATION

KLC Multigrade Hydraulic & Lubricating Oils KLC-5/20 · KLC-10/30 · KLC-20/50

KLC Multigrade Oils are premium antiwear oils for use in multiple applications. They are USDA "H-2"* products.

Features:	Benefits:
• Ashless	• Low sludge and deposit formation; ease of disposal
• Antiwear; Rust and Oxidation Inhibited	• Longer pump life, increased oil life
• High Viscosity Index	• Less change in viscosity with temperature
• Low Pour Points	• Easier cold operation
• Shear Stable	• Stays in viscosity grade in most applications

* USDA "H-2", Lubricants with no incidental contact: may be used in food processing plants where there is no possible contact of the lubricant with edible products.

Applications:
KLC Multigrade Oils are designed for use in low and high pressure hydraulic systems. They are inhibited against rust and oxidation, antiwear treated, and are ashless to prevent sludge. They can be used in plain and antifriction bearings.

KLC-5/20, Keystone's most popular multigrade hydraulic oil, is dyed green for identification. It is approved by Racine and other manufacturers. Call Keystone for details.

Typical Properties

Property	Test	KLC 5/20	KLC-10/30	KLC-20/50
SAE Grade		5W-20	10W-30	20W-50
Viscosity @ 210°F, SUS:	D-2161	50	65	95
@ 100°F, SUS:	D-2161	180	320	650
@ 100°C, cst:	D-445	7.1	10.9	18.5
@ 40°C, cst:	D-445	35.4	63.1	129
Viscosity Index:	D-2270	170	165	160
Pour Point, °F:	D-97	−40	−25	−10
Specific Gravity, 20°C/20°C:	D-1298	0.864	0.874	0.885
Flash Point, C.O.C. °F:	D-1298	400	430	460
Ash, %:	D-874	Nil	Nil	Nil
4-Ball Wear Test, 40 kg.,mm:	D-2266	0.45	0.45	0.45
Oxidation, Hours to NN=2:	D-943	3000	2800	2500
Rust Protection:	D-665-A,B	Pass	Pass	Pass
Demulsibility, Minutes to 40/37/3	D-1401	10	20	40
Copper Corrosion:	D-130	1b	1b	1b
Neutralization Number:	D-664	0.3	0.3	0.3
Vickers Vane Pump Wear, mg:	D-2882	20	25	30

Keystone Division of Pennwalt Corporation, with a policy of continuous improvement reserves the right to change specifications as our technology progresses. We are not responsible for misuse and/or misapplication of our products.

700 Third Line
Oakville, Ontario L6J5A3
Canada: 416-827-9841

21st and Lippincott Streets
Philadelphia, PA 19132

Call Toll Free 800-344-2241
in Pennsylvania: (215) 225-7473

KEYSTONE
PENNWALT
CHEMICALS ■ EQUIPMENT
HEALTH PRODUCTS

TS-006A
8/84C

LUBRICANTS

Keystone/Pennwalt Corporation
21st and Lippincott Streets
Philadelphia, PA 19132
(215) 225-7473

PRODUCT INFORMATION

KEYSTONE PENNWALT
PRECISION LUBRICANTS
21st and LIPPINCOTT STREETS, PHILADELPHIA, PA. 19132 U.S.A.

KLC® LUBRICATING AND HYDRAULIC OILS

KLC oils are formulated to meet many equipment builder's specification. KLC oils are manufactured in 6 viscosity grades and meet USDA requirements for non-food contact applications. They are suitable for machine tools, vacuum pumps and gear boxes.

Features	Benefits
• Non-Detergent	• Eliminates Deposits Due to Moisture
• Rust, Oxidation & Anti-Wear Additives	• Protects Components, Maximizes Oil Life
• High Viscosity Index	• Minimizes Viscosity Change with Temperature

TYPICAL PROPERTIES	KLC LUBRICATING OILS					
	10	15	20	30	40	50
ISO Visosity Grade:	32	46	68	100	150	220
Viscosity SUS @ 100°F:	160	225	345	530	800	1300
Viscosity SUS @ 210°F:	42	48	59	65	77	97
Viscosity Index:	100	100	100	98	96	95
Viscosity, cs at 40°C:	31.4	42.0	68.2	99.03	144.7	237.5
Viscosity, cs at 100°C:	5.31	6.24	8.79	11.16	14.21	19.73
Color-ASTM-D-1500:	2	3	4.0	5.0	5.0	6
Flash Point °F:	420	420	420	425	435	475
Pour Point - °F:	- 10	0	+ 10	+ 10	+ 15	+ 15
Rust Test ASTM-D-665 (Proc. A):	Pass	Pass	Pass	Pass	Pass	Pass
Oxidation Test ASTM-D-943 hours:	2000	2000	2000	2000	2000	2000
SAE No.:	10	20	20	30	40	50
AGMA Rating:	--	1	2	3	4	6

REPLACEMENT KEY

KLC (New Name)	KLC 10	KLC 15	KLC 20	KLC 30	KLC 40	KLC 50
Replaces (Old Name)	KLC-6	KLC-5	KLC-4A	KLC-4	KLC-3	KLC-2

Keystone/Pennwalt Corporation
21st and Lippincott Streets
Philadelphia, PA 19132
(215) 225-7473

KEYSTONE **PRODUCT INFORMATION**

KEYGEAR® Enclosed Gear Oils

The products listed provide efficient lubrication to a wide variety of industrial gearing. All of the lubricants meet various manufacturer's specifications. Keygear 80/140 for example is a multi-graded "State of the Art" product with a high Timken rating (60#) and low pour point (−25°F). All grades are non-corrosive to usual metals like bronze and steel.

Features:	**Benefits:**
• USDA H-2*	• Acceptable in non-food contact areas
• Versatile	• Satisfies majority of industrial gear applications

* Authorized by USDA for use in Federally inspected meat and poultry plants where no contact with edible product is possible.

Typical Properties KEYGEAR Lubricants	**K-600**	**80/140**	**90**	**110**	**140**
ISO Viscosity Grade:		—	150	—	460
SAE Viscosity Grade:	140	80W-140	90	90	140
AGMA Grade:	7(C)	6 EP	4 EP	6 EP	7 EP
Saybolt Viscosity, SUS, 100°F:	2140	1400	720	1320	2240
Saybolt Viscosity, SUS, 210°F:	145	124	81	111	147
Viscosity Index:	100	120	100	100	100
Pour Point, °F:	45	−25	5	10	20
Flash Point, C.O.C. °F:	550	475	470	500	520
Timken OK Load, lbs.:	—	60	65	65	65
4-Ball Weld Load, kg.:	160	280	320	320	320
4-Ball Wear, 40 kg., Scar diam. mm.:	.70	0.50	.60	.60	.60
Kinematic Viscosity, cSt, 40°C:	402	267	138	250	420
Kinematic Viscosity, cSt. 100°C:	29.5	25.1	15.3	22.2	30
Kinematic Viscosity, cSt. 0°F(−18°C):	—	40,000	—	—	—
Rust Protection:	—	pass	pass	pass	pass
Copper Corrosion:	—	lb	lb	lb	lb
Demulsibility, 180°F, ml. oil/ml. water/ ml. emulsion (mins.). ASTM D-1401	—	42/38/0 (5)	42/38/0 (5)	42/38/0 (5)	42./38/0 (5)

NOTE: KEYGEAR Products are not recommended for Hypoid Gears or Automotive Applications like differentials.

Replacement Key
The KEYGEAR Products Replace Keystone Products Noted

KEYGEAR (New name)	K-600	80/140	90	110	140
Replaces (Old name)	K-600	KEYGEAR EP-6	WG-3 & 1790	WG-1	WG-A, K-610, & 1791

700 Third Line
Oakville, Ontario L6J5A3
Canada: 416-827-9841

21st and Lippincott Streets
Philadelphia, PA 19132

Call Toll Free 800-344-2241

KEYSTONE
PENNWALT
CHEMICALS ■ EQUIPMENT
HEALTH PRODUCTS

LUBRICANTS

Keystone/Pennwalt Corporation
21st and Lippincott Streets
Philadelphia, PA 19132
(215) 225-7473

KEYSTONE

PRODUCT INFORMATION

KEYSTONE® GP Oils
GP-20, GP-30, GP-50

Keystone GP Oils are tacky, extreme-pressure oils for use on ways, chains, bearings and rock drills. GP oils are rust and oxidation inhibited, and are water-displacing. GP oils meet various equipment manufacturers' specifications. Consult Keystone for details.

Features:	**Benefits:**
• Tackiness	• Adhesion to ways and chains; less drip, less housekeeping
• Extreme-pressure treated	• Carries heavy loads, shock loading
• Rust-inhibited, water-displacing	• Suitable for wet environments

Applications:
Keystone GP oils may be used on ways and slides where tacky, stick-slip-resistant oils are required. They may also be used in bearings, chains, conveyors and other applications were minimal oil drippage is desired.

Typical Properties

	GP-20	GP-30	GP-50
SAE Viscosity Grade:	20	30	50
ISO Viscosity Grade:	68	150	320
Viscosity, SUS, 100°F:	320	720	1800
Viscosity, SUS, 210°F:	50	64	95
Pour Point, °F:	0	5	10
Flash Point, C.O.C., °F:	375	385	430
Rust Test, ASTM D-665:	pass	pass	pass
Timken OK Load, lbs.:	43	43	43
U.S.D.A. Rating*	H2	H2	H2

*H2. Lubricants with no contact. These compounds may be used as a lubricant on equipment and machine parts in locations in which there is no possibility of the lubricant or lubricated part contacting edible products.

Keystone Division of Pennwalt Corporation, with a policy of continuous improvement, reserves the right to change specifications as our technology progresses. We are not responsible for misuse and/or misapplication of our products.

700 Third Line
Oakville, Ontario L6J5A3
Canada: 416-827-9841

21st and Lippincott Streets
Philadelphia, PA 19132

Call Toll Free 800-344-2241
in Pennsylvania: (215) 225-7473

KEYSTONE
PENNWALT
CHEMICALS ■ EQUIPMENT
HEALTH PRODUCTS

Keystone/Pennwalt Corporation
21st and Lippincott Streets
Philadelphia, PA 19132
(215) 225-7473

KEYSTONE PRODUCT INFORMATION

ZENIPLEX®
Aluminum Complex EP Grease

ZENIPLEX greases are high performance greases for industrial applications. ZENIPLEX greases are U.S.D.A. H-2* rated for use in food processing applications. ZENIPLEX greases meet many manufacturers' specifications. Consult Keystone for specific information.

Features	Benefits
• Aluminum Complex	• Stable to shear and high temperature
• Extreme Pressure	• Lubricates heavy loads.
• Water Resistant	• Reduced washout in wet applications
• U.S.D.A. H-2*	• Broad range of applications

* H-2. Lubricants with no contact. These compounds may be used as a lubricant where there is no possibility of the lubricant or lubricated part contacting edible products.

Applications:
General purpose and heavy duty applications up to 400°F in plain and antifriction bearings, slides and way, automatic systems (NLGI 0).

ZENIPLEX 2 and ZENIPLEX 1 may be used intermittently to 450°F with frequent relubrication. ZENIPLEX greases are compatible with most industrial greases. Some softening can occur when grease types are mixed. Old greases should be flushed from the bearings.

LUBRICANTS

Typical Properties

	ZENIPLEX 0	ZENIPLEX 1	ZENIPLEX 2
Operating Range, °F:	0-400	0-400	0-400
Penetration, worked:	375	325	285
NLGI Grade:	0	1	2
Dropping Point, °F:	470	480	490
Appearance:	Smooth, Brown	Smooth, Brown	Smooth, Brown
Water Washout, % loss, 175°F:	N/A	5	2
Rust Test (D-1743):	Pass	Pass	Pass
Oxidation, psi loss, 100 hrs.:	2	2	2
Copper Corrosion (D-130):	Pass	Pass	Pass
Timken OK Load, lbs.:	50	50	50
4-Ball Weld Load, kg.:	280	280	280
4-Ball Wear, 40 kg., (D-2266), mm:	0.55	0.55	0.55
Base Oil Properties:			
Viscosity at 100°F, SUS:	1100	1100	1100
Viscosity at 210°F, SUS:	85	85	85
Flash Point, C.O.C., °F:	450	450	450
Pour Point, °F:	10	10	10

Keystone Division of Pennwalt Corporation, with a policy of continuous improvement, reserves the right to change specifications as our technology progresses. We are not responsible for misues.and/or misapplication of our products.

700 Third Line
Oakville, Ontario L6J5A3
Canada: 416-827-9841

21st and Lippincott Streets
Philadelphia, PA 19132

Call Toll Free 800-344-2241

KEYSTONE
PENNWALT
CHEMICALS ■ EQUIPMENT
HEALTH PRODUCTS

TS-227C
8/83

Keystone/Pennwalt Corporation
21st and Lippincott Streets
Philadelphia, PA 19132
(215) 225-7473

KEYSTONE

PRODUCT INFORMATION

KSL® Synthetic Air Compressor Lubricants

KSL synthetics are diester-based air compressor lubricants designed for high-efficiency operation in most air compressors. Available in six viscosities to meet individual compressor requirements. Many major compressor manufacturers approve the use of KSL Synthetic Lubricants. Copies of specific OEM recommendations and approval letters are available upon request.

Features	**Benefits**
• Fully synthetic	• No mineral oil; lower oil consumption.
• Diester-based	• Carbon-free, low maintenance costs; low friction, low energy costs.
• Ashless	• Free of sludge and deposits for higher efficiency.
• Inhibited	• Long oil life, less down time for changes.

Applications

• Rotary Screw Compressors	• KSL-220, KSL-224 *
• Reciprocating Compressors	• KSL-219, KSL-221, KSL-222
• Rotary Vane Compressors	• KSL-214, KSL-224 *

*Some manufacturers recommend other viscosity grades, such as SAE 10W for rotary screw or SAE 40 for rotary vane compressors.

U.S.D.A. Rating

KSL synthetics are U.S.D.A. rated "H2" (BB). These compounds may be used as a lubricant on equipment and machine parts in locations in which there is no possibility of the lubricant or lubricated part contacting edible products.

KSL® Synthetic Lubricants Typical Properties

	KSL® 214	KSL® 219	KSL® 220	KSL® 221	KSL® 222	KSL® 224
SAE Viscosity Grade:	10W	20	20	30	40	15W-20
ISO Viscosity Grade:	32	80 *	68	100	150	46
Viscosity, SUS, 100 °F:	145	430	320	525	750	208
Viscosity, SUS, 210 °F:	44	54	51	60	72	48
Viscosity Index:	115	50	70	75	80	110
Flash Point, C.O.C., °F:	470	490	485	525	525	480
Autoignition Temperature, °F:	770	775	770	780	780	775
Pour point, °F:	−50	−30	−40	−30	−20	−50

*** This is viscosity in centistokes at 40 °C. No I.S.O. Grade exists.**

LUBRICANTS

(See other side for additional information)

Keystone/Pennwalt Corporation
21st and Lippincott Streets
Philadelphia, PA 19132
(215) 225-7473

Materials Compatibility

COMPATIBILITY	SEAL MATERIAL	PLASTICS	PAINTS
Suitable	Fluorocarbons *(Teflon®, Viton®) Nitrile rubber (with high nitrile, content over 36% Buna N) Polysulfide Fluorosilicone	Fluorocarbon *Delrin® *Celcon® Nylon	Epoxy 2 Component Urethane Oil Resistant Alkyd
Not Recommended	Natural rubber Neoprene Nitrile rubber (with low nitrile content) SBR rubber	Polystyrene PVC ABS Polycarbonate	Varnish PVC Acrylic

***Trademarks of DuPont Corporation**

Recommendations for changeover to KSL® Synthetic Lubricants:

- Select proper product according to OEM; check compatibility.
- Drain reciprocating compressor hot; clean as well as possible.
- Inspect and clean valves and downstream equipment.
- Drain rotary compressors hot; turn drive manually for complete drain; drain accessory equipment and lines.
- Flushing is recommended following severe service or irregular oil changes.
- Contact Keystone for specific recommendations.

KSL® Synthetic Lubricants Additional Typical Properties

	KSL®-214	KSL®-219	KSL®-220	KSL®-221	KSL®-222	KSL®-224	ASTM Test
Kinematic Viscosity, cSt, 100 °C:	5.2	8.1	7.3	10.0	12.9	6.4	D-445
Kinematic Viscosity, cSt, 40 °C:	28.1	80.9	61.1	100	142	40.4	D-445
Kinematic Viscosity, cSt, 0 °C:	1,250	2,100	1,150	2,350	3,600	460	D-445
Carbon Residue, Conradson, %:	nil	nil	nil	0.01	0.01	nil	D-189
Evaporation, 6½ hrs., 400 °F, %:	8	4	5	3	3	7	D-972
Demulsibility, 130 °F, ml oil/ ml water/ ml emulsion (minutes)	40/40/0(5)	39/39/2(60)	40/39/1(15)	38/38/4(60)	38/37/5(60)	40/39/1(10)	D-1401
4-Ball Wear, 40 kg., 1200 rpm, 75 °C, 1 hour, scar dia.,mm.:	0.74	0.71	0.72	0.68	0.67	0.73	D-2266
Coefficient of friction:	0.075	0.078	0.072	0.075	0.077	0.074	
Surface Tension, dynes/cm, 20 °C:	31.5	31.0	31.2	30.8	30.5	31.6	
100 °C:	29.8	28.9	29.3	28.8	28.5	29.9	
Thermal Conductivity, Cal/hr/cm²	1.26	1.33	1.30	1.35	1.37	1.28	
Specific Heat, cal/gm/ °C,100 °F:	.47	.50	.48	.51	.52	.48	
200 °F:	.51	.54	.52	.55	.56	.51	
300 °F:	,55	.58	.56	.59	.60	.55	
Bulk Coefficient of Expansion:	0.00043	0.00042	0.00042	0.00041	0.00040	.00042	
Vapor Presssure, mm Hg, 100 °F:	0.00003	0.00002	0.000027	0.000018	0.000014	.00003	
200 °F:	0.00075	0.00047	0.00061	0.00041	0.00033	.0007	
300 °F:	0.019	0.012	0.014	0.0095	0.0077	.017	

Keystone Division of Pennwalt Corporation, with a policy of continuous improvement, reserves the right to change specifications as our technology progresses. We are not responsible for misuse and/or misapplication of our products.

700 Third Line
Oakville, Ontario L6J5A3
Canada: 416-827-9841

21st and Lippincott Streets
Philadelphia, PA 19132

Call Toll Free 800-344-2241
in Pennsylvania: (215) 225-7473

KEYSTONE
PENNWALT
CHEMICALS ■ EQUIPMENT
HEALTH PRODUCTS

LUBRICANTS

Kluber Lubrication Corporation
Grenier Industrial Airpark
Manchester, NH 03103
(603) 669-7789

ISOFLEX Greases

Delivery Program ISOFLEX ALLTIME, PDB, PDL, and TEL Types

ISOFLEX ALLTIME greases

With an operating temperature from -70 to 150 °C, these are high-quality, wide-temperature range greases and special purpose lubricant greases for high temperature, and are highly water resistant, in conformance with DIN 51 807.

They also offer extremely long life in case of relatively high, speed-oriented friction and, thus, heat development, since they are best capable of withstanding the resultant localized temperature peaks.

ISOFLEX ALLTIME greases
present new possibilities in automotive engineering for starters, generators and general high-speed rolling bearings in electric motors, turbines as well as in applications in the aerospace industry.

ISOFLEX PDB Greases

Dynamic, very light greases. Proven for miniature bearings, encased small bearings, gear units, sliding and porous bearings, control motors, segmental rings and other sliding points in precision engineering. Where very low torque is present in sound and film equipment, e. g. for motor-driven lens adjustment, as well as for measuring instruments and devices and friction clutches.

ISOFLEX PDL and ISOFLEX TEL Greases

ISOFLEX PDL greases are highly water resistant in conformance with DIN 51 807. They are especially suitable in aerospace, telecommunications, precision, electrical and refrigeration applications. The most favorable low temperature behavior in the ISOFLEX grease series is demonstrated by type ISOFLEX PDL 300 A.

ISOFLEX SUPER TEL has long been applied in high speed grinding spindles. This type is still prescribed today for the main spindles of metal cutting machine tools.

ISOFLEX NBU 15 and LDS 18 greases are preferable for more recent designs.

LUBRICANTS

ISOFLEX Types	Colour	Drop point DIN 51 801/1 (°C)	Operating temperature range (°C)	Worked penetration DIN 51 804 (0.1 mm) approx.	Consistency class DIN 51 818	Dynamic viscosity (mPa s) approx.	Speed parameter ($n \cdot d_m$)	Applications
ISOFLEX ALLTIME ML 2	yellow-brown	190	-40 . . . 150	280	2	2500	10^6	**Rolling bearing grease** for very high speeds and/or high temperatures.
ISOFLEX ALLTIME MSV 2 MF	metallic black	>220	-60 . . . 180	280	2	3000	$5 \cdot 10^5$	**Rolling and sliding bearing grease** for broad temperature range. Emergency running properties through UNIMOLY.
ISOFLEX ALLTIME SL 2	beige	>180	-70 . . . 150	280	2	2500	10^6	**Rolling bearing grease** for very low and/or high temperatures or high speeds.
ISOFLEX ALLTIME SL 1	yellowish	>180	-70 . . . 150	325	1	2000	$1{,}3 \cdot 10^6$	**Rolling bearing grease** for extremely high speeds.
ISOFLEX PDB 38/3000	natural, beige	>250	-65 . . . 100	300	1/2	1500	10^6	**Instrument grease** for segmental rings and general use for sliding points in precision engineering.
ISOFLEX PDB 38/CX 1000 EP	yellow-brown	–	-70 . . . 120	1000 (acc. to Klein)	–	500	$1{,}3 \cdot 10^6$	**Low temperature fluid grease** for grease lubricated gear units or for sump lubrication. Meets Bauer Special Test standards.
ISOFLEX PDB 38/CX 1000	yellow-brown	–	-70 . . . 120	1000 (acc. to Klein)	–	500	$1{,}3 \cdot 10^6$	**Fluid grease** and impregnating grease for porous bearings with ≤3 mm bore.
ISOFLEX PDB 38/CX 2000	yellow-brown	–	-70 . . . 120	liquid	–	–	–	**Impregnating grease** for porous bearings. Reduces friction and temperature.
ISOFLEX PDL 250	whitish	195	-60 . . . 125	275	2	2500	10^6	**Instrument grease** for telecommunications and aerospace industries.
ISOFLEX PDL 300 A	green-yellow	195	-73 . . . 125	300	1/2	2500	$1{,}3 \cdot 10^6$	**High performance grease** meets MIL-G-7421 A, MIL-G-23827 A, DTD 5598.
ISOFLEX SUPER TEL	beige	180	-65 . . . 60	320	1	2000	$1{,}3 \cdot 10^6$	**Spindle bearing grease** for extremely high speeds.
ISOFLEX TEL 3000 ALTEMP	beige	140	-50 . . . 60	320	1	2700	–	**Sliding grease** for friction surfaces in optical equipment and textile machinery.

Kluber Lubrication Corporation
Grenier Industrial Airpark
Manchester, NH 03103
(603) 669-7789

ISOFLEX Greases

Delivery Program – Problem Solution with ISOFLEX LDS 18 Types

ISOFLEX LDS 18 greases
are water resistant to 90 °C, in conformance with DIN 51 807.

Proven in high-speed grinding spindles, main spindles in metal-cutting machine tools, in spindle bearings, roller clutches, freewheels, starters and encased bearings (2Z / 2 RS) in motors for generators, textile spindles, laying rollers, false twist tubes, turbines, noise-tested bearings.

ISOFLEX LDS 18/200 serves as sealing and lubricating grease for felt, felt gaskets and couplings as well as for the fine threads of adjusting units in optical equipment.

Problem: Lifetime lubrication of freewheels under open air weather conditions.

Solution: Functioning or force-locking through **ISOFLEX LDS 18 SPECIAL A.** is guaranteed at any possible temperature.

Problem: High-speed spindle bearings in an automatic filling bobbin. Operation for several years without relubrication desireable.

Solution: ISOFLEX LDS 18 SPECIAL A. Over three years in operation at 12,000 rpm.

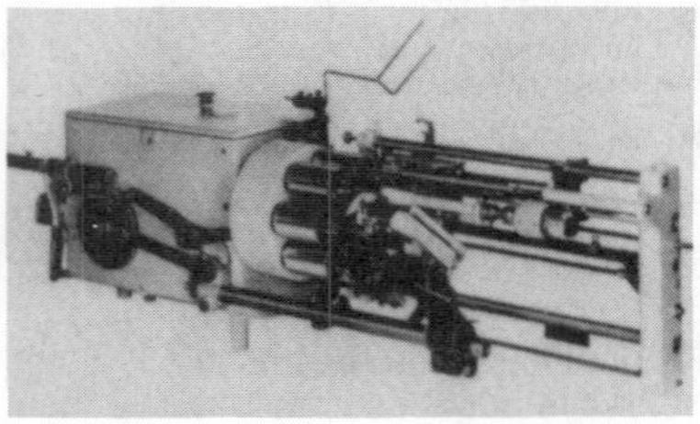

Problem: Permanent grease lubrication at maximum speeds for grinding spindles. Maintenance-free operation and long service life required at speeds of 55,000 rpm or more.

Solution: ISOFLEX LDS 18 SPECIAL A. This type offers excellent one time grease lubrication for long periods and guarantees maximum possible tool life.

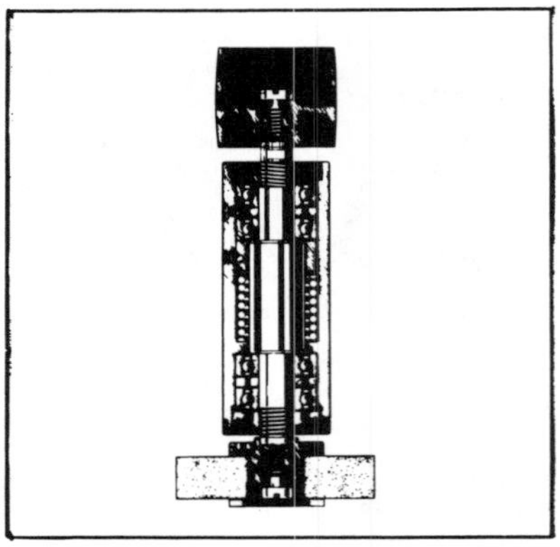

ISOFLEX LDS 18 Types	Colour	Drop point DIN 51 801/1 (°C)	Operating temperature range (°C)	Worked penetration DIN 51 804 (0.1 mm) approx.	Consistency class DIN 51 818	Dynamic viscosity (mPa s) approx.	Speed parameter $(n \cdot d_m)$	Applications
ISOFLEX LDS 18 SPECIAL A	green-yellow	>185	-60 . . . 130	280	2	3000	10^6	**Rolling bearing grease** for high speed spindle bearings, conforms to MIL-G-7118 A, MIL-G-3278 A, MIL-G-23827.
ISOFLEX SUPER LDS 18	green-yellow	>185	-60 . . . 130	280	2	3000	$1{,}3 \cdot 10^6$	**Rolling bearing grease** for extremely high speeds. Test run B at 120 °C acc. to DIN 51 806 conform.
ISOFLEX LDS 18/250	green-yellow	>185	-60 . . . 130	250	2/3	2800	$5 \cdot 10^5$	**Lubricating and sealing grease** for the optical industry and precision engineering.
ISOFLEX LDS 18/200	yellowish	>185	-50 . . . 130	200	3/4	8500	$5 \cdot 10^5$	**Seal and thread grease** for felt seals and fine threads in optical equipment.

ISOFLEX Greases
Problem Solution with ISOFLEX LDS 18 Types

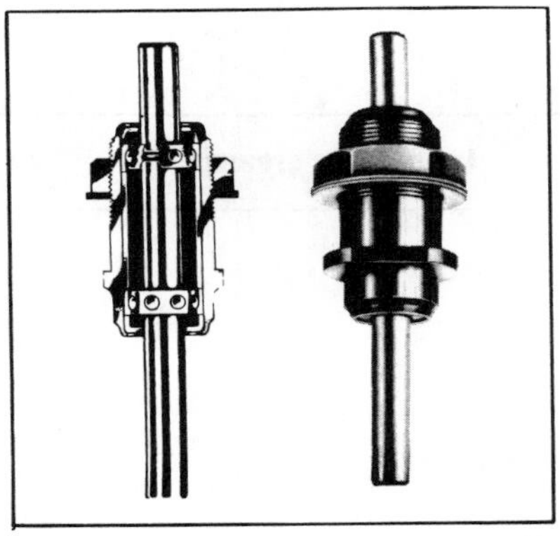

Problem. Life-time lubrication of false-twist tubes for twisting and texturing. A grease for 80,000 rpm which cuts power needs and energy costs at the same time.

Solution: ISOFLEX SUPER LDS 18
At $n \cdot d_m$ values, in part $>10^6$, this type meets the designer's requirements. Service life corresponds to that of the bearings.

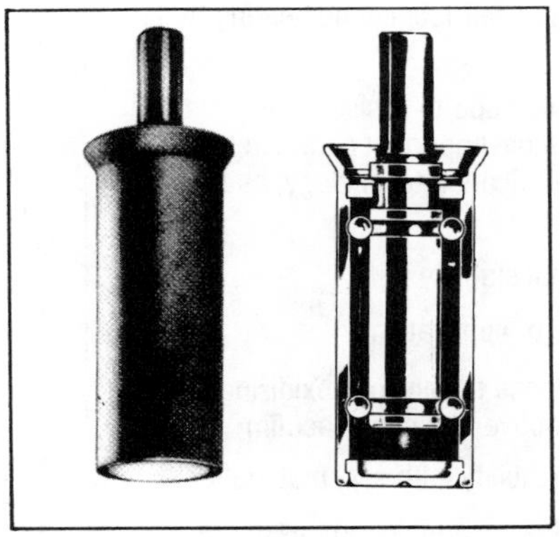

Problem: Lubrication of laying rollers in draw twisters and draw spooling machines. Minimum starting resistance to prevent thread breakage, lowest power requirement during dynamic loading, high speeds to 30,000 rpm ($n \cdot d_m$ = approx. 600,000).

Solution: ISOFLEX SUPER LDS 18 DISPERSION 15 F
The rollers work 18,000 hours and longer in three-shift-operation without relubrication. Previously spindle oil had to be applied and relubrication was carried out monthly.

Problem: Change-over from oil to grease lubrication of spindle units in metal cutting machine tools. Bearing temperature of 53 °C must be reduced and maximum maintenance intervals attained.

Solution: ISOFLEX SUPER LDS 18
Bearing temperature was reduced by 18 K to 35 °C. The first spindle series has run trouble-free for seven years in three-shift-operation.
n = 1300 to 1800, concentricity = 1 μm.

Duplicating Milling Machine

High-speed Milling spindle
Bearing: inclined roller bearing

Bearing temperature: 37 °C
Speed: 9,000 rpm

Problem solution:
ISOFLEX SUPER LDS 18
The initial spindle series, produced three years ago, is still operating trouble-free.

Universal Milling Machine

Milling spindle
Bearing: cylindrical roller bearing

Bearing temperature: 36 °C
Speed. 2,000 rpm

Problem solution:
ISOFLEX LDS 18 SPECIAL A
Bearing temperature was reduced to 36 °C as opposed to 48 °C with the grease used previously.

Lathe

Spindle head
Bearing: Needle bearing/ inclined roller bearing

Bearing temperature 35/36 °C
Speed. 12,000 rpm

Problem solution:
ISOFLEX LDS 18 SPECIAL A
Bearing temperature (with the use of low-viscosity spindle oils at times over 70 °C) was reduced significantly.

Kluber Lubrication Corporation
Grenier Industrial Airpark
Manchester, NH 03103
(603) 669-7789

This survey is a brief classification of the Klüber Lubrication programme. We have, however, limited ourselves to the more essential products. It is not possible here to enter into areas such as thermo electrically conductive greases, separating agents for plastic and rubbers, surface treatment for selflubricating faces, auxiliaries for M. I. G. welding and self-lubricating bearing material etc.

Instead, we hope to assist in the selection from the more important parameters which bear on lubrication technology, as follows:

- thermal stability
- range of application
- resistance to solvents, oxidizing agents, radioactive radiation, vacuum
- compatibility with seal materials

In the following product groups, there is an overlap in the properties. Detailed information is available in our individual leaflets and special prints.

As far as service temperature is concerned, the deciding factor is the suitability at low and high temperatures. Since most of the lubricants listed are suitable within a temperature range of -20 . . . 100/120 °C, the normal temperature range has been omitted.

Certain lubricants are available in different physical forms and these are listed under the heading "type of lubricant", e. g! UNIMOLY is available as a suspension, a paste, a bonded coating or simply as a powder.

Valve greases
Steel rope lubricants
Sealing greases
Precision engineering lubricants
Fluid greases
Gear lubricants
Plain bearing lubricants
Sliding agents
High temperature greases
Hydraulic oils
Instrument lubricants
Electrical contact lubricants
Long-term lubricants
Sea water-resistant lubricants
Silicone lubricants
Sintered bearing lubricants
Spindle oils
Radioactive radiation resistant lubricants
Low temperature lubricants
Clock and watch greases
Vacuum lubricants
Rolling bearing greases
Thermo conductive lubricants
Gear wheel lubricants

⇨ **Lubricating oils**

⇨ **Lubricating greases**

Running-in compounds
Bonded coatings
Graphited lubricants
Graphite-water suspensions
Hot screw compounds
High performance lubricants
Metallic pastes
Molybdenum disulphide lubricants
Assembly pastes
Lubricating powders
Screw compounds
Separating coatings

⇨ **Solid lubricants**

Emulsions
Drop forging lubricants
Plunger lubricants
Coolants
Needle oils
Needle bar oils
Ring and traveller lubricants
Cutting oils
Stamping oils
Extrusion lubricants
Hot pressing lubricants
Drawing lubricants

⇨ **Special lubricants**

Diecasting separating agents
Ladle dressings
Glass moulding separating agents
Gravity Die Dressings
Preservatives
Coolants
Plastic release agents
Solvents
Cleansing agents
Rust protection agent
Rust solvents
Selflubricating materials
Diluting agents

⇨ **Special products**

Kluber Lubrication Corporation
Grenier Industrial Airpark
Manchester, NH 03103
(603) 669-7789

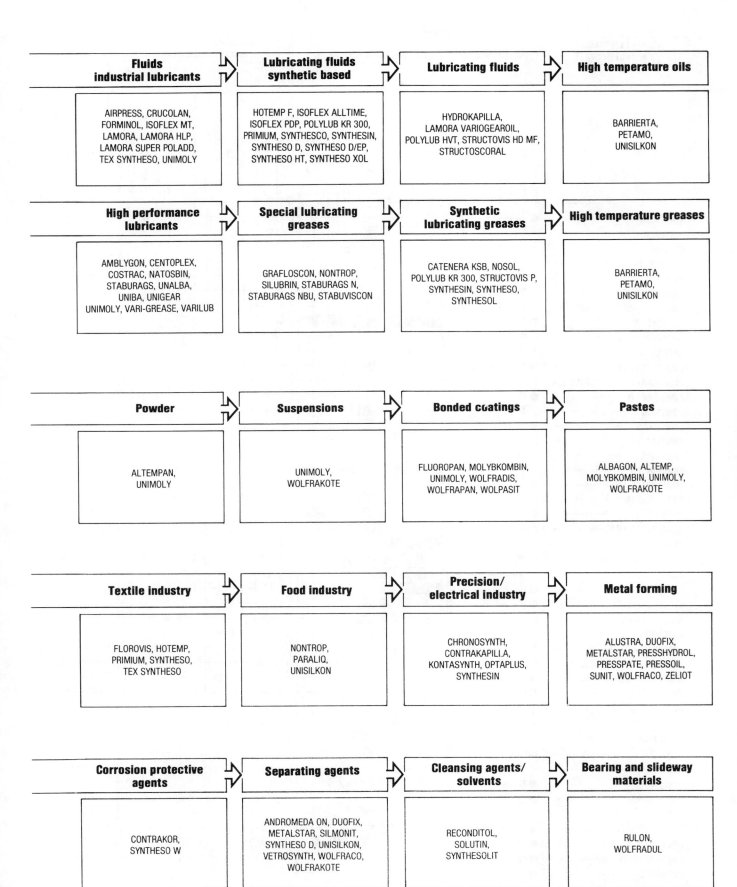

Fluids industrial lubricants	Lubricating fluids synthetic based	Lubricating fluids	High temperature oils
AIRPRESS, CRUCOLAN, FORMINOL, ISOFLEX MT, LAMORA, LAMORA HLP, LAMORA SUPER POLADD, TEX SYNTHESO, UNIMOLY	HOTEMP F, ISOFLEX ALLTIME, ISOFLEX PDP, POLYLUB KR 300, PRIMIUM, SYNTHESCO, SYNTHESIN, SYNTHESO D, SYNTHESO D/EP, SYNTHESO HT, SYNTHESO XOL	HYDROKAPILLA, LAMORA VARIOGEAROIL, POLYLUB HVT, STRUCTOVIS HD MF, STRUCTOSCORAL	BARRIERTA, PETAMO, UNISILKON

High performance lubricants	Special lubricating greases	Synthetic lubricating greases	High temperature greases
AMBLYGON, CENTOPLEX, COSTRAC, NATOSBIN, STABURAGS, UNALBA, UNIBA, UNIGEAR UNIMOLY, VARI-GREASE, VARILUB	GRAFLOSCON, NONTROP, SILUBRIN, STABURAGS N, STABURAGS NBU, STABUVISCON	CATENERA KSB, NOSOL, POLYLUB KR 300, STRUCTOVIS P, SYNTHESIN, SYNTHESO, SYNTHESOL	BARRIERTA, PETAMO, UNISILKON

Powder	Suspensions	Bonded coatings	Pastes
ALTEMPAN, UNIMOLY	UNIMOLY, WOLFRAKOTE	FLUOROPAN, MOLYBKOMBIN, UNIMOLY, WOLFRADIS, WOLFRAPAN, WOLPASIT	ALBAGON, ALTEMP, MOLYBKOMBIN, UNIMOLY, WOLFRAKOTE

Textile industry	Food industry	Precision/ electrical industry	Metal forming
FLOROVIS, HOTEMP, PRIMIUM, SYNTHESO, TEX SYNTHESO	NONTROP, PARALIQ, UNISILKON	CHRONOSYNTH, CONTRAKAPILLA, KONTASYNTH, OPTAPLUS, SYNTHESIN	ALUSTRA, DUOFIX, METALSTAR, PRESSHYDROL, PRESSPATE, PRESSOIL, SUNIT, WOLFRACO, ZELIOT

Corrosion protective agents	Separating agents	Cleansing agents/ solvents	Bearing and slideway materials
CONTRAKOR, SYNTHESO W	ANDROMEDA ON, DUOFIX, METALSTAR, SILMONIT, SYNTHESO D, UNISILKON, VETROSYNTH, WOLFRACO, WOLFRAKOTE	RECONDITOL, SOLUTIN, SYNTHESOLIT	RULON, WOLFRADUL

Kluber Lubrication Corporation
Grenier Industrial Airpark
Manchester, NH 03103
(603) 669-7789

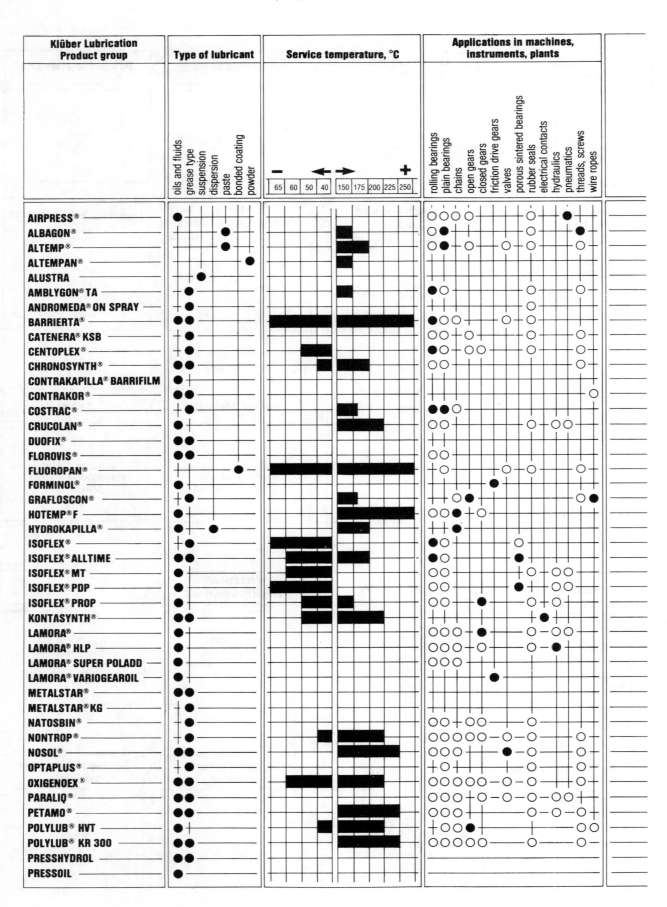

LUBRICANTS

Klüber Lubrication Product group	Type of lubricant							Service temperature, °C									Applications in machines, instruments, plants													
	oils and fluids	grease type	suspension	dispersion	paste	bonded coating	powder	− 65	60	50	40 ←	→ 150	175	200	225	250 +	rolling bearings	plain bearings	chains	open gears	closed gears	friction drive gears	valves	porous sintered bearings	rubber seals	electrical contacts	hydraulics	pneumatics	threads, screws	wire ropes
AIRPRESS®	●																○	○	○	○					○			●		
ALBAGON®					●							■					○	●							○				●	
ALTEMP®					●							■	■				○	●		○			○		○				○	
ALTEMPAN®							●					■																		
ALUSTRA			●																											
AMBLYGON® TA		●										■					●	○							○				○	
ANDROMEDA® ON SPRAY		●																							○					
BARRIERTA®	●	●						■	■	■	■	■	■	■	■	■	●	○	○				○		○					
CATENERA® KSB		●															○	○		○					○				○	
CENTOPLEX®		●								■	■						●	○		○	○				○				○	
CHRONOSYNTH®	●	●									■	■	■				○	○											○	
CONTRAKAPILLA® BARRIFILM	●																													
CONTRAKOR®	●	●																												○
COSTRAC®		●										■	■				●	●	○											
CRUCOLAN®	●											■	■	■			○	○							○		○	○		
DUOFIX®	●	●																												
FLOROVIS®	●	●															○	○												
FLUOROPAN®						●		■	■	■	■	■	■	■	■	■		○					○		○				○	
FORMINOL®	●																					●								
GRAFLOSCON®		●										■	■						○	●									○	●
HOTEMP® F	●											■	■	■	■	■	○	○	●		○									
HYDROKAPILLA®	●			●								■	■						●											
ISOFLEX®		●						■	■	■	■						●	○						○						
ISOFLEX® ALLTIME	●	●							■	■	■	■	■				●	○						●						
ISOFLEX® MT	●								■	■	■						○	○							○		○	○		
ISOFLEX® PDP	●							■	■	■	■						○	○						●			○	○		
ISOFLEX® PROP	●									■	■	■					○	○			●				○		○			
KONTASYNTH®	●	●								■	■	■	■													●				
LAMORA®	●																○	○	○		●				○		○	○		
LAMORA® HLP	●																○	○	○		○				○		●			
LAMORA® SUPER POLADD	●																○	○	○											
LAMORA® VARIOGEAROIL	●																					●								
METALSTAR®	●	●																												
METALSTAR® KG		●																												
NATOSBIN®		●															○	○		○	○				○					
NONTROP®		●									■	■	■				○	○	○	○	○		○		○				○	
NOSOL®	●	●										■	■	■	■		○	○	○				●		○				○	
OPTAPLUS®		●																○							○				○	
OXIGENOEX®	●	●							■	■	■	■	■				○	○	○	○	○		○		○				○	
PARALIQ®	●	●															○	○	○		○		○		○		○	○		
PETAMO®	●	●										■	■	■	■		○	○	○						○		○		○	
POLYLUB® HVT	●										■	■	■					○	○	●									○	○
POLYLUB® KR 300	●	●										■	■	■	■		○	○	○	○	○				○				○	
PRESSHYDROL	●	●																												
PRESSOIL	●																													

Kluber Lubrication Corporation
Grenier Industrial Airpark
Manchester, NH 03103
(603) 669-7789

Indications

○ properties and general field of application
● principal use

service temperature
■ dry or emergency lubrication by solid lubricants, in case of exceeding service temperatures

Effect of external media
Tests acc. to DIN 51 807. No lubricant is absolutely resistant to mechanical running-in or to being worked. Dissolution in the excess of the fluid media occurs.

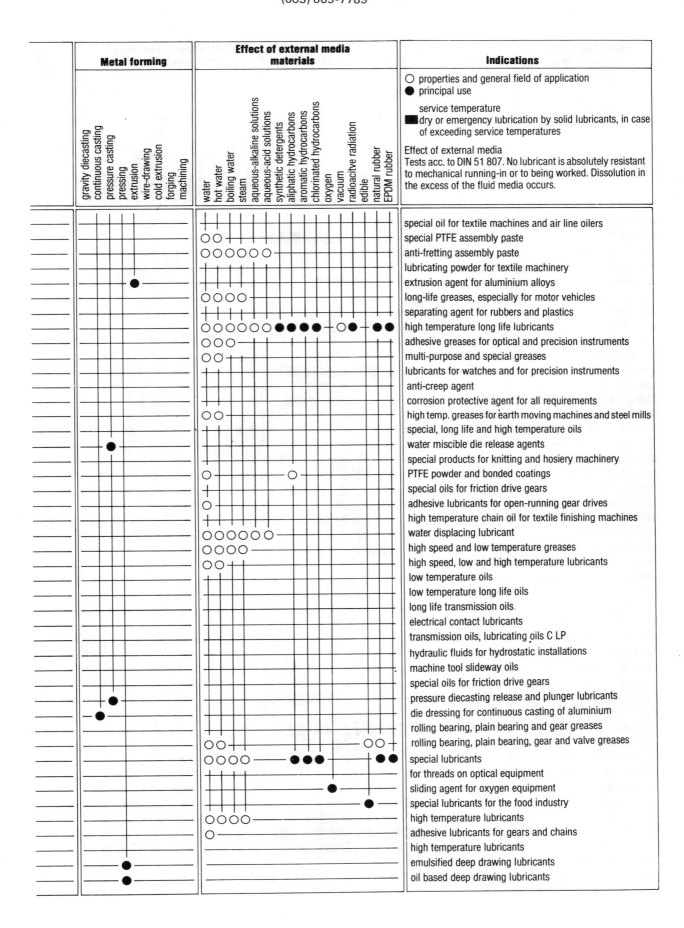

Metal forming									Effect of external media materials																Indications
gravity diecasting	continuous casting	pressure casting	pressing	extrusion	wire-drawing	cold extrusion	forging	machining	water	hot water	boiling water	steam	aqueous-alkaline solutions	aqueous-acid solutions	synthetic detergents	aliphatic hydrocarbons	aromatic hydrocarbons	chlorinated hydrocarbons	oxygen	vacuum	radioacitve radiation	edible	natural rubber	EPDM rubber	
																									special oil for textile machines and air line oilers
									○	○															special PTFE assembly paste
									○	○	○	○	○	○											anti-fretting assembly paste
																									lubricating powder for textile machinery
				●																					extrusion agent for aluminium alloys
									○	○	○	○													long-life greases, especially for motor vehicles
																									separating agent for rubbers and plastics
									○	○	○	○	○	○	●	●	●	●		○	●		●	●	high temperature long life lubricants
									○	○	○														adhesive greases for optical and precision instruments
									○	○															multi-purpose and special greases
																									lubricants for watches and for precision instruments
																									anti-creep agent
																									corrosion protective agent for all requirements
									○	○															high temp. greases for earth moving machines and steel mills
																									special, long life and high temperature oils
		●																							water miscible die release agents
																									special products for knitting and hosiery machinery
									○							○									PTFE powder and bonded coatings
																									special oils for friction drive gears
									○																adhesive lubricants for open-running gear drives
																									high temperature chain oil for textile finishing machines
									○	○	○	○	○	○											water displacing lubricant
									○	○	○	○													high speed and low temperature greases
									○	○															high speed, low and high temperature lubricants
																									low temperature oils
																									low temperature long life oils
																									long life transmission oils
																									electrical contact lubricants
																									transmission oils, lubricating oils C LP
																									hydraulic fluids for hydrostatic installations
																									machine tool slideway oils
																									special oils for friction drive gears
		●																							pressure diecasting release and plunger lubricants
	●																								die dressing for continuous casting of aluminium
																									rolling bearing, plain bearing and gear greases
									○	○												○	○		rolling bearing, plain bearing, gear and valve greases
									○	○	○	○				●	●	●					●	●	special lubricants
																									for threads on optical equipment
																			●						sliding agent for oxygen equipment
																						●			special lubricants for the food industry
									○	○	○	○													high temperature lubricants
									○																adhesive lubricants for gears and chains
																									high temperature lubricants
			●																						emulsified deep drawing lubricants
			●																						oil based deep drawing lubricants

Kluber Lubrication Corporation
Grenier Industrial Airpark
Manchester, NH 03103
(603) 669-7789

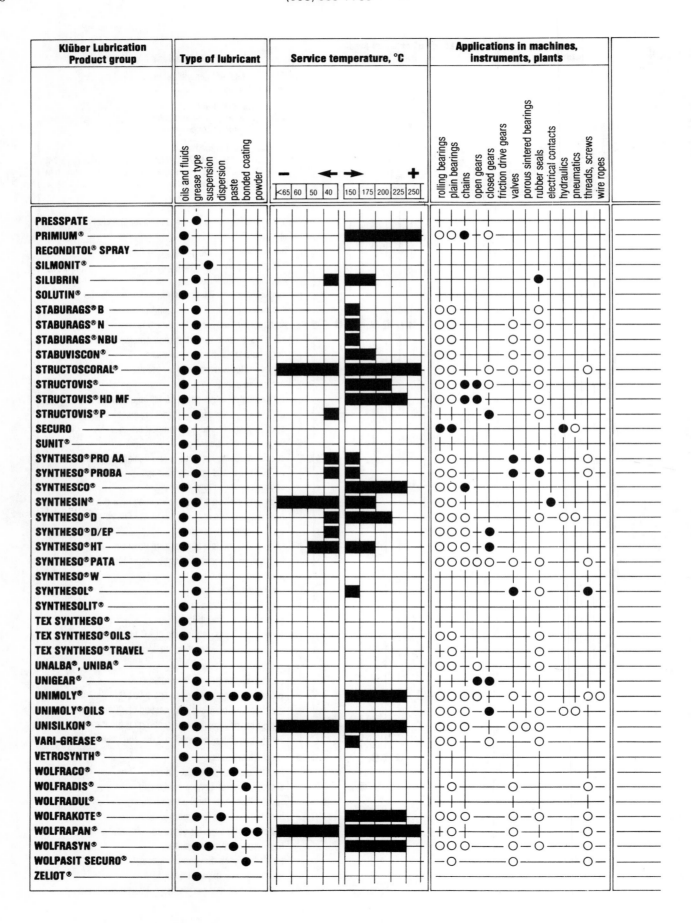

Klüber Lubrication Product group	Type of lubricant: oils and fluids	grease type	suspension	dispersion	paste	bonded coating	powder	Service temperature, °C (–): <65	60	50	40	Service temperature, °C (+): 150	175	200	225	250	Applications in machines, instruments, plants: rolling bearings	plain bearings	chains	open gears	closed gears	friction drive gears	valves	porous sintered bearings	rubber seals	electrical contacts	hydraulics	pneumatics	threads, screws	wire ropes
PRESSPATE		●																												
PRIMIUM®	●											■	■	■	■	■	○	○	●		○									
RECONDITOL® SPRAY	●																													
SILMONIT®			●																											
SILUBRIN		●									■	■	■												●					
SOLUTIN®	●																													
STABURAGS® B		●										■					○	○							○					
STABURAGS® N		●										■					○	○					○		○					
STABURAGS® NBU		●										■					○	○					○		○					
STABUVISCON®		●										■	■				○	○					○		○					
STRUCTOSCORAL®	●	●						■	■	■	■	■	■	■	■	■	○	○			○		○		○				○	
STRUCTOVIS®	●											■	■	■			○	○	●	●	○				○					
STRUCTOVIS® HD MF	●											■	■	■	■		○	○	●	●					○					
STRUCTOVIS® P		●									■										●				○					
SECURO	●																●	●									●	○		
SUNIT®	●																													
SYNTHESO® PRO AA		●									■	■					○	○					●		●				○	
SYNTHESO® PROBA		●									■	■					○	○					●		●				○	
SYNTHESCO®	●											■	■	■	■		○	○	●											
SYNTHESIN®	●	●						■	■	■	■	■	■				○	○								●				
SYNTHESO® D	●										■	■	■	■			○	○	○						○		○	○		
SYNTHESO® D/EP	●										■						○	○	○		●									
SYNTHESO® HT	●									■	■	■	■				○	○	○		●									
SYNTHESO® PATA	●	●															○	○	○	○	○		○		○				○	
SYNTHESO® W		●																												
SYNTHESOL®		●										■											●		○				●	
SYNTHESOLIT®	●																													
TEX SYNTHESO®	●																													
TEX SYNTHESO® OILS	●																○	○							○					
TEX SYNTHESO® TRAVEL		●																○							○					
UNALBA®, UNIBA®		●															○	○		○					○					
UNIGEAR®		●																		●	●									
UNIMOLY®		●	●		●	●	●					■	■	■	■		○	○	○	○			○		○				○	○
UNIMOLY® OILS	●																○	○	○		●				○		○	○		
UNISILKON®	●	●						■	■	■	■	■	■	■	■		○	○	○				○	○	○					
VARI-GREASE®		●										■					○	○			○				○					
VETROSYNTH®	●																													
WOLFRACO®		●	●		●																									
WOLFRADIS®						●												○					○						○	
WOLFRADUL®																														
WOLFRAKOTE®		●		●								■	■	■	■		○	○	○				○		○				○	
WOLFRAPAN®						●	●	■	■	■	■	■	■	■	■	■		○					○						○	
WOLFRASYN®		●	●		●							■	■	■	■		○	○	○				○		○				○	
WOLPASIT SECURO®						●												○					○						○	
ZELIOT®		●																												

Kluber Lubrication Corporation
Grenier Industrial Airpark
Manchester, NH 03103
(603) 669-7789

Indications

○ properties and general field of application
● principal use

service temperature
■ dry or emergency lubrication by solid lubricants, in case of exceeding service temperatures

Effect of external media
Tests acc. to DIN 51 807. No lubricant is absolutely resistant to mechanical running-in or to being worked. Dissolution in the excess of the fluid media occurs.

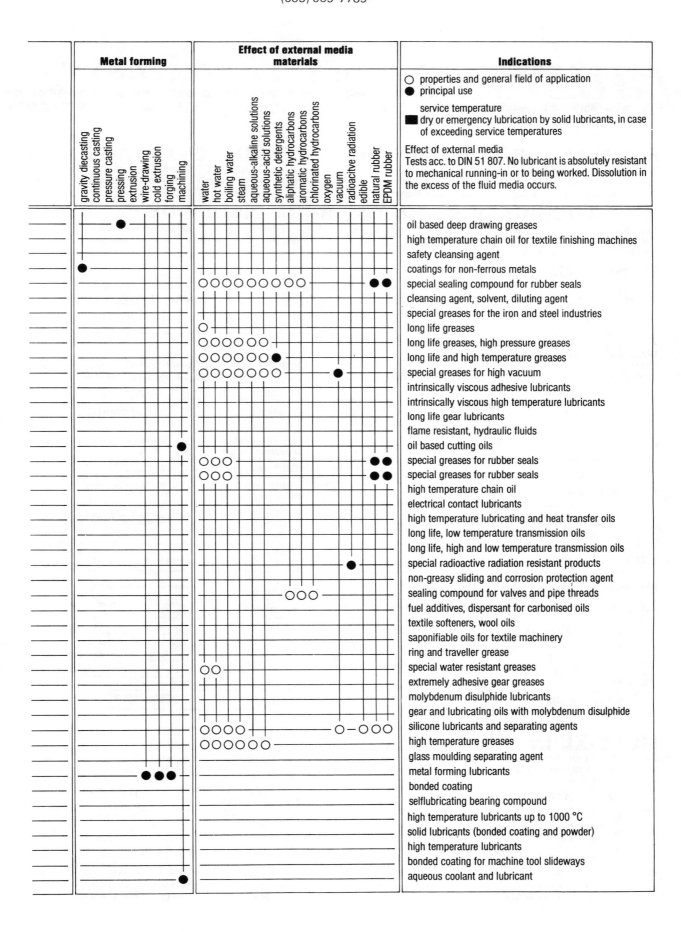

Metal forming									Effect of external media materials																Indications
gravity diecasting	continuous casting	pressure casting	pressing	extrusion	wire-drawing	cold extrusion	forging	machining	water	hot water	boiling water	steam	aqueous-alkaline solutions	aqueous-acid solutions	synthetic detergents	aliphatic hydrocarbons	aromatic hydrocarbons	chlorinated hydrocarbons	oxygen	vacuum	radioacitve radiation	edible	natural rubber	EPDM rubber	
			●																						oil based deep drawing greases
																									high temperature chain oil for textile finishing machines
																									safety cleansing agent
●																									coatings for non-ferrous metals
									○	○	○	○	○	○	○	○	○						●	●	special sealing compound for rubber seals
																									cleansing agent, solvent, diluting agent
																									special greases for the iron and steel industries
									○																long life greases
									○	○	○	○	○	○											long life greases, high pressure greases
									○	○	○	○	○	○	●										long life and high temperature greases
									○	○	○	○	○	○	○					●					special greases for high vacuum
																									intrinsically viscous adhesive lubricants
																									intrinsically viscous high temperature lubricants
																									long life gear lubricants
																									flame resistant, hydraulic fluids
								●																	oil based cutting oils
									○	○	○												●	●	special greases for rubber seals
									○	○	○												●	●	special greases for rubber seals
																									high temperature chain oil
																									electrical contact lubricants
																									high temperature lubricating and heat transfer oils
																									long life, low temperature transmission oils
																									long life, high and low temperature transmission oils
																					●				special radioactive radiation resistant products
																									non-greasy sliding and corrosion protection agent
																○	○	○							sealing compound for valves and pipe threads
																									fuel additives, dispersant for carbonised oils
																									textile softeners, wool oils
																									saponifiable oils for textile machinery
																									ring and traveller grease
									○	○															special water resistant greases
																									extremely adhesive gear greases
																									molybdenum disulphide lubricants
																									gear and lubricating oils with molybdenum disulphide
									○	○	○	○								○		○	○	○	silicone lubricants and separating agents
									○	○	○	○	○	○											high temperature greases
																									glass moulding separating agent
					●	●	●																		metal forming lubricants
																									bonded coating
																									selflubricating bearing compound
																									high temperature lubricants up to 1000 °C
																									solid lubricants (bonded coating and powder)
																									high temperature lubricants
																									bonded coating for machine tool slideways
								●																	aqueous coolant and lubricant

Moly-XL Company
P.O. Box 219
Vienna, OH 44473
(216) 394-2181

MOLY-XL®

MOLY - XL COMPANY, Inc.

THE COMPLETE LINE OF QUALITY MOLY-LUBRICANTS

MOLY-XL LUBRICANTS PROVIDE THESE ADVANTAGES . . .

Reduce Friction, Heat, Wear . . . Extend Equipment Life . . . Reduce Energy Consumption . . . Extend Lubrication Intervals . . . Simplify Oil House Stores . . . Batch Tested Production . . . Fewer Environmental Pollutants . . . Engineered Products to Meet Your Requirements . . . Excellent Break-in Lubricants

SELF-LUBRICATING MOLYBDENUM DISULFIDE

Reduces Friction, Heat & Energy Consumption . . . Lubricates to 750°F . . . Sustains Loads to 500,000 PSI . . . Lubricates After Carrier is Gone . . . Plates Out on Bearing Surface . . . Stays Longer: — Molecular Adherence . . . Builds Up Lubricating Film . . . An Ideal Lubricant by Itself.

RESULTS:

Higher Production . . . Less Down Time . . . Overall Costs are Reduced . . . Fewer Maintenance Repairs . . . Minimizes Environmental Pollutants . . . Lubrication Labor Costs are Reduced . . . Solves Problem Applications. . . Reduces Fretting, Galling & Seizing.

MOLY XL GREASES

These greases were designed to meet all requirements for plant maintenance and provide lubrication protection for new equipment warranties. These longer life moly-disulfide greases permit longer intervals between applications and extend service life.

MOLY XL	N.L.G.I. GRADE	BASE	DROP POINT MIN.	ADDITIVES	DESCRIPTION
1A	1½	Lithium	360°F	Extreme Pressure Oxidation Inhibitors	High performance, versatile grease with good pumpability.
47-F2-75	1½	Lithium	350°F	Extreme Pressure	Excellent all purpose grease. Hi and low speed and wide temperature uses.
Valve & Anti-Seize	2	Lithium	350°F	Extreme Pressure Oxidation Inhibitors	High % of Moly Solids used where seizing must be prevented.
2AP	2	Lithium	350°F	Extreme Pressure R&O Inhibitor Polyethylene	Excellent multi purpose grease. Can be used for most industrial uses, truck and construction equipment. Good pumpability.
202AC	2	Aluminum Complex	475°F	Rust & Oxidation Inhibitors	Has many uses in industry where varying temperatures are encountered.
HI SPEED	1½	Calcium	275°F	Rust & Oxidation Inhibitors	Excellent anti-water-washout characteristics. Operates at −65°F. to over 200°F.
BOTTLING PLANT	2½	Calcium	200°F	Stringiness Agents	A special grease for bottling machine uses subjected to water spray.
FOOD MACHINERY USDA	2	Calcium	260°F	Rust & Oxidation Inhibitors	A white, colorless, edible, non contaminant grease, meeting F.D.A. standards. (No Moly)
FOOD MACHINERY EP	2	Aluminum Complex	450°F	Rust & Oxidation Inhibitors	A white, colorless, non contaminant grease, USDA approved. (No Moly)
703 BENTONE	1½	Bentone	None	Extreme Pressure	A tacky, no drop grease for smaller open gear applications & hi-temp work.
OPEN GEAR BARIUM	2	Barium	320°F	Extreme Pressure	An extremely tacky, cohesive, no drop lubricant for all open gear uses.

MOLY-XL INDUSTRIAL OILS

Manufactured from highest quality base materials. Microscopic self lubricating Molybdenum Disulfide particles are homogenized to produce a lubricant that provides excellent boundary lubrication.

MOLY XL	SAE	API	AGMA	VISCOSITY SSU @ 210°F	VISCOSITY SSU @ 100°F	FLASH POINT MIN.	ADDITIVES
10	10	10	—	39/45	100/150	330°F	E.P.—R&O—Anti Foam—Dis.
20	20	20	2EP	45/58	300/500	330°F	E.P.—R&O—Anti Foam—Dis.
30	30	30	3EP	55/65	500/600	400°F	E.P.—R&O—Anti Foam—Dis.
40	40	40	4EP	70/85	500/600	350°F	E.P.—R&O—Anti Foam

E.P.—Extreme Pressure.
R&O—Rust & Oxidation Inhibitors.
Dis.—Dispersant.

N.L.G.I.—National Lubricating Grease Institute.
A.G.M.A.—American Gear Manufacturers Association.

S.A.E.—Society Automotive Engineers.
A.P.I.—American Petroleum Institute.

Moly-XL Company
P.O. Box 219
Vienna, OH 44473
(216) 394-2181

MOLY-XL®

MOLY - XL COMPANY, Inc.

THE COMPLETE LINE OF QUALITY MOLY-LUBRICANTS

MOLY-XL GEAR OILS

These are from highest stocks, highly refined, balanced both Chemically and Metallically to provide the utmost protection for internal gears. Used in transmissions and gears of over the road truck fleets and heavy construction equipment. Also as the universal speed reducer oil. In many instances drops friction heat to a safe margin and extends gear life. Retards leakage. Moly especially helpful on worn, noisy gears.

MOLY XL	SAE	API	AGMA	VISCOSITY SSU @ 210°F	VISCOSITY SSU @ 100°F	FLASH POINT MIN.	ADDITIVES
85W-140	85W-140	—	—	135/140	1750/2000	410°F	E.P.—R&O—Anti Foam
90	90	90	5EP	80/90	1000/1250	360°F	E.P.—R&O—Anti Foam
140	140	140	7EP	130/160	2500/3500	590°F	E.P.—R&O—Anti Foam
250	250	250	8EP	200/220	3500/4000	600°F	E.P.—R&O—Anti Foam

MOLY-XL OPEN GEAR COMPOUND (GEAR TRAK)

Moly-XL Gear Trak is a very adhesive, tacky fluid type compound to be brushed on, sprayed on or used through central systems at any temperature. Designed to protect against wear. We offer this product as the perfect all purpose open gear compound.

4000/5000	800/1000	600°F	E.P.—R&O

Spraymatic Gear Trak, Heavy Duty (Aerosol). An open gear compound. Easy to apply. Goes direct to spot intended. No Waste. Packed in 13 oz. cans, 12 to a case. This is a cutback lube with a tacky, adhesive character and long life. No fluorocarbons.

MOLY-XL SPINDLE OILS

These are the FINEST Spindle Oils Produced. Ideal for High Speed Spindles: High Temperature: Precision machinery application and Conveyor system lubrication.

No.1	10	10	—	40/45	100/150	375°F	E.P.—R&O—Anti Foam
No.3	10	10	—	39/42	100/125	300°F	E.P.—R&O—Anti Foam

MOLY-XL HEAVY DUTY PRESS & WAY OILS

MOLY XL	SAE	SAE GEAR	VISCOSITY SSU @ 210°F	VISCOSITY SSU @ 100°F	FLASH POINT	DESCRIPTION E.P. HEAVY DUTY LUBE OIL
Press Oil 351	20	80	50/60	325/350	340°F	A mild E.P. oil. R&O. For use on presses and press brakes. Anti Foam. No lead.
Press Oil 1401	50	90	100/110	1350/1400	425°F	E.P. in character for use on heavy duty press operations. R&O, Anti Foam. No lead.
E. P. Heavy Duty Lube Oil	50	90	100/110	1050/1150	425°F	E.P. in character for use on heavy duty press operations. R&O, Anti Foam. No lead.
Way Oil 20	20	80	40/50	210/240	330°F	A good adhesive extreme pressure oil with exceptional lubricating qualities. E.P. anti-stick slip and plating action.

MOLY-XLS HI TEMP LUBRICANTS

MOLY XL	NGLI	BASE	DROP POINT	VISCOSITY SSU @ 210°F	VISCOSITY SSU @ 100°F	DESCRIPTION
XLS Synthetic Grease	1	Synthetic	500°F	93	550	For furnace, and oven conveyor bearings, hinges, hoist sheaves, etc.
XLS Synthetic Grease	2	Synthetic	550°F	103	550	For furnace, and oven conveyor bearings, hinges, hoist sheaves, etc.
XLS Bentone Grease	2	Bentone	None	250/275	1650/1800	The ultimate grease for high temperature applications when a conventionally formulated grease cannot provide lubricity.

	SAE	SAE GEAR	VISCOSITY SUU @ 210°F	VISCOSITY SUU @ 100°F	FLASH POINT	
XLS 1600 Hi Temp. Fluid	40	90	1000/1100	215/235	395°F	Synthetic Hydrocarbons Boron Nitride; Lubricating Range to 1600°F.
XLS Light Hi Temp. Oil	40	80	93.1	550	490°F	The ultimate lubricant for enameling and core oven conveyors, lazer and lehr bearings, speed reducers, etc. Where temperatures eliminate the use of conventional lubricants.

INDUSTRIAL OILS • GREASES • PASTES • DRY FILM • HI TEMP & SYNTHETIC LUBRICANTS

ORB Industries Incorporated
2 Race Street/P.O. Box 1067
Upland, PA 19015
(215) 874-2537

ORB aerosol bulletin

ORB #510 LUBRICATING OIL

What It Is

ORB #510 Lubricating Oil is an aerosol spray lubricant containing a superior rust inhibiting oil of medium viscosity.

What It Does

Provides a quick easy method of applying lubrication and rust protection to various mechanisms which can best be done by aerosol spray and which cannot be reached by usual methods of application. Can also be used for pinpoint application of lubricant by means of the unique extension nozzle provided.

Uses

For use wherever a medium viscosity oil is required. May be used on open gears, eccentrics, machine ways, slides, chains, guides, shafting, springs, latches, toggles, hinges, nuts and bolts, tools and dies and similar mechanisms requiring light lubrication, particularly where both lubrication and rust prevention are desired.

Advantages

Quick and easy to use. Saves time and labor. Reaches spots that are otherwise inaccessible. Always ready for use - no oil can to fill - no waste - no spillage. Can be carried in the tool box by every mechanic and maintenance man. Suitable both as a lubricant and as a protective agent because of rust inhibiting qualities.

Price

See price list enclosed (subject to applicable trade discounts).

Packing

12 --- 14 oz. cans per case --- shipping weight 15 lbs. per case.

ORB aerosol bulletin

ORB #571 MOLY DRY FILM LUBRICANT

What It Is

ORB #571 is an extreme pressure lubricant based on colloidal molybdenum disulfide. Supplied in convenient aerosol but also available in bulk form for standard spray, brush or dip methods of application. Predominately used on metals, but effective on all substrates not affected by the solvent carrier.

What It Does

Provides effective high pressure lubrication over a wide temperature range. Advantages include: fast air drying; superior extreme-pressure lubrication, excellent release characteristics; excellent adhesion to most substrates; little or no surface pre-treatment required; and effective in thin films - .003 to .005 inch.

Uses

Use when oil or liquid lubricants are not effective or desired. Useful in assembly of close fitting or press fit parts, high temperature lubrication, mating surfaces of business machine parts; lubrication of automotive and industrial gaskets, chutes and bins, tumbler locks, rubber components, sliding surfaces and gears, internal combustion engine assembly and break-in, vending machines, parking meters, etc.

Physical Properties

Lubricant: Molybdenum Disulfide
Binder: Thermoplastic cellulosic resin
Solvent Carrier: Predominately nonflammable
Color: Dark Gray
Density: 1.184 (9.88 lbs./gal.)
Coefficient of Friction: 0.23 (static)
Service Temp.: ∠ 0 to 390° F (200° C)
Intermittent to 660° F (350° C)

Directions

Mix thoroughly before use. Substrates should be clean and dry. Adhesion and wear properties are normally excellent, but can be improved by sandblasting, phosphatizing or chemical etching. Apply thin uniform coating and allow film to air dry approximately five minutes before use. When applied during periods of high humidity, color may become mottled or blushed. This does not affect performance. Provide adequate ventilation when using. Do not use around open flame or ignition sources, since decomposition products are hazardous. Not recommended for use in oxygen service.

Pricing

See enclosed price list. Subject to quantity discounts.

Packing

Aerosol: 12 - 16 ounce cans per case - shipping weight 16 lbs. per case.

Bulk: One (4 gal. min.), five and 55 gallon containers.

LUBRICANTS

ORB Industries Incorporated
2 Race Street/P.O. Box 1067
Upland, PA 19015
(215) 874-2537

ORB aerosol bulletin

ORB #572 GRAPHITE DRY FILM LUBRICANT

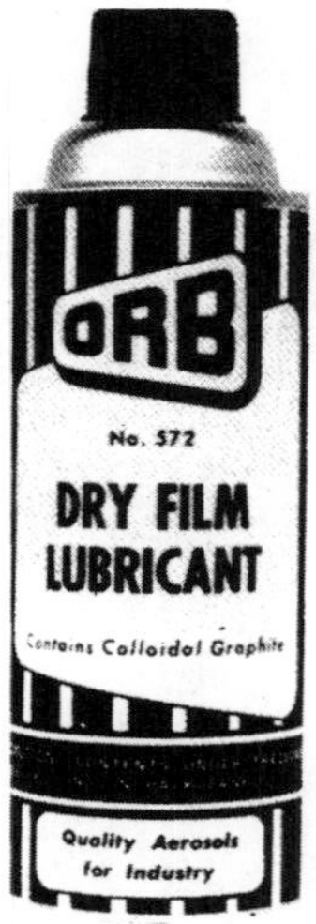

What It Is

ORB #572 is a specially processed graphite dry film lubricant, electrically conductive with excellent adhesion to most substrates with minimum surface preparation. Supplied in convenient aerosol form but also available in bulk for standard spray, brush or dip methods of application.

What It Does

Provides excellent, long wearing, conductive, high temperature lubrication at low film thickness. Advantages include: fast air drying; superior lubricity; excellent adhesion to most substrates; simple solvent wipe pretreatment usually sufficient pretreatment; and effective lubrication in film down to .0005 inch.

Uses

Use where effective, high temperature, long lasting lubrication is required. Typical applications include: internal combustion engine components for assembly and break-in; mating surfaces in business machines; vending machines, parking meters and tumbler locks; automotive and industrial gaskets, rubber component assembly and break-in; thread anti-seize lubricant; opaque coating for negatives; oven chains; etc.

Physical Properties

Lubricant: Processed micro-graphite
Binder: Thermoplastic resin
Solvent Carrier: Predominately nonflammable
Color: Black
Density: 1.179 (9.84 lbs./gal.
Coefficient of Friction: 0.19 (static)
Service Temp: $<$ 0 to 850° F (450° C)*

*The small amount binding resin present for application slowly decomposes above 200° F.

Directions

Mix thoroughly before use. Substrates should be clean and dry. A simple solvent wipe is usually sufficient pretreatment, but adhesion can be improved by sandblasting, phosphatizing or chemical etching. Apply thin uniform coating and allow to air dry approximately five minutes before use. Provide adequate ventilation when using. Do not use around open flame or ignition sources since decomposition products are hazardous.

Pricing

See enclosed price list. Subject to quantity discounts.

Packing

Aerosol: 12 - 16 ounce cans per case. Shipping weight 16 lbs. per case.

Bulk: One (4 gal. min.), five and 55 gallon containers.

Renite Company
2500 East Fifth Avenue/P.O. Box 19235
Columbus, OH 43219
(614) 253-5509

office and factory
2500 East Fifth Avenue
Phone (area 614) 253-5509

RENITE ®
Company *Lubrication Engineers*

P.O. BOX 19235 • COLUMBUS, OHIO 43219 • U.S.A.

RENITE S-26

NEW WATER-BASED METAL-WORKING LUBRICANT

Renite S-26 is a dispersion of graphite in water, recommended as lubricant and release agent for steel, brass, copper and related forging operations, some extrusions, permanent mold casting of brass or iron (and certain aluminum and brass die castings) where die temperatures (at the moment Renite is sprayed) are between 300 and 800°F (150 and 425°C).

When this lubricant is sprayed on working surfaces of tools and dies, the water vehicle evaporates, leaving a solid film lubricant coating well suited to high temperature and high pressure applications. Lubricity is quite high (comparable to, if not better than, oil/graphite lubricants), and the metal-working operation is carried out without smoke, fumes and flames.

Renite S-26 is a versatile product, supplied as a concentrate, capable of being adjusted to suit a variety of different applications by varying the dilution ratio. Richer mixtures are used for hotter and heavier jobs, typical ratios being 10-20 to 1 for die casting, 5-10 to 1 for forging and extrusion, and 2-3 to 1 for the hottest and heaviest jobs. The concentrate is stable and uniform. After dilution, done with clean tap water, the mixture should be provided with a little agitation to maintain graphite suspension. Light stirring several times during a shift and after any prolonged shutdown is generally sufficient. A recirculating system providing continuous flow through lines and back to the lubricant reservoir tank is ideal—ask us about our Renite Model 40 Tank and Pump System.

Special features of Renite S-26 are its ability to coat at temperatures often considered above the range for water-based products, and relative freedom from problems with nozzle clogging and build-up on dies. Clogging and build-up generally are the result of dried lubricant residues that are hard and water-resistant. Dried Renite S-26 residue is soft and redisperses readily when wetted with water. Recycling and reusing dried residue and "over-spray" is even possible.

Proper application is the key to success in using a water-based lubricant. We have many years of experience with such products and also have a full line of manual and automatic spray equipment especially suited to their application.

We would be happy to provide a sample of Renite S-26 and to consult with you on application details.

Renite Company
2500 East Fifth Avenue/P.O. Box 19235
Columbus, OH 43219
(614) 253-5509

office and factory
2500 East Fifth Avenue
Phone (area 614) 253-5509

RENITE® Company *Lubrication Engineers*

P.O. BOX 19235 • COLUMBUS, OHIO 43219 • U.S.A.

RENITE S-28

NEW WATER-BASED METAL-WORKING LUBRICANT FOR FORGING AND EXTRUSION OF ALUMINUM

Renite S-28 is a dispersion of graphite in water, recommended as lubricant and release agent for aluminum forging and extrusion operations where die and tool temperatures, at the time of spraying, are between 300 and 600°F (150 and 315°C). (External heating of dies may be provided when needed.)

When this lubricant is sprayed on working surfaces of tools and dies, the water vehicle evaporates, leaving a solid film lubricant coating well suited to high temperature and high pressure applications. Lubricity is quite high (comparable to, if not better than, oil/graphite lubricants), and the metal-working operation is carried out without smoke, fumes and flames.

Renite S-28 is a versatile product, supplied as a concentrate, capable of being adjusted to suit a variety of different applications by varying dilution ratio. Richer mixtures are used for hotter and heavier jobs, typical ratios being on the order of 5-20 to 1. The concentrate is stable and uniform. After dilution, done with clean tap water, the mixture should be provided with a little agitation to maintain graphite suspension. Light stirring several times during a shift and after any prolonged shutdown is generally sufficient. A recirculating system providing continuous flow through lines and back to the lubricant reservoir tank is ideal--ask us about our Renite Model 40 Tank and Pump System.

Special features of Renite S-28 are the extra metal movement it gives to aluminum and relative freedom from build-up problems, even on narrow draft dies. Lubricant residues are soft and flaky--tend to be knocked off and blown away when the forging or extrusion is removed and the coating for the next piece applied.

Proper application is the key to success in using a water-based lubricant. We have many years of experience with such products and also have a full line of manual and automatic spray equipment especially suited to their application.

We would be happy to provide a sample of Renite S-28 and to consult with you on application details.

Renite Company
2500 East Fifth Avenue/P.O. Box 19235
Columbus, OH 43219
(614) 253-5509

office and factory
2500 East Fifth Avenue
Phone (area 614) 253-5509

RENITE® Company Lubrication Engineers

P.O. BOX 19235 • COLUMBUS, OHIO 43219 • U.S.A.

RENITE IF

NEW EMULSION-TYPE HAMMER DIE LUBRICANT FOR IMPACT FORGING

Recently developed as a hammer die lubricant for impact forging, Renite IF is a graphited oil-in-water emulsion. It combines in a single product the lubricity and release characteristics of oil, bolstered by graphite so as to stand up under extremely high temperatures and pressures, with the cooling effect, safety, environmental acceptability and economy of water.

The product is supplied as a concentrate, designed for dilution with ordinary tap water. Typical dilution ratio is six parts water to one of Renite IF. More concentrated mixtures may be used for more difficult jobs, but to keep the release effect from becoming overly vigorous, there should be no less than two parts of water to one of Renite IF.

Renite IF is normally swab-applied, though it is also suited to spray application. The customary hammer die temperature range for application is $200\text{-}500^{\circ}$F. (For unusually hot dies, try stirring in a little flour and apply by swabbing - lab tests indicate that this permits swabbing at temperatures well above the above range.) Complex die shapes are no problem, as the steam generated by the water contacting a hot die serves to kick around the oil and graphite. Because of this spattering effect the operator should, naturally, wear safety glasses and suitable protective clothing.

The oil droplets in the product are surrounded by a water jacket which serves to reduce if not eliminate fire problems. Smoke is minimal since water is the diluent for the product. Having a water base, the product should be kept from freezing, though accidentally-frozen material is still useable if thawed.

The concentrate is quite stable - needs no stirring up before dilution - and dilution is easily accomplished with only light hand stirring. Stability is also quite good after dilution - occasional hand stirring every few hours is sufficient to maintain suspension.

No preservatives have been added, thus avoiding the allergy problems associated with such materials. There are no caustic, acidic, or especially toxic additives, and the emulsifiers in the product are biodegradable.

Samples are readily available - write or phone for prompt shipment.

Sprayway, Incorporated
484 Vista Avenue
Addison, IL 60101
(312) 628-0998

Sprayway

No. 714

dry graphite film lubricant

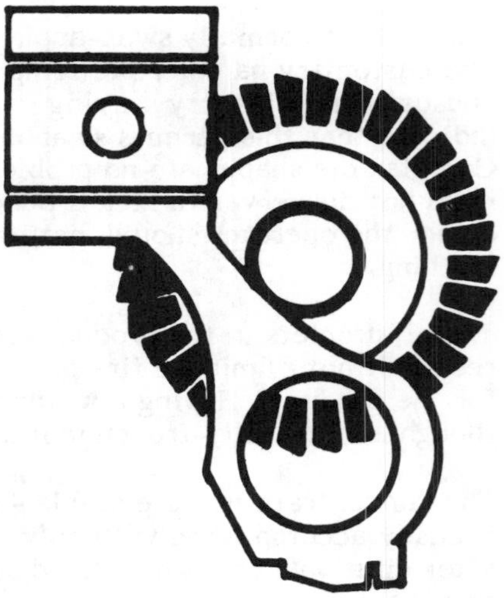

NET WT. 16 OZ. (1 LB.) (453.6 GRAMS)

PRE-ASSEMBLY LUBRICATION FOR ENGINE AND MACHINERY PARTS. PREVENTS SCORING, SCUFFING, SEIZING

PROVIDES EXCELLENT LUBRICATION FROM 800°F. to −75°F.

Sprayway, Incorporated
484 Vista Avenue
Addison, IL 60101
(312) 628-0998

Sprayway®

WHITE *Lithium* GREASE

LUBRICANT SPRAY

For anything that squeaks, sticks or rusts

A highly concentrated, all purpose white lubricant that leaves a long lasting film that will not wash away, run off, or dry out.

Sprayway

NO. 715

NET WT.

16 OZ. (1 LB.)

INDUSTRIAL: Cams and slides, chains, couplings, plain bearings, open gears (small), fork lift truck masts.

MARINE: Excellent for general marine use because of high resistance to rusting and corrosive action.

AUTOMOTIVE: Use on accelerator linkage, bearings, brake service linkage, ash tray slides, radio antenna, cams, gears, hinges, hood locks, door locks, springs, striker plates, window regulators and all body hardware.

HOME & OFFICE: Use wherever metal meets metal. Windows, garage doors, hinges, faucets, lawn mowers, bicycles, toys, garden and household tools, fishing reels, guns, filing cabinets, office equipment, etc.

All Purpose Heavy Duty Lubricant for

AUTOMOTIVE
MARINE
INDUSTRY
HOME
OFFICE

Sprayway, Incorporated
484 Vista Avenue
Addison, IL 60101
(312) 628-0998

NET WT. 16 OZ. (1 LB.)
453.6 GRAMS

Sprayway®

No. 718

CHAIN DRIVE SPRAY

OPEN GEAR LUBE

PROTECTS ROLLERS, SLIDES AND GUIDES
INDUSTRIAL MACHINERY FARM EQUIPMENT
EXTRA TOUGH-HEAVY BODIED

FOR INDUSTRIAL USE

Recommended For

Open Gears, chains, cables, sprockets, rollers, racks pinions, guides, slides, wear plates, shafts, sockets.

Ultrachem Incorporated
1400 North Walnut Street/P.O. Box 2053
Wilmington, DE 19899
(302) 571-8520

QUICK-REFERENCE CHART

OILS

APPLICATIONS	Chemlube Gear Oils	C-201	C-207	C-209	C-215	C-217	C-225	C-226	C-228	C-230	C-400	C-501	C-530	C-645	C-650	C-680	C-751	C-1000	C-1102	S-220	S-330	S-440	O-100, 300	O-350, 550	O-Non-Tox	BG-105, 108	PM-325	AF#1
INDUSTRIAL & PRODUCTION APPLICATIONS																												
Air Compressors (Reciprocating)											★					★												
Air Compressors (Reciprocating - PAO Oil)												★																
Air Compressors (Rotary)					★					★																				
Air Compressors (Rotary - PAO Oil)									★																				
Process Gas Compressors											★					★												
Refrigeration Compressors											★																		
Steam Turbines											★					★												
Ball & Roller Bearings			★																									
Cables, Chains, Wire Ropes							★	★																				
Cams, Slides																												
Conveyors																												
Damping Fluid																												
Gear Boxes |★|																											
Heat Transfer Fluid					★	★																						
High Temperature Bath Oil					★	★																						
Hydraulic Oil					★	★																						
Open Gears																												
Worm Gears																												
Rust Prevention																											★		
Steel Mills - Casters																												
Textile Equipment			★				★	★				★					★												
O.E.M. APPLICATIONS																												
Cables																												
Dials & Potentiometers																												
Electric Connectors																			★										
Impregnating Oils for Powder Metal Bearings:																												
High Temperature				★		★								★	★	★				★	★	★	★	★	★				
Low Temperature		★	★	★		★								★	★	★				★	★	★	★	★	★				
Plastic Compatible																							★	★	★				
USDA Approved - H1																									★				
Instrument Oils		★	★															★	★										
Watches, Clocks & Timing Devices																		★	★										
GOVERNMENT APPLICATIONS																												
Mil-L-3918A																		★											
Mil-L-6085A		★																											
Mil-L46000B																												
Mil-L-46150																												
Mil-L-46152											★																		
Federal Stock #9150-00-159-4012																													★

GREASES

APPLICATIONS	V-310	V-350, 350M	V-352	V-373	V-375	V-501	V-502	V-507	V-4076	V-4076SS	C-AY4	C-303	C-303T-25
INDUSTRIAL & PRODUCTION APPLICATIONS												
Air Compressors (Reciprocating)												
Air Compressors (Reciprocating - PAO Oil)												
Air Compressors (Rotary)												
Air Compressors (Rotary - PAO Oil)												
Process Gas Compressors												
Refrigeration Compressors												
Steam Turbines												
Ball & Roller Bearings			★					★	★				
Cables, Chains, Wire Ropes				★									
Cams, Slides	★			★		★	★						
Conveyors	★			★		★	★						
Damping Fluid	★			★		★	★						
Gear Boxes					★								
Heat Transfer Fluid												
High Temperature Bath Oil												
Hydraulic Oil												
Open Gears		★		★									
Worm Gears					★								
Rust Prevention												
Steel Mills - Casters										★			
Textile Equipment		★												
O.E.M. APPLICATIONS												
Cables	★	★	★										
Dials & Potentiometers	★		★			★	★						
Electric Connectors											★		
Impregnating Oils for Powder Metal Bearings:												
High Temperature												
Low Temperature												
Plastic Compatible												
USDA Approved - H1												
Instrument Oils												
Watches, Clocks & Timing Devices												
GOVERNMENT APPLICATIONS												
Mil-L-3918A												
Mil-L-6085A												
Mil-L46000B												★	
Mil-L-46150													★
Mil-L-46152												
Federal Stock #9150-00-159-4012												

Ultrachem Incorporated
1400 North Walnut Street/P.O. Box 2053
Wilmington, DE 19899
(302) 571-8520

VERSATILE LUBRICANTS SATISFYING

O.E.M. LUBRICANTS

- *Fractional H.P. Electric Motors*
- *Powder Metal Bearings*
- *Anti-Friction Bearings*
- *Speedometer & Tachometer Cables*
- *Instruments*
- *Permanent Gear Box Lubricants*
- *Appliances*

Ultrachem's synthetic lubricants offer versatility for manufacturers of a wide range of consumer products. From impregnating oils for powder metal sleeve bearings to a broad selection of greases, O.E.M. applications are met with confidence and cost efficiency.

AUTOMOTIVE LUBRICANTS

- *Premium Motor Oil*
- *2-Cycle Oil*
- *Race Oil*
- *Gear Oil*
- *Engine Additive*

Ultrachem offers a proven track record of success in the manufacture of synthetic lubricants for the automotive after-market. Now the technical expertise gathered through years of successful service to industry and government is applied to automotive requirements. The result is better performance and longer life for your cars.

Ultrachem Incorporated
1400 North Walnut Street/P.O. Box 2053
Wilmington, DE 19899
(302) 571-8520

MANY DIFFICULT APPLICATIONS

INDUSTRIAL MAINTENANCE LUBRICANTS

- *Compressors*
- *Chains*
- *Conveyors*
- *Ovens*
- *High Temperature Applications*
- *Oil Baths*
- *Textile Machinery*
- *Motors*

Roller bearings on a steel mill's continuous caster are prime examples of a high temperature grease application.

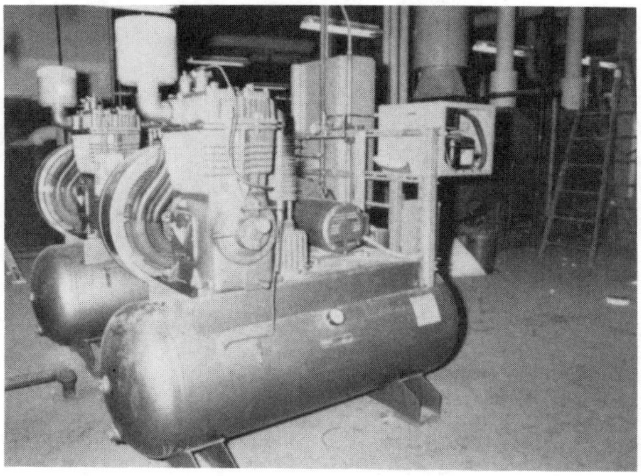

Ultrachem's synthetic lubricants are the perfect answer for industrial maintenance needs. Available on a direct basis or through our worldwide network of distributors, these products are specially designed to solve costly plant lubrication problems. This means reduced relubrication intervals, longer equipment life, less down-time, and reduced maintenance costs. Most importantly, synthetic lubricants can be used in applications where petroleum lubricants simply won't function effectively.

GOVERNMENT LUBRICANTS

- *Military Specifications*
- *Federal Stock Items*

Ultrachem expertise is reflected in many ways. One of the most significant is our presence on the Federal government's Qualified Products List as a primary manufacturer of military specified lubricants. Ultrachem's ability to formulate oils and greases for demanding military specifications demonstrates our commitment to excellence in the production of synthetic lubricants.

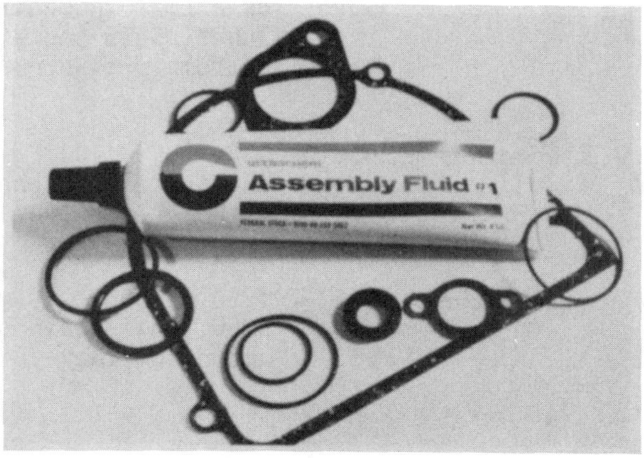

Ultrachem Incorporated
1400 North Walnut Street/P.O. Box 2053
Wilmington, DE 19899
(302) 571-8520

TECHNICAL INFORMATION

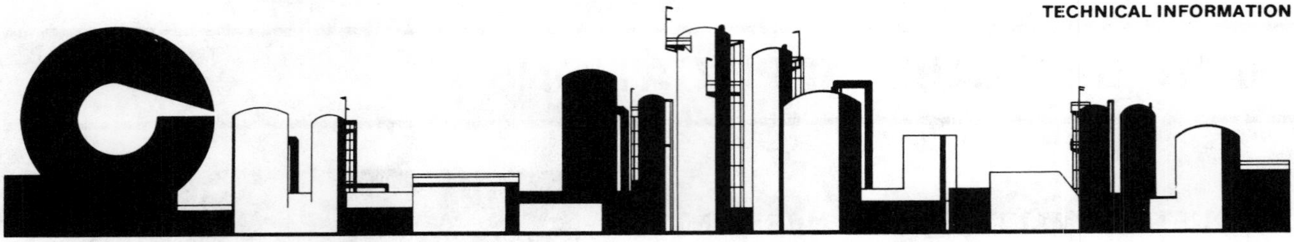

SYNTHETIC HIGH TEMPERATURE CHAIN LUBRICANTS

FROM ULTRACHEM

Chemlube ® 225	850 SUS @ 100°F. The original tacky heavy chain lubricant with a diester base fluid and a high temperature organo-moly extreme pressure anti-wear additive. Use with a drip or hand applied application system.
Chemlube ® 226	223 SUS @ 100°F. A thinner version of 225 with no tack for higher temperatures. Excellent for application through a mister or atomizer type lubricator.
Chemlube ® 5051	635 SUS @ 100°F. This is a pure diester chain oil with no tack added. High flash point of 590°F indicates its high temp characteristics.

* * * * * * * * * * * *

Chemlube ® 5045	1200 SUS @ 100°F. This is one of three truly heavy chain lubricants made from a proprietary blend of two synthetic base stocks. No tack added. Use with a forced feed lubricator such as a Lincoln Centromatic. Like our other chain lubes, 5045 contains a high temp organo-moly extreme pressure anti-wear additive.
Chemlube ® 5050	1320 SUS @ 100°F. This uses the same base fluid and additive technology as 5045 with the addition of heavy tack for adherence to chains.
Chemlube ® 5055	1250 @ 100°F. Similar in make up to the 5045 and 5050, this chain oil contains a greatly reduced amount of tack in comparison to 5050.

ULTRACHEM® INC.

1400 N. WALNUT ST./P.O. BOX 2053
WILMINGTON, DE 19899
(302) 571-8520

Ultrachem Incorporated
1400 North Walnut Street/P.O. Box 2053
Wilmington, DE 19899
(302) 571-8520

	BASE FLUID	ORGANO-MOLY EXTREME PRESSURE ANTI-WEAR ADDITIVE	VISCOSITY	FLASH POINT	POUR POINT	SAE	TACK	APPLICATION METHOD
Chemlube 225	Diester	Yes	850 SUS @ 100°F	470°F	-20°F	50	Yes	Drip or Hand
Chemlube 226	Diester	Yes	223 SUS @ 100°F	500°F	-50°F	20	No	Mist
Chemlube 5051	Diester	Yes	635 SUS @ 100°F	590°F	-50°F	50	No	Forced Feed
Chemlube 5045	Proprietary Synthetic	Yes	1200 SUS @ 100°F	530°F	-35°F	N.A.	No	Forced Feed
Chemlube 5050	Proprietary Synthetic	Yes	1320 SUS @ 100°F	525°F	-30°F	N.A.	Yes	Forced Feed
Chemlube 5055	Proprietary Synthetic	Yes	1250 SUS @ 100°F	525°F	-30°F	N.A.	Reduced Tack	Forced Feed

Ultrachem Incorporated
1400 North Walnut Street/P.O. Box 2053
Wilmington, DE 19899
(302) 571-8520

TECHNICAL INFORMATION

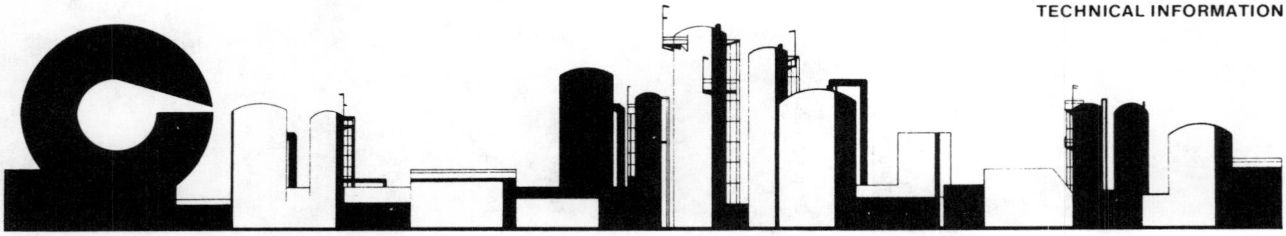

CHEMLUBE ® 80W, 85W-90, 140 and 250

SYNTHETIC GEAR OILS

Applications:

The Chemlube 80W, 85W-90, 140 and 250 Series consists of 4 SAE rated, fully synthetic gear oils composed of synthetic ester base fluids for a wide variety of extreme pressure industrial applications. Areas of special interest are extreme environmental applications, as well as, those where high heat resistance, anti-wear and superior lubricity of enclosed gear boxes is a necessity.

Characteristics:

Extremely low pour points.
High viscosity index.
Low evaporation rates.
Savings in energy costs - reduced utility bills.
High polarity with metal surfaces.
Reduced downtime due to extended change intervals.
Longer life than conventional mineral gear oils.
Reduction in gearbox lubricant temperature - gearboxes actually run cooler due to reduced friction.

Typical Properties:

		80W	85W-90	140	250
Viscosity 210°F SUS	(ASTM D-445)	64.2	90	140	200
100°F SUS	(ASTM D-445)	332	760	1130	1880
210°F Cs.	(ASTM D-445)	11.4	18.0	29.4	42.8
100°F Cs.	(ASTM D-445)	71.5	164	244	406
-15°F Cps.		3400			
AGMA Lubricant Number		2 EP	4 EP	5 EP	7 EP
Viscosity Index	(ASTM D-2270)	150	140	135	130
SAE Viscosity Number		80W	85W-90	140	250
ISO Viscosity Grade		68	150	220	460
Pour Point °F	(ASTM D-97)	-50	-50	-50	-35
Pour Point °C	(ASTM D-97)	-46	-46	-46	-37
Flash Point °F	(ASTM D-92)	450	450	450	450
Flash Point °C	(ASTM D-92)	232	232	232	232
Timken O.K. Load (Lbs.)		70	70	70	70
Copper Corrosion	(ASTM D-130)	1B	1B	1B	1B
Rust Prevention	(ASTM D-665)	Pass	Pass	Pass	Pass
Carbon Residue	(ASTM D-189)	Nil	Nil	Nil	Nil
Color	(ASTM D-1500)	Amber	Amber	Amber	Amber

2/83

ULTRACHEM® INC.
1400 N. WALNUT ST./P.O. BOX 2053
WILMINGTON, DE 19899
(302) 571-8520

Ultrachem Incorporated
1400 North Walnut Street/P.O. Box 2053
Wilmington, DE 19899
(302) 571-8520

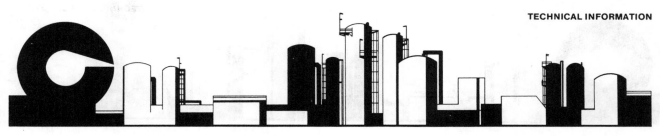

TECHNICAL INFORMATION

VISCHEM ® 4076

SYNTHETIC HIGH TEMPERATURE GREASE

Applications:

Ball and roller bearings - cams - slides - gears for appliance, automotive, business machines, electric motor and instrument applications. Especially useful where unusual thermal and oxidative stability are required.

Characteristics:

Extreme long life.
Wide useful temperature range, -40°F to 450°F.
Excellent oxidation stability, no deposits.
Outstanding thermal stability.
Good load carrying capability.
Resists moisture.
Versatile, for heavy or light loads.

Typical Properties:

Penetration (Worked 60 strokes)	265-295
NLGI Grade	2
Norma Hoffman Bomb Oxidation	
100 hrs. PSI loss	0.0
500 hrs. PSI loss	0.3
Dropping Point, min	500°F
Oil Separation Rate, max	3%
Base Oil	Synthetic
Viscosity, 210°F ASTM D-445 (49.9 SUS)	7.25 cSt.
100°F ASTM D-445 (186.8 SUS)	40 cSt.
-40°F	22,500 cSt.
Evaporation Rate, 400°F, max. ASTM D-972	5%
Pour Point ASTM D-97 (-53.89° C)	-65°F
Flash Point ASTM D-92 (265.56° C)	510°F

Also available as Vischem 4076SS in a softer penetration.

11/82

ULTRACHEM® INC.
1400 N. WALNUT ST./P.O. BOX 2053
WILMINGTON, DE 19899
(302) 571-8520

Ultrachem Incorporated
1400 North Walnut Street/P.O. Box 2053
Wilmington, DE 19899
(302) 571-8520

TECHNICAL INFORMATION

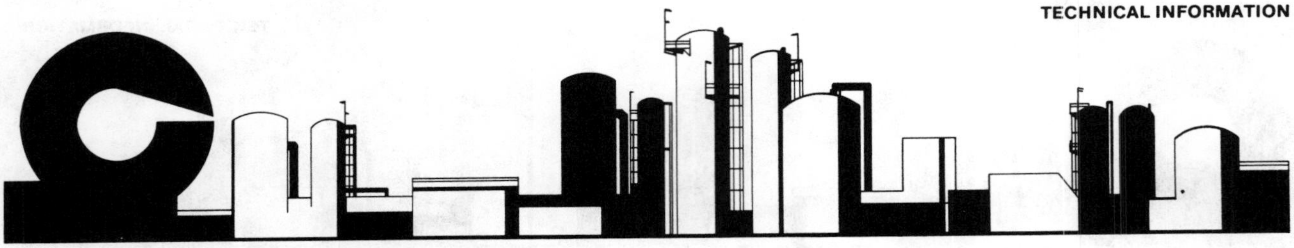

LUBRICANTS

OMNILUBE ® 1

PLASTICS COMPATIBLE SYNTHETIC GREASE

Applications:

This multi-purpose PAO based grease is excellent for a wide variety of applications, especially where plastic compatibility is desired with the protection and wear prevention characteristics of a synthetic. Long life, wide temperature range, and waterproof characteristics. Use in sliding bearings, anti-friction bearings, etc.

Characteristics:

Excellent stability. Prevents rust in bearings.
Wide useful temperature range. Pumpable at -40°F and useful to 300°F continuous and 400°F intermittently.
Superior for wet conditions.
Tacky for better adherence.

Typical Properties:

NLGI Grade		1
Penetration (worked)		310-340
Dropping Point	ASTM D-2265	500°F
Rust Test	ASTM D-1743	#1 (No Rust)
Shell 4-Ball EP	ASTM D-2569	
Weld Load		315 kg.
Load Wear Index		56.6 kg.
Water Wash		Nil
Thickener		Aluminum Complex
Base Oil		Synthetic Hydrocarbon (Polyalphaolefin)
Viscosity 210°F	ASTM D-445 (73.4 SUS)	13.97 cs.
Viscosity 100°F	ASTM D-445 (512 SUS)	110.4 cs.
Viscosity 0°F	ASTM D-445 (7000 SUS)	1512 cs.
Viscosity Index	ASTM D-2270	135
Flash Point	ASTM D-92 (246°C)	475°C
Pour Point	ASTM D-97 (-57°C)	-70°F

5/84

ULTRACHEM® INC.
1400 N. WALNUT ST./P.O. BOX 2053
WILMINGTON, DE 19899
(302) 571-8520

Ultrachem Incorporated
1400 North Walnut Street/P.O. Box 2053
Wilmington, DE 19899
(302) 571-8520

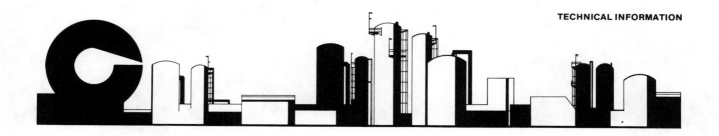

TECHNICAL INFORMATION

OMNILUBE ® 2

PLASTICS COMPATIBLE SYNTHETIC GREASE

Applications:

This multi-purpose PAO based grease is excellent for a wide variety of applications, especially where plastic compatibility is desired with the protection and wear prevention characteristics of a synthetic. Long life, wide temperature range, and waterproof characteristics. Use in sliding bearings, anti-friction bearings, etc.

Characteristics:

Excellent stability. Prevents rust in bearings.
Wide useful temperature range. Pumpable at -40°F and useful to 300°F continuous and 400°F intermittently.
Superior for wet conditions.

Typical Properties:

NLGI Grade			2
Penetration (worked)			265-295
Change after 10,000 strokes	ASTM D-217		310-315
Dropping Point	ASTM D-2265		500°F
Rust Test	ASTM D-1743		#1 (No Rust)
Shell 4-Ball EP	ASTM D-2569		
Weld Load			315 kg.
Load Wear Index			56.6 kg.
Water Wash			Nil
Thickener			Aluminum Complex Gel
Base Oil			Synthetic Hydrocarbon (Polyalphaolefin)
Viscosity 210°F	ASTM D-445	(73.4 SUS)	13.97 cs.
Viscosity 100°F	ASTM D-445	(512 SUS)	110.4 cs.
Viscosity 0°F	ASTM D-445	(7000 SUS)	1512 cs.
Viscosity Index	ASTM D-2270		135
Flash Point	ASTM D-92	(246°C)	475°F
Pour Point	ASTM D-97	(-57°C)	-70°F

2/84

ULTRACHEM® INC.
1400 N. WALNUT ST./P.O. BOX 2053
WILMINGTON, DE 19899
(302) 571-8520

Van Straaten Chemical Company
630 West Washington Boulevard
Chicago, IL 60606
(312) 454-1000

VAN STRAATEN
Chemical Company

Metal Forming Lubricants

Cold metal forming—stamping, drawing, cold heading, cold forging, and cold extrusion—is growing in importance among metalworking operations as the technology becomes more advanced. The obvious advantages these operations offer occur in the areas of energy and raw material savings. But to appreciate full productivity from any forming operation, it is essential to choose a specific lubricant based on the difficulty of the task, realistic die life, and economies achieved through the effective use of time, equipment and work flow. Van Straaten metal forming lubricants represent state-of-the-science thinking based on these considerations.

Basic Products

Oils and Oil Soluble Products
Many of these are superior to the traditional lubricating products that provide good adhesion, effective lubrication, enhanced tool life, and corrosion protection between operations.

Water Solubles
Water soluble lubricants provide all the basic advantages of oil-based products with the additional benefit of ease of handling and cleaning.

Powders
Powdered lubricants are easy to store and provide good lubrication and release in a number of otherwise difficult operations.

Applications

The applications of cold forming processes occur in almost any industry. Some of the areas in which Van Straaten products are being used include:

Sheet Metal Forming
This area covers the range from appliances to automotive and truck bodies, off-highway equipment, and industrial and farm equipment.

Tube Drawing
These operations are found in tube mills, hydraulics, building equipment, heat exchanger manufacturing, and especially in the manufacture of chairs, tables, and related furniture.

Stamping
Some of the parts currently being cold stamped with the use of Van Straaten lubricants include automotive, truck and off-highway equipment components such as rocker arms, relay covers, valve covers and a myriad of other parts; domestic appliances; metal furniture; components for garden equipment such as lawn mowers and edge trimmers.

Deep Drawing and Extrusion
Bearing cups, piston pins, disc brake pistons, and shell and missile casings are all made by cold extrusion using Van Straaten lubricants.

Cold Heading
Cold heading applications produce a variety of metal fasteners, spark plug shells, valve guides, shock absorber components, piston pins, engine valves, etc.

Westmont Products
P.O. Box 224049
Dallas, TX 75222
(214) 438-0588

BIO-GLIDE™

WESTMONT PRODUCTS
P.O. Box 224049
Dallas, TX 75222, 214/438-0588

Provides a tough, even film of lubrication for machine tool slideways

BIO-GLIDE is a compounded oil that provides an even film of lubrication between ways and riders. Anti-stick-slip characteristics ensure smoother machining operations and fewer rejects. BIO-GLIDE resists forceful squeezing and wiping action without building up and is tacky enough to withstand washoff by metal cutting fluids. It extends machine tool life by protecting metal surfaces from friction and wear. BIO-GLIDE meets or exceeds strict specifications for adhesiveness, oiliness and film strength for a premium, all-purpose way lubricant.

Benefits of BIO-GLIDE

- **REDUCES REJECTS**—Helps prevent workpiece disfigurement from stick-slip conditions caused by poor lubrication of machine tool ways.
- **EXTENDS MACHINE TOOL LIFE**—Protects metal surfaces from friction and wear.
- **WON'T BUILD UP**—Resists buildup under riders, so workpiece tolerances can be held precisely.
- **WITHSTANDS WASHOFF**—Tacky enough to withstand washoff by water-soluble metal cutting fluids.
- **HIGH QUALITY**—Meets or exceeds major machine manufacturers requirements for adhesiveness, oiliness and film strength.

Application

Use BIO-GLIDE on ways supporting the motion of worktables, carriages, saddles, tool holders and various types of heads for all kinds of production machines and machine tools. BIO-GLIDE is excellent for use on all slides, gibs, crosshead guides, worm gears, feed screws and columns.

BIO-GLIDE is available in bulk or convenient aerosol form. For best results, remove old lubricant with a solvent that evaporates quickly, then apply a thin, even coat of BIO-GLIDE. BIO-GLIDE can also be used in automatic lubricators. Reapply every two or three months or as needed.

Consult label for complete directions before using.

DISTRIBUTED BY:

Wynn Oil Company
P.O. Box 4370/2600 East Nutwood Avenue
Fullerton, CA 92634
(714) 992-2000

WYNN'S
INDUSTRIAL PRODUCTS DIVISION

WYNN'S ULTRA-KUT
727

WYNN'S® ULTRA-KUT 727

A HEAVY DUTY METALWORKING LUBRICANT DESIGNED AND FORMULATED FOR THE DIFFICULT TO MACHINE ALLOYS IN THE METALWORKING INDUSTRY.

DESCRIPTION

WYNN'S ULTRA-KUT 727 IS A TRANSPARENT, NON EMULSIFIABLE METAL CUTTING AND FORMING OIL. IT IS FORTIFIED WITH EXTREME PRESSURE, ANTIWELD AND LUBRICITY ADDITIVES FOR SEVERE MACHINING OPERATIONS ON FERROUS ALLOYS, WHERE STAINING IS NOT IMPORTANT.

ADVANTAGES & BENEFITS

- GENERAL PURPOSE - CAN BE USED FOR BOTH HEAVY DUTY GRINDING AND CUTTING OPERATIONS.
- SUPERIOR FINISH - ON ALL METALS. EXCELLENT LUBRICITY.
- INCREASES FEED & SPEED - LOW VISCOSITY LUBRICANT ALLOWS FASTER COOLING OF WORK PIECE AND TOOL.
- HIGH PERFORMANCE - WILL PROVIDE A HIGH PRODUCTION RATE, OPTIMUM TOOL LIFE AND QUALITY FINISH.
- CHEMICAL ACTIVITY - WHERE TOUGH ALLOYS ARE CONCERNED.
- RECYCLABLE - PETROLEUM BASE ALLOWS EASY RECYCLING OF PRODUCT TO EXTEND SERVICE.
- CHLORINE FREE - EXCELLENT E.P. WITHOUT THE USE OF CHLORINE.

Wynn Oil Company
P.O. Box 4370/2600 East Nutwood Avenue
Fullerton, CA 92634
(714) 992-2000

TYPICAL SPECIFICATIONS

APPEARANCE	CLEAR LIQUID
ODOR	PERFUMED
COLOR	DARK BROWN
API GRAVITY @ 60°F	27.1
VISCOSITY @ 40°C	34.0 - 37.0 CST
VISCOSITY @ 100°F	160 - 175 SUS
FLASH POINT COC, °F	380 MIN.
TOTAL SULFUR	1.5% WT
SULFURIZED FAT	4.0% WT
TOTAL PHOSPHOROUS	0.05% WT
LB/GAL @ 60°F	7.4

ASK YOUR WYNN'S REPRESENTATIVE ABOUT OTHER QUALITY PRODUCTS FOR THE METALWORKING INDUSTRY.

WYNN OIL COMPANY
2600 EAST NUTWOOD AVENUE . P.O. BOX 4370, FULLERTON, CALIFORNIA 92634

ITEM 7409 US (9/83)

Wynn Oil Company
P.O. Box 4370/2600 East Nutwood Avenue
Fullerton, CA 92634
(714) 992-2000

WYNN'S
INDUSTRIAL PRODUCTS DIVISION

TECHNICAL DATA
WYNN'S SUPERKON 604
Ref. 861

WYNN'S SUPERKON 604 is a heavy-bodied lubricant designed to be used as an anti-weld improver for cutting fluids and as a lubricant for severe metal forming operations. This product, when added to pale oils, performs exceedingly well in machining the stainless steels and other tough, ferrous metals.

BENEFITS

FUNCTIONS AS AN EFFECTIVE EXTREME PRESSURE LUBRICANT WHEN ADDED TO CUTTING OILS.

WILL NOT STAIN NON-FERROUS METALS.

MULTI-PURPOSE METALWORKING LUBRICANT.

HOT WATER RINSEABLE.

TYPICAL DATA

Form:	Viscous Fluid	**Flash Point** °F	500
Color:	Brown	**% Chlorine:**	46
Odor:	Typical	**lb./gal.:**	10.4
Spec. Gravity @ 60° F:	1.25		

Wynn Oil Company
P.O. Box 4370/2600 East Nutwood Avenue
Fullerton, CA 92634
(714) 992-2000

GENERAL APPLICATION AND CONCENTRATION TABLE

MACHINING METHOD	TOOL MATERIAL	STEELS: FREE MACHINING MEDIUM CARBON MALLEABLE CAST	STEELS: LOW CARBON HIGH CARBON ALLOY STEELS	STEELS: STAINLESS STEELS TOUGH CARBON TOUGH ALLOY	CAST IRON: ALLOYED CHILLED	CAST IRON: SOFT GRAY CLOSE GRAIN	NON-FERROUS: ALUMINUM SOFT BRASS COPPER	NON-FERROUS: ALUMINUM ALLOY HARD BRASS BRONZE	TITANIUM & OTHER SULFUR & CHLORINE SENSITIVE ALLOYS
Broaching	HSS	20%	20%	20%	20%		10%	10%	5%
	CARBIDE	20%	20%	20%	20%		10%	10%	5%
Threading, Pipe Reaming	HSS	20%	20%	20%	20%		10%	10%	
	CARBIDE	20%	20%	20%	20%		10%	10%	
Threading, Tapping	HSS	25%	25%	25%	25%		10%	10%	
	CARBIDE	25%	25%	25%	25%		10%	10%	
Gears: Hob, Cut, Shape, Shave	HSS						10%	10%	
Drilling, Deep Hole	HSS	10%	10%	15%	15%		10%	10%	
	CARBIDE	10%	10%	15%	15%		10%	10%	
Milling	HSS	5%	5%	10%	10%		5%	5%	
	CARBIDE	5%	5%	10%	10%		5%	5%	
Drilling	HSS	5%	5%	10%			5%	5%	
	CARBIDE	5%	5%	10%			5%	5%	
Turret Lathes, Automatic Lathes, Single and Multiple Spindle Automatics, Drill Forming	HSS	5%	5%	10%	10%		5%	5%	
	CARBIDE	5%	5%	10%	10%		5%	5%	
Turning, Forming Boring Lathes	HSS	5%	5%	10%	10%		5%	5%	
	CARBIDE	5%	5%	10%	10%		5%	5%	
Sawing, Circular & Hack	HSS	10%	10%	10%	10%		10%	10%	
	CARBIDE	10%	10%	10%	10%		10%	10%	
Grinding, Plain & Surface							5%	5%	
							5%	5%	
Grinding, Centerless Thread							10%	10%	
							10%	10%	
Form Rolling		10%					10%	10%	
Stamping		10%					15%	15%	
Drawing		20%					20%	20%	

The above percentage concentrations of WYNN'S SUPERKON 604 should be added to mineral oil of proper viscosity; generally 100 SUS @ 100° F pale oil.

NOTE: These recommendations are necessarily general. Factors of speed, feeds, finish requirements, and tool materials used will be considered by your WYNN'S representative in making custom recommendations.

For SPECIFIC RECOMMENDATIONS, see your Wynn Oil Company Representative.

SECTION 3

CUTTING FLUID SYSTEMS

Blaser Swisslube Incorporated
1 Holland Avenue
White Plains, NY 10603
(914) 997-6931

Jetmix

Art. Nr. 9261

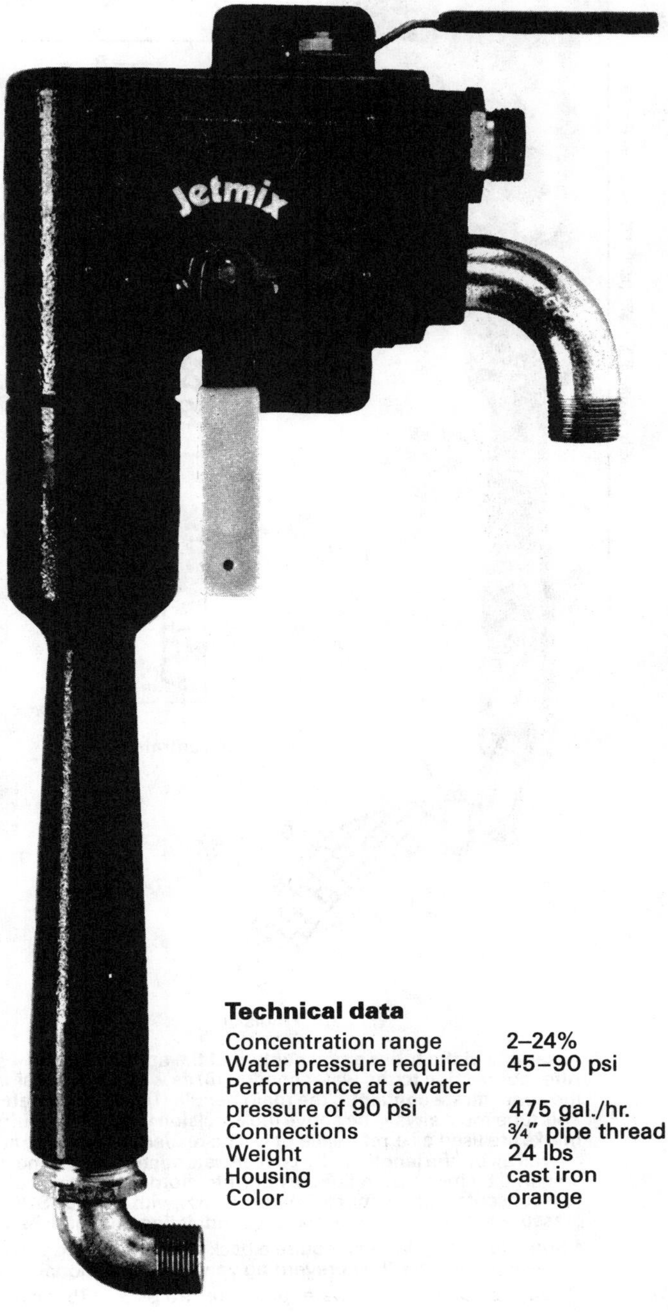

Mixing apparatus for water soluble metal working liquids

The use of our mixing apparatus offers you considerable advantages such as:

- **Saving of time**
  on mixing and distribution of the emulsion. You will always dispose of the necessary quantity within a short time
- **Easy and correct mixing**
  providing the best possible emulsion quality – therefore better performance
- **Central control**
  of the metal working liquid means rationalization
- **Guarantee**
  for an always correct concentration, therefore lower corrosion risks
- **No loss of concentrate**
  since the effectively used quantity of concentrate will be mixed only

The Jetmix mixing apparatus is
easy to handle – quick – reliable – economical!

Product description

Compact and solid design, no moving parts in the mixing zone. New mixing system; the effective mixing does not take place by simple confluence of water and emulsion-concentrate, but by shock waves (cavitation). This procedure guarantees absolutely homogeneous emulsions. Water pressure is the only energy required.

The supply valves for concentrate and water are operated by one single lever. The concentrate is delivered to the mixing zone by means of the venturi principle.

A valve in the concentrate suction pipe prevents a backflow of the concentrate. The mixing ratio can be adjusted by a removable handle. The repeatability of the concentration is extremely good, provided the water pressure is between 45–90 psi.

Technical data

Concentration range	2–24%
Water pressure required	45–90 psi
Performance at a water pressure of 90 psi	475 gal./hr.
Connections	¾″ pipe thread
Weight	24 lbs
Housing	cast iron
Color	orange

Blaser Swisslube Incorporated
1 Holland Avenue
White Plains, NY 10603
(914) 997-6931

Assembly- and service-manual

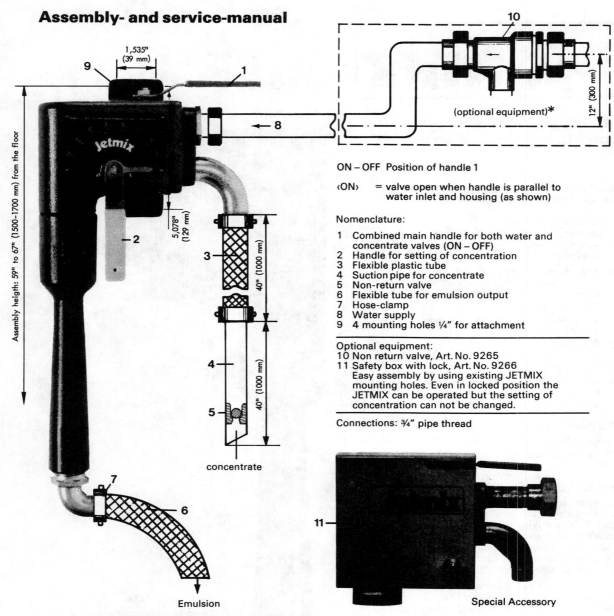

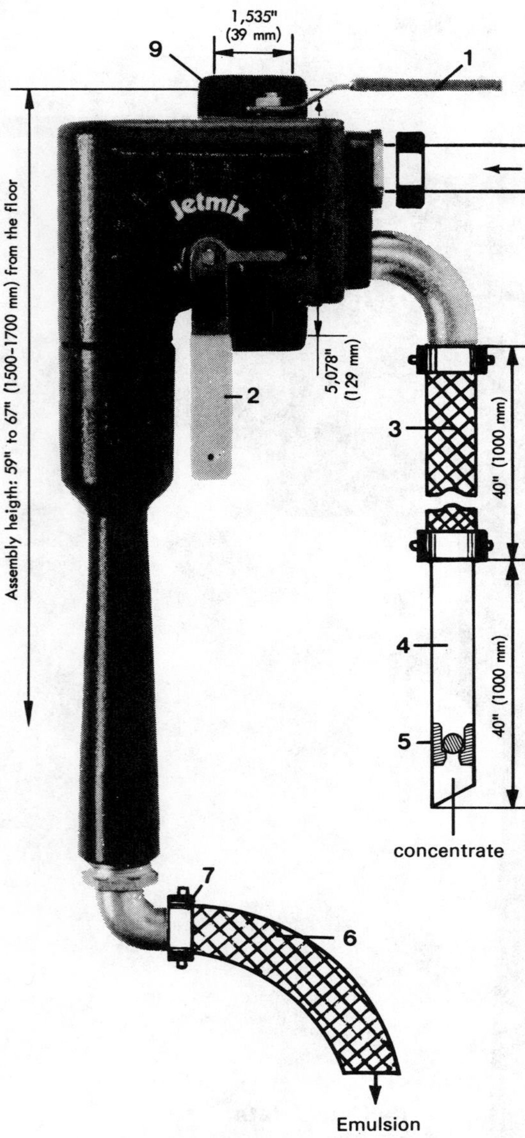

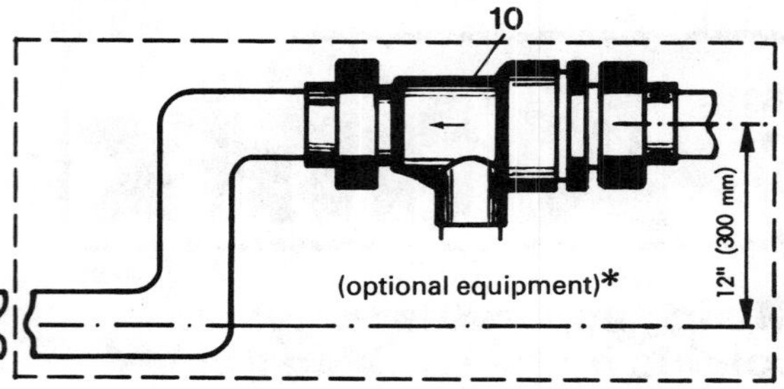

ON – OFF Position of handle 1

‹ON› = valve open when handle is parallel to water inlet and housing (as shown)

Nomenclature:

1 Combined main handle for both water and concentrate valves (ON – OFF)
2 Handle for setting of concentration
3 Flexible plastic tube
4 Suction pipe for concentrate
5 Non-return valve
6 Flexible tube for emulsion output
7 Hose-clamp
8 Water supply
9 4 mounting holes ¼" for attachment

Optional equipment:
10 Non return valve, Art. No. 9265
11 Safety box with lock, Art. No. 9266
Easy assembly by using existing JETMIX mounting holes. Even in locked position the JETMIX can be operated but the setting of concentration can not be changed.

Connections: ¾" pipe thread

11

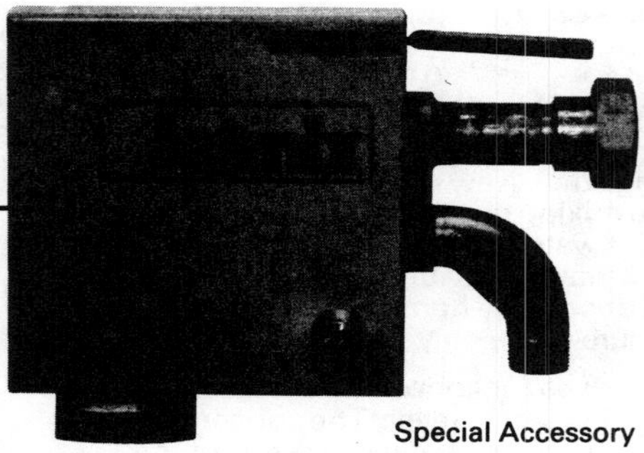

Special Accessory

Assembly of the tubes and mounting of the apparatus with 4 screws according to above picture.
After connection to the water conduit (8) as well as immersion of the suction pipe for concentrate (4) into the concentrate container, the main handle (1) can be operated. Emulsion output is from the plastic tube (6). This tube must always be above the emulsion level. The required concentration is set by handle (2). The scale marks are used as a reference only. The respective concentration depends on the viscosity of the concentrate as well as on the length of the concentrate suction pipe. The Jetmix will therefore have to be adjusted for the first time by means of a fluid tester (refractometer). Scale position III at 75 psi water pressure corresponds to a concentration of approximately 5–6% with our BLASOCUT 2000 Universal concentrate. In order to avoid pressure shock waves in the water conduit line, the main handle (1) always has to be closed slowly.

* Some local regulations require a back flow preventing valve if the water conduit is not protected yet. We deliver the back flow preventing valve (10) as optional equipment.

Throughput depends on water pressure: 90 psi = 475 gal./hr.; 45 psi = 380 gal./hr.

Blaser Swisslube Incorporated
1 Holland Avenue
White Plains, NY 10603
(914) 997-6931

Jetmix

Drum Mounting Assembly

The JETMIX drum mounting assembly facilitates the supply of coolant to individual machines. It allows the drum of coolant concentrate, with the necessary length of high pressure water hose, to be brought directly to the machine to be filled. The drum mounting assembly fits every JETMIX already delivered.

Advantages:

- Time savings; the correct amount of concentrate at the desired place, fast and easy mixing and distribution of coolant.
- Cost savings; no spillage of concentrate, no spillage of emulsion, the correct mixture, no guesswork, less machine down time.
- Maintenance free operation:
  rugged design, fast and easy to change over from one drum to the next.

No. 9264

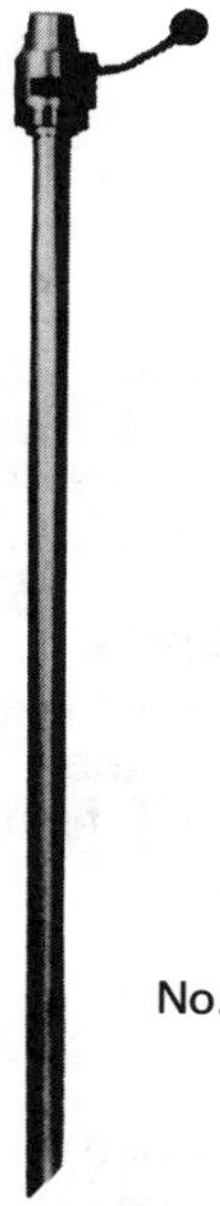

No. 9267

Order No. 9267

Drum mounting pipe and eccentric clamp for existing JETMIX.

Order No. 9264

JETMIX with drum mounting assembly complete

Blaser Swisslube Incorporated
1 Holland Avenue
White Plains, NY 10603
(914) 997-6931

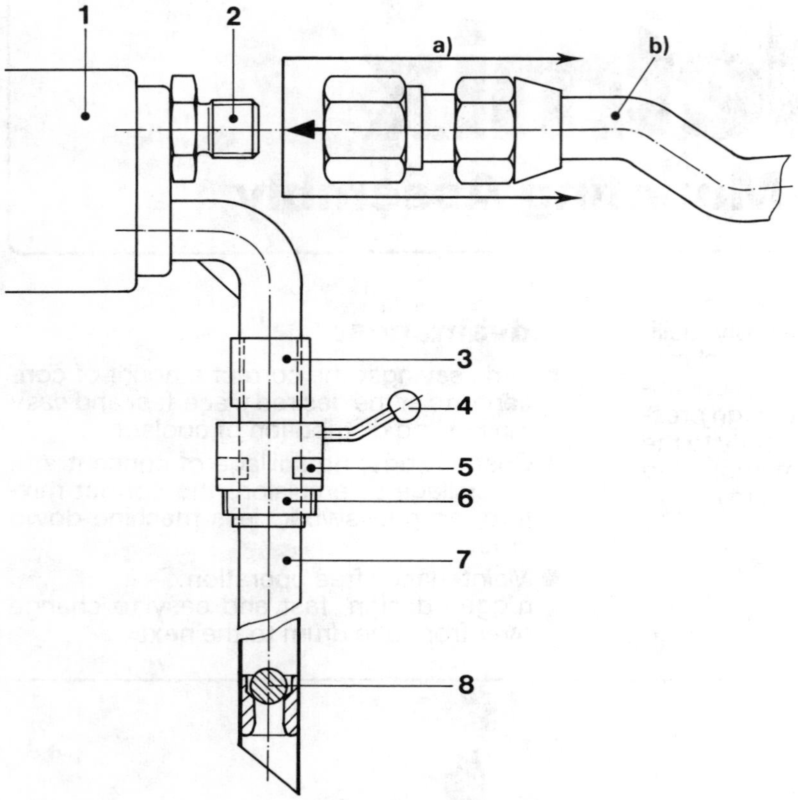

NOMENCLATURE:

1. JETMIX – Mixing Apparatus
2. Water connection: 3/4" pipe thread
3. Pipe connection: JETMIX – Suction Pipe
4. Eccentric Lever Assembly: clamping of JETMIX suction pipe
5. Two parallel flats for open-end wrench (55 mm = 2.165") to tighten clamping assembly #4 into drum bund hole
6. 2" thread for drum bung hole mounting
7. Concentrate suction pipe
8. Check valve inside suction pipe to prevent siphoning of water or coolant back into the drum.

Mounting Instructions:

1. When fitting the drum mounting to the JETMIX, the connection pipe #3 must be sealed with teflon tape. 2. Then the clamping assembly (part Nos. 4,5,6) is threaded into the bung hole of the drum. 3. The JETMIX with mounted suction pipe (#3, #7) is inserted into the clamping assembly until the pipe reaches the bottom of the drum. 4. Clamp the pipe with JETMIX firmly by turning the eccentric lever (#4).

Water Connection, a:

Connect high pressure water hose to JETMIX 3/4" coupling (part #2) with swivel type or quick release connection. High pressure hose and connector is not supplied with JETMIX. IMPORTANT: Make sure that no water leaks on top or inside the drum. When JETMIX is not in use, disconnect the hose and release water pressure. Minimum dynamic pressure required is 45 psi.

Water Hose, b:

Use high pressure water hose with I.D. 20.6 mm (.8" dia.). We strongly recommend a 0–100 psi pressure gage to be hooked up between JETMIX and hose connection to allow monitoring of flow pressure. Do not use JETMIX if dynamic pressure drops below 45 psi (insufficient agitation).

DoALL Company
254 North Laurel Avenue
Des Plaines, IL 60016-4398
(312) 824-1122

NEW! NEW! NEW!

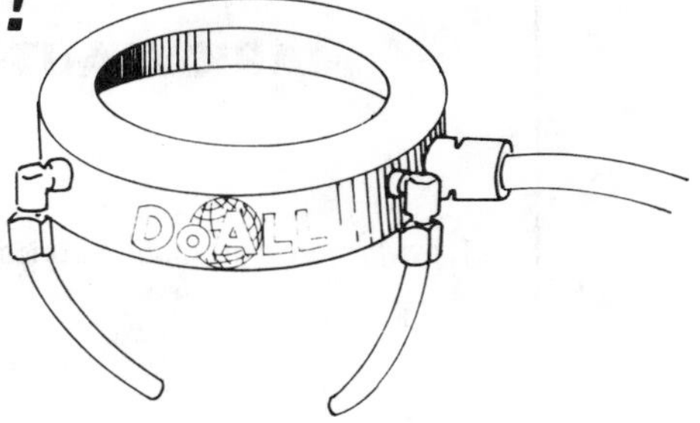

DoALL's CON-TROL-FLO
Coolant Applicator

The exclusive DoALL CON-TROL-FLO Coolant Applicator puts the cutting fluid stream where it's needed on Series 1 vertical milling machines. It fits around the spindle and mounts on the quill allowing the fluid nozzles to follow the tool in both the horizontal and vertical planes. There is no need for the constant nozzle adjustment found with other applicators. This is a flood unit not a misting unit (fluid and air mixture).

Can be installed in a few minutes and connected to the mill's existing coolant system (Model L-19, Cat. No. 789-140019) or is available with its own cutting fluid pump unit (Model L-5, Cat. No. 789-140118) for those mills lacking a coolant system. Fittings and tubing supplied in both kits. Model L-5 includes shutoff valve.

Coolant flow is constant to the exact point desired. Additional nozzles can be installed by the user.

Assures longer tool life, better finish, reduced operator time, and more continuous work flow.

Designed to fit Series 1 DoALL, Bridgeport, Moog, and other mills having spindles and quills that will accept coolant manifold with 1.9 in. I.D. and 3.2 in. O.D. Mounting screw circle 2.9 in.

Call your nearest DoALL Industrial Supply Center.

DoALL

For the whole shop and everything to keep it running.

968-105247 1/83

Printed in U.S.A.

Dri-Slide, Incorporated
411 North Darling
Fremont, MI 49412
(616) 924-3950

MARK VI AUTOMATIC LUBRICATOR APPLICATION SHEET

Mark VI Automatic Lubricators are used by nearly every one of the conveyor users on the Fortune 500 lists.

"I" Beam or enclosed track systems, overhead or underground systems -- all of these types are used and lubricated automatically in General Motors, Ford, Westinghouse, General Electric, Brillion Iron Works, SAAB, etc.

The Mark VI is very adaptable. Automatic chain lubrication in electro-plating operations where one solenoid was at ground level and the second solenoid was twenty-five feet higher was successfully accomplished, even though the control box and drum were twenty feet away, with no problem.

The Mark VI Automatic Lubricator will sense chain stop condition and shut itself down if that condition exists for more than one minute. Upon resumption of chain travel, our lubricator will turn itself back on and continue to lubricate. If a counter model of the Mark VI is used, the lubrication cycle of the conveyor can be counted and Dri-Slide applied only when needed. This feature counteracts today's driving tendency to over-lubricate. A "normal" situation with a chain and trolley wheels in bad condition might be: set cycle counter to inject Dri-Slide on the chain pins and trolley bearings on every rotation of the chain for two weeks. After visual inspection to insure that the pins and bearing are free, set cycle counter to lubricate once every 4th or 5th chain cycle. Again inspect the chain and bearings. Chain pins should be free and no evidence of trolley bearing binding or dragging of trolley wheels should be noted. At this point, in consultation with maintenance people at plant, set counter to function where desired.

ENERGY NEWSLETTER — Consumers Power

Published by Energy Consulting Services
Consumers Power Company, Room 585
212 W. Michigan Ave., Jackson, Mi 49201

ENERGY NEWS BRIEFS

• Dri-Slide, Inc. announces Mark VI Automatic Power Conveyor Lubricator. Conveyor system energy savings of 10 to 70 percent have been documented with the use of this new system. The Mark VI Automatic Lubricator replaces conventional units usually equipped with a fluid tank tied to air pressure to spray lubricating fluid on conveying equipment. The Dri-Slide lubricant, MoS_2 (molybdenum disulfide), dry film also reduces maintenance and downtime. For more information, contact Dri-Slide, Inc., Fremont, Michigan 49412.

• The US Department of Commerce/National Bureau of Standards in cooperation with the Federal Energy Administration/Conservation and Environment has recently published NBS Handbook 115, "Energy Conservation Program Guide for Industry and Commerce." This is a comprehensive guide to setting up an energy conservation program in commercial and industrial operations. A copy can be purchased for $2.50 from: Superintendent of Documents
US Government Printing Office
Washington, DC 20402
SD Cat No C13.11:115

• The 1975 National Electrical Code, NFPA No. 70-1975 has been published and is presently available from the National Fire Protection Association, 470 Atlantic Avenue, Boston, Massachusetts 02210.

Dri-Slide, Incorporated
411 North Darling
Fremont, MI 49412
(616) 924-3950

MARK VI AUTOMATIC LUBRICATOR

FEATURES:	**BENEFITS:**
Can be used with 100 Lb. or 55 Gal. Drums	Requires less operator maintenance
Fully Automatic	Requires no hands-on operator
110 V. Power	Easily Powered
Extreme Engineering Flexibility	Can be fitted to very unusual circumstances
Drum Empty Light	Signals maintenance
Counter or Timer	Will deliver Dri-Slide on a counted rotation of assembly line or will lubricate periodically on continuous line basis.
Solid State Electronics	Totally reliable

	Code No.	Shipping Weight
Mark VI	03120	87 Lbs.

DRI-SLIDE, INC. • 411 North Darling • Fremont, Michigan 49412
Watts: 800-647-1206 • In Michigan 616-924-3950

Dri-Slide, Incorporated
411 North Darling
Fremont, MI 49412
(616) 924-3950

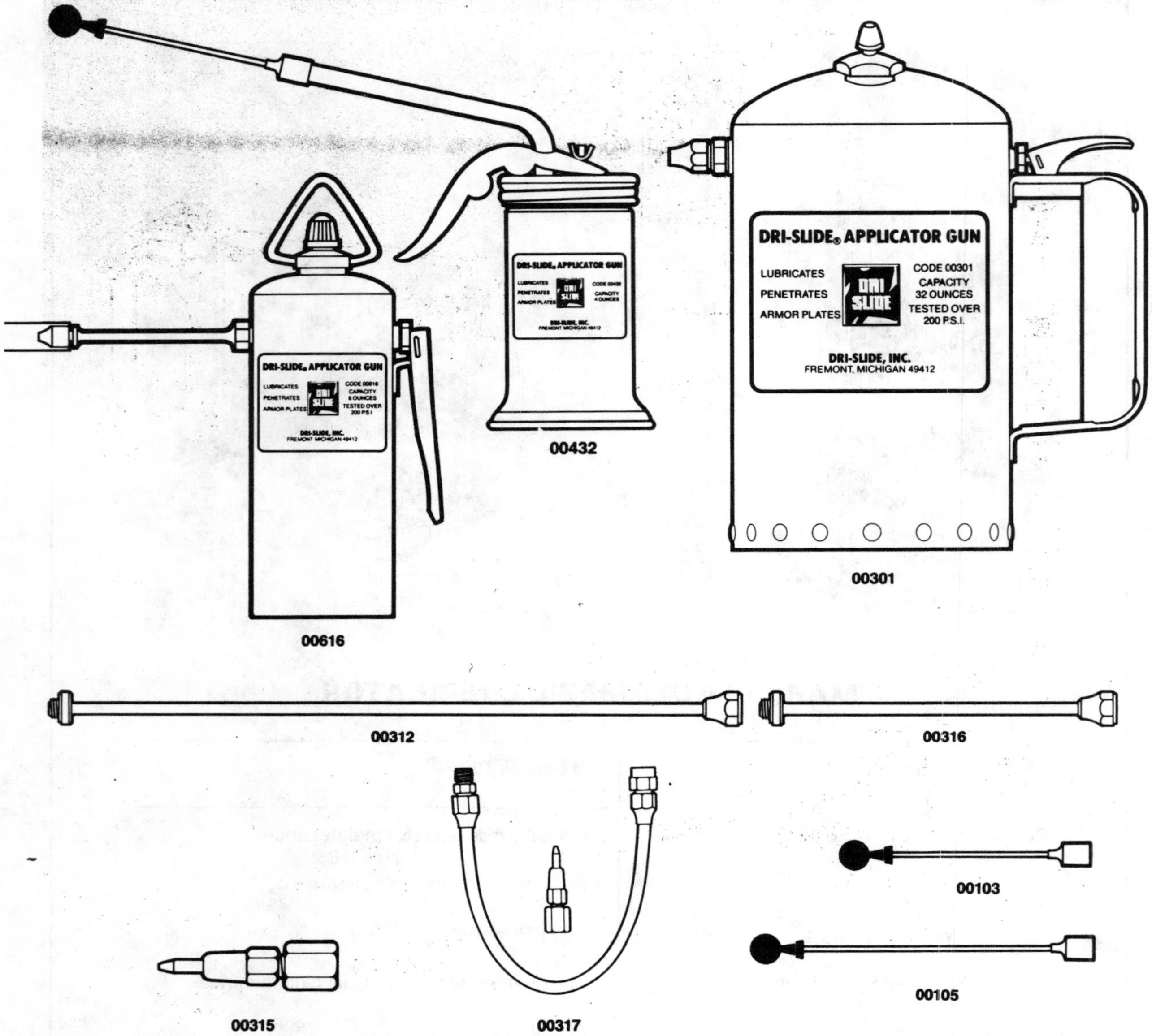

Hand and Air Sprayers
00432 4 oz. Hand Sprayer with 3½″ needle
00616 6 oz. Air Sprayer with jet & zerk
00301 32 oz. Air Sprayer

Convenient refillable sprayers. Enamel painted surface means long life. Needle/nozzle for hard to reach places. Air Sprayers can be pressurized from shop air.

Sprayer Accessories
00312 12″ chrome extension
00315 Grease fitting nozzle
00316 6″ chrome extension
00317 Revitalizer Kit

Adapt your sprayer with an extension for those places that your arm won't reach. Grease fitting nozzle easily allows you to apply liquids to grease fittings.

Revitalizer Kit designed to revitalize old grease by applying liquid through grease fitting nozzle.

Needle Applicators
00103 3½″ stainless steel needle
00105 5½″ stainless steel needle

Fits any ¼″ — 28 thread or any plastic spout such as on Dri-Slide 4 oz. and 8 oz. cans. Protective Tip.

Dri-Slide, Incorporated
411 North Darling
Fremont, MI 49412
(616) 924-3950

00533

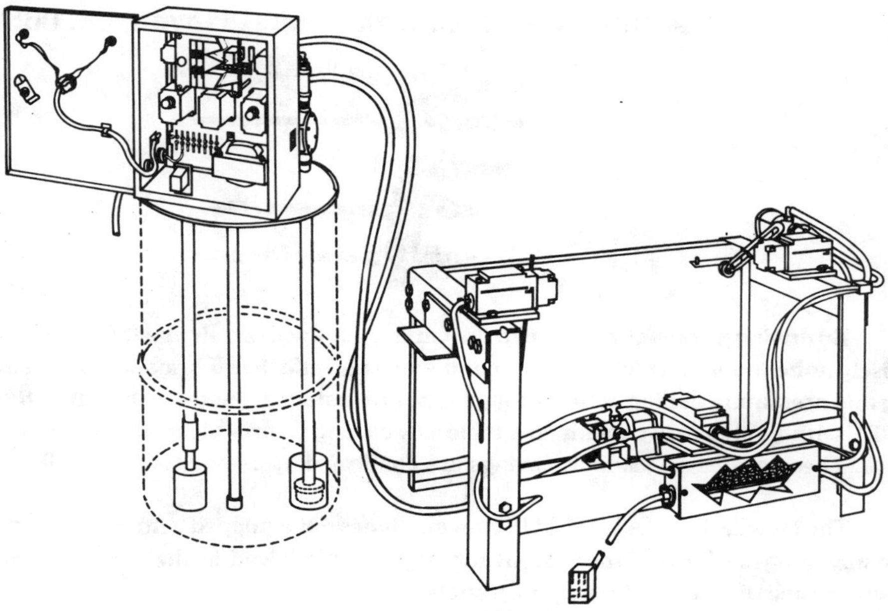

03120

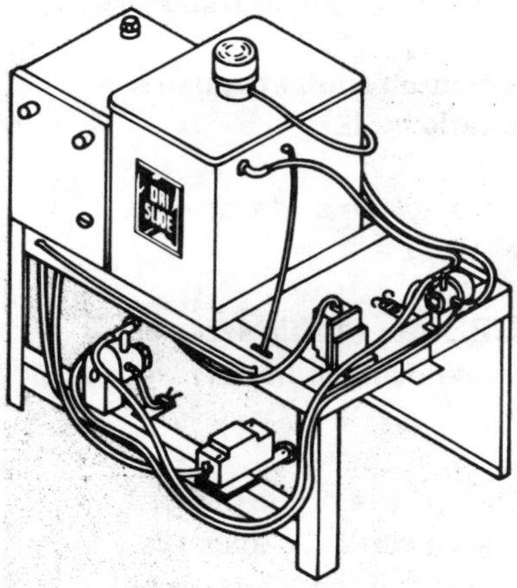

03150

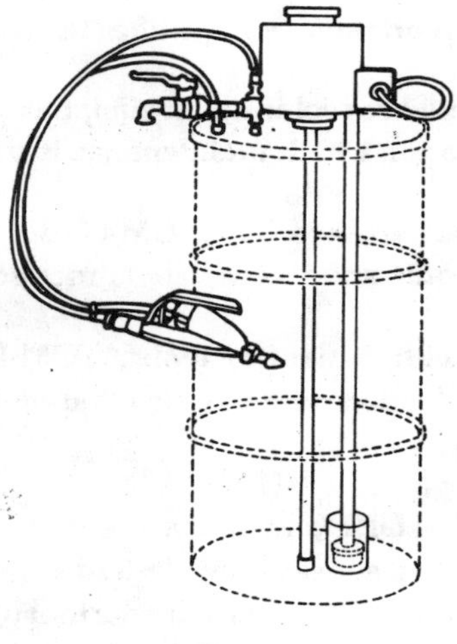

03140

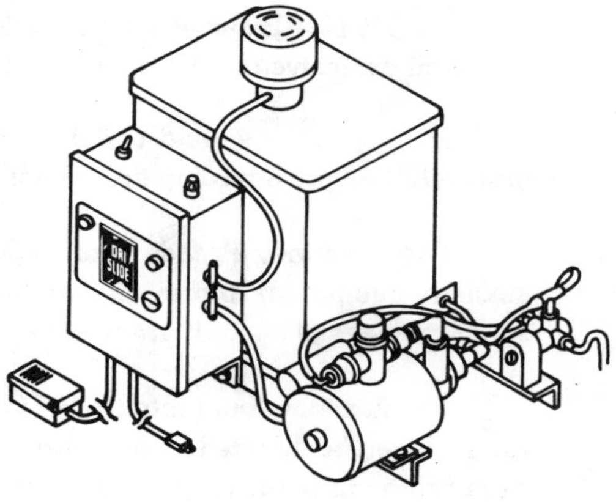

03170

Shirt Pocket Applicator
00533 Empty Shirt Pocket Applicator

Handy and easy to use applicator clips to shirt pocket. Fill with any liquid. Has 2″ needle/nozzle to reach problem areas. Apply a drop or a squirt. Protective Tip.

Automatic Lubricators
03120 Standard Unit

Automatically provides the right amount of lubrication at the right time and place. Can lubricate according to revolutions or time. Will shut down and start up when necessary.

03150 Mini-Mark

Scaled down model of standard unit, Mini-Mark has 2 gallon reservoir for lubricating short conveyor line. Unit turned on & off by operator.

03140 E-4 Agitator with draw off valve and hand gun

Fits standard 16 gal. drum and mixes with recirculating action of pump. Hose has 20 ft. length. Standard hand gun. Pump uses 110 volt. Ideal for lubricating chain, cables, dies or anything that needs to be sprayed.

03170 Parts Assembly Lubricator

Foot switch allows operator to apply lubricant on line during small parts assembly. Adjustable volume.

Garsons, Incorporated
P.O. Box 6
East Haven, CT 06512-0006
(203) 469-0723

ATOM LUBE NON-CIRCULATING MIST COOLANT SYSTEMS

Garsons, Inc., P.O. Box 6, East Haven, Ct. 06512

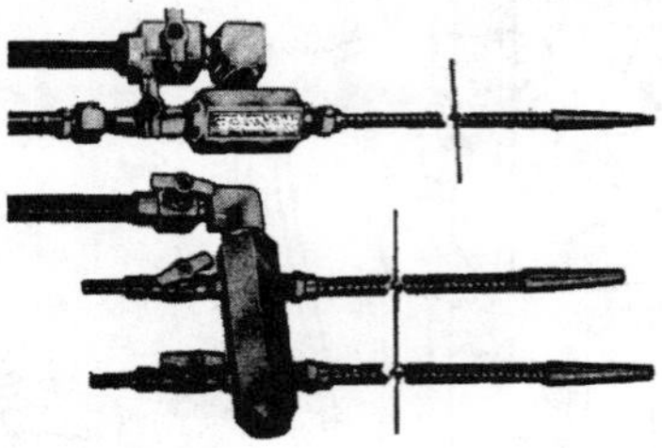

By driving a coolant vapor on to the cutting edge with an air blast, ATOM-LUBE produces a closely adhering film which provides continuous and effective lubrication, thus eliminating burning and premature dulling of the cutting tools. For instance, our tests show that the use of ATOM-LUBE with a soluble oil coolant on a table saw cutting Cupro-Nickel tubes increased blade life by 30%. These results and similar savings are confirmed by users of ATOM-LUBE

The installation of ATOM-LUBE on machines not equipped with a cooling system allows all the advantages of a coolant without the high cost involved in the purchase, installation, and maintenance of a conventional pump coolant system.

For machines which cannot be modified for a pump coolant system ATOM-LUBE is alone the only effective means of cooling and lubricating and justifies its installation without question.

ATOM-LUBE can also be considered as replacement equipment for pump coolant systems on which extensive repairs are needed or on which maintenance is regularly high.

Drip systems should definitely be replaced by ATOM-LUBE. Drip systems are usually make-shift affairs affording only spasmodic and incomplete lubrication.

For machines already equipped with coolant systems ATOM-LUBE can readily serve as auxiliary equipment to prevent lost production time in case of pump breakdowns or in case frequent changes of coolant are necessary.

The flexibility built into the ATOM-LUBE unit allows the most effective use of the coolant vapor. It can be directed against the cutting edges at any desired angle, even vertically upwards. At the same time the installation does not interfere with the normal operation of the machine or with the comfort of the operator. This flexibility is missing from either pump or drip coolant systems.

ATOM-LUBE handles a wide variety of liquids, from pure water, water and soluble oil mixtures and machine oils. Rapid change-over of coolants is possible since ATOM-LUBE is self-cleaning and the coolant feed line is simply moved from one container to another on the floor. Note: The coolant container must be elevated if air pressure used is below 50 PSI.

A simple needle valve allows any desired concentration of the coolant in the air blast. This setting remains constant and is not lost when the ATOM-LUBE unit is shut off.

Air blast blow chips away from area of cut allowing operator full vision and increasing accuracy of the work. This is particularly helpful in the dry cutting of cast iron.

Because of its effectiveness and flexibility ATOM-LUBE uses a minimum amount of coolant, resulting in economy of operation, cleaner machines, and improved working conditions.

Garsons, Incorporated
P.O. Box 6
East Haven, CT 06512-0006
(203) 469-0723

ATOM LUBE NON-CIRCULATING MIST COOLA[illegible]

Garsons, Inc., P.O. Box 6, East Haven, CT [illegible]

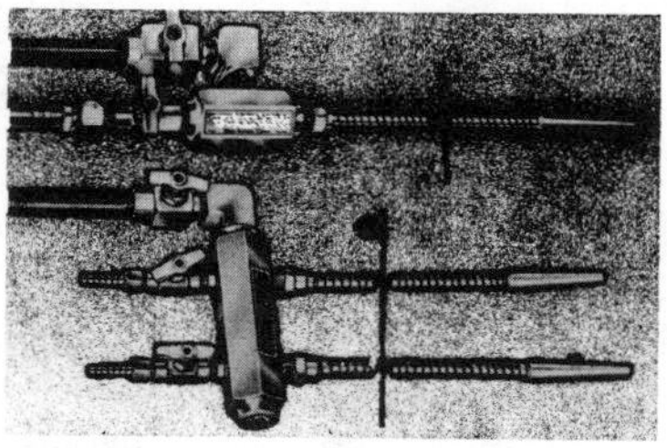

Over 31 years of dependable service to the aluminum, plastic and metal cutting industries.

Increases life of tools, grinding wheels, drills, abrasive belts, cutters, tool sharpeners.

Improves cutting quality and finish.

Dissipates heat to keep cutting tools cool.

Easily installed and maintained on any machine including those with complicated systems or those that formerly cut dry.

Any liquid from plain water to heavy machine oils can be used.

Coolant supplied from any portable container or coolant line.

Nozzles easily adjustable from any angle.

Required Pressure 50 to 60 psi. Average comsumption of air is 6 to 8 cubic feet per minute in normal operation. Equipped with 12″ tubes, (special lengths available) a mounting bracket, a ¼″ SPS nipple for air connection, and four feet of ¼″ ID flexible, clear plastic hose for the coolant supply.

Grinding

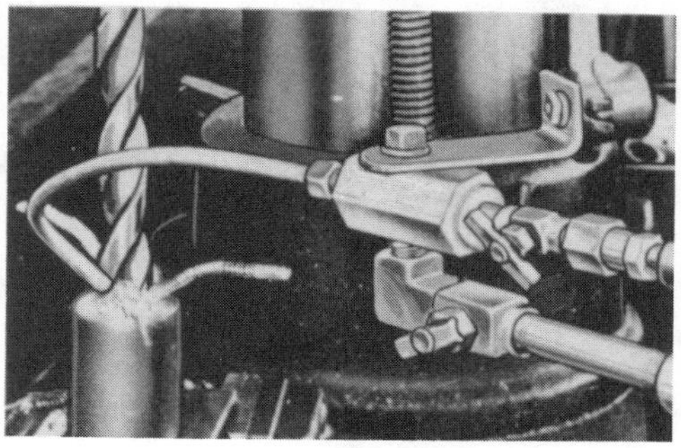

Drilling

Milling

COMPLETE UNIT (Packed one unit per box)

Stock No.	Description	Net Each
114	Single 1/4″ nozzle	$77.00
118	Single 1/8″ nozzle	77.00
214	Double 1/4″ nozzle	103.00
218	Double 1/8″ nozzle	103.00

REPLACEMENT SETS (Packed two sets per box)

Consisting of 1 outer flexible tube and nozzle and 1 inner tube.

Stock No.	Description	Net Each
R 14	Single 1/4″ nozzle	$11.50
R 18	Single 1/8″ nozzle	11.50
R 14D	Double 1/4″ nozzle	11.50
R 18D	Double 1/8″ nozzle	11.50

Madison-Kipp Products Corporation
201 Waubesa Street/P.O. Box 3037
Madison, WI 53704
(608) 244-3511

Madison-Kipp Lubricators...

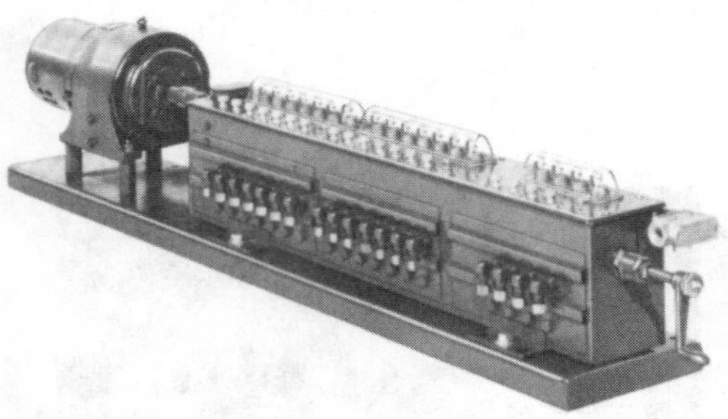

MODEL 50 LUBRICATOR

The Model 50 Lubricator is ideal for a multitude of applications including diesel, gas or steam engines, steam hammers, air compressors, cranes, presses, shears, machine tools, wood working equipment—wherever a closely controlled volume of oil is required. Each pump will deliver up to .023 in.^{3}per impulse against up to 200 PSI back pressure. From one to twenty-five impulses per minute attainable depending on drive speed and internal ratio. For more details request bulletin L-3401.

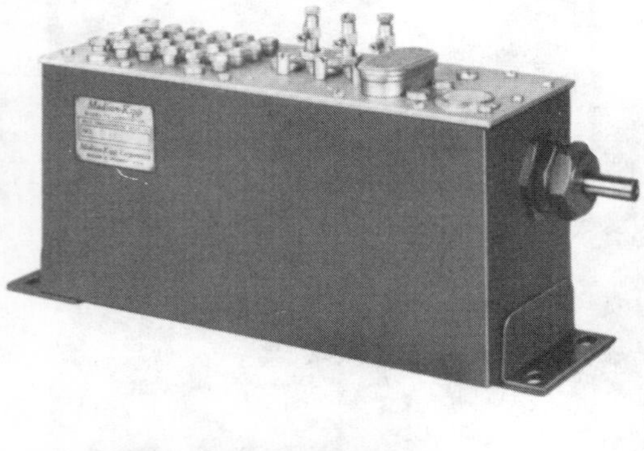

MODEL FD LUBRICATOR

The Model FD was designed for applications requiring less oil per lubrication point. Pump output is the same as for the Model 50 (.023 in^3) but it is delivered sequentially to eight outlet ports by means of an internal indexing distributor assembly. This lubricator has been used extensively on shears, brake presses, screw machines, diesel engines, labeling machinery and many other applications. For more detailed information request bulletin L-3611.

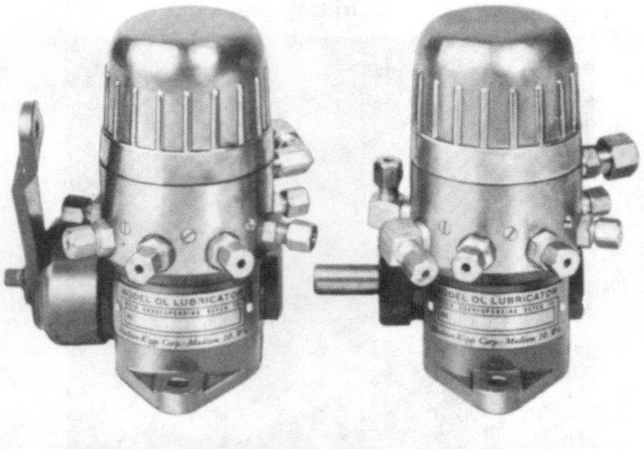

MODEL OL LUBRICATOR

The OL Lubricator was designed to fill the need for the smallest possible lubricator. Up to eight plungers in one pump body each deliver up to .004 in.^{3} per impulse at up to 150 PSI., and up to 27 impulses per minute. Requires separate reservoir. Applications include injection molding machines, screw machines, grinders, woodworking machinery, machine tools and many others. For more information request bulletin L-3361.

Madison-Kipp Products Corporation
201 Waubesa Street/P.O. Box 3037
Madison, WI 53704
(608) 244-3511

a size and type for every need

MODEL DSL LUBRICATOR

The DSL Lubricator was designed primarily for air and process gas compressor lubrication but is not limited to these applications. The pump features a vacuum type dry sight feed, and delivers up to .008 in^3 per impulse against up to 3000 PSI back pressure. Features weatherproof construction, and is suitable for either petroleum or synthetic lubricants. For additional information request bulletin L-6000.

MODEL SVH LUBRICATOR

The Model SVH Lubricator was also designed primarily, but not exclusively, for compressor applications. Three different liquid filled sight feeds are available for 250, 1400 or 5000 PSI service. Each pump delivers .015 in^3 maximum per impulse. Available in watertight construction. Ideal for very heavy oil applications; up to 5000 SSU. can be used with synthetic lubricants with appropriate sight feed fluid. For more details request bulletin L-3502.

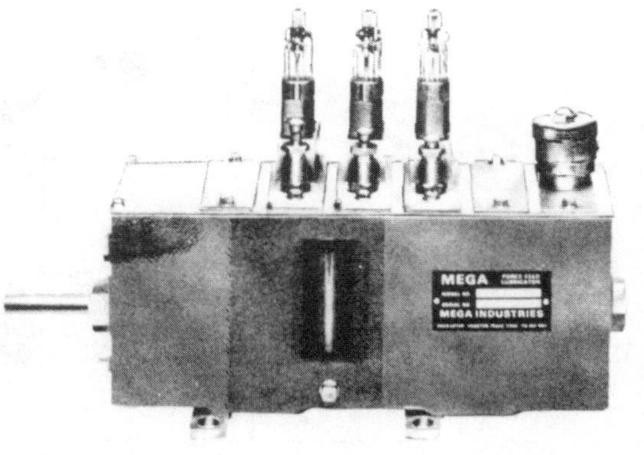

MODEL M55 LUBRICATOR

The "MEGA" M55 Lubricator features a vacuum type dry sight feed pump with two plunger sizes available; 1/4" for up to .018 in^3 per impulse at 6000 PSI maximum pressure, and 3/8" for up to .041 in^3 per impulse at 3500 PSI. Wide range of drive ratios available. Complete lubricators are physically interchangeable with McCord and Lincoln units, and M55 pumping elements will interchange in McCord or Lincoln reservoirs. Write or call for more details.

Madison-Kipp Products Corporation
201 Waubesa Street/P.O. Box 3037
Madison, WI 53704
(608) 244-3511

Conveyor maintenance products—for every conveyor style, includes lubricators for automatic grease or oil delivery, automatic oilers and motor driven conveyor cleaners.

Extend conveyor life—automatic conveyor maintenance will extend the life of your conveyors. It will also eliminate conveyor surging and will significantly reduce drive motor strain and power consumption.

Economical—due to the precise method used to deliver lubricants to the wear points of your conveyors, very small amounts are needed to keep them properly lubricated. The result is a clean, economical operation.

Conveyor industry standard—ask any of the worlds leading conveyor manufacturers. Opco's product superiority coupled with a reputation for cooperation and service is your guarantee of satisfaction.

Simple installation—units can easily be installed by your conveyor supplier or your own factory maintenance personnel.

Automatic lubrication—units come equipped with a timer for selected cycle control or with an optional counter control module (CCM) when more precise lubrication cycles are required.

Chain Pin Oiler:

Model OP-4—Automatic chain pin oiler for rivetless-type chain on overhead monorail conveyors. It is air operated and electrically controlled for precise delivery of oil directly to the pin. The result is an efficient, dripless, economical operation.

Combination Oiler:

Model OP-47—Automatic chain pin and open race trolley wheel oiler with two pumps activated by non-contacting proximity switches. Each pump dispenses .0045 cubic inches of oil per nozzle, per trip. It is air operated and electrically controlled for precise delivery of oil directly to the chain pins and trolley wheel bearings.

Model OP-47E—All electric version of the OP-47 (no air required). It has a single pump with adjustable oil volume and velocity control.

Free Trolley Oiler:

Model OP-9—Automatic oiler for open race free trolley wheels and guide rollers of power and free conveyor system. OP-9 is also used to oil free trolley pivot points.

Model OP-9E—Totally electric version of the OP-9 (no air required).

Floor Conveyor Oilers:

Models OP-10, OP-10E, OP-43—Automatic oilers for a variety of in-floor conveyors to lubricate chain pins and rider plates, or chain only. The OP-43 is provided with a dual reservoir to allow delivery of different weight oils to chain pins and rider plates.

High Speed, Small Pitch Roller Chain Oiler

Model OP-34—Specifically designed for High Speed Roller Chain as used in the container industry with RC-60 and similar chains. You control amount of oil used and frequency of application. Oil is delivered directly to chain in solid drop form. No possibility of drippage results in properly lubricated chain.

Multiple Use Pneumatic Oiler

Model OP-139—Its modular concept allows it to be used for a variety of applications. The reservoir can be removed for remote mounting. The lower module is sealed for use in dusty and dirty areas. The nozzles and trip mechanism are mounted by the customer at any desired location. It is de-

Application	Roller Chain		
	Grease	Oil	w/Grease Fitting
Function	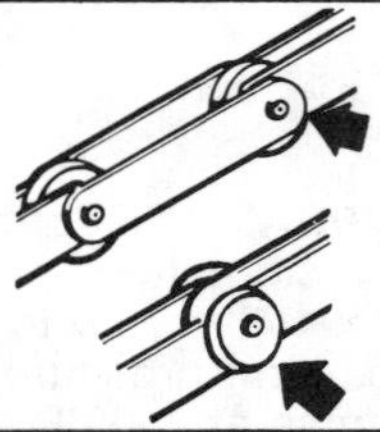	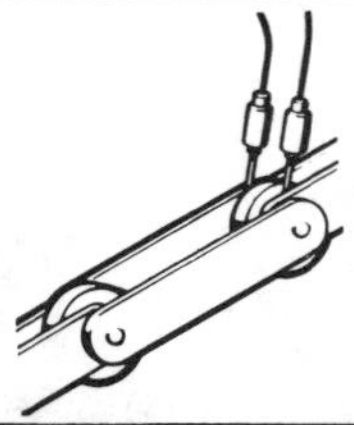	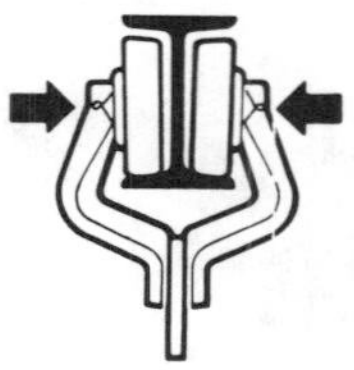
Type	OP-331 OP-311	OP-10, OP-139, OP-40	OP-201
Size			2″, 3″, 4″, 6″/Std. metrics
Grease	X		X
Oil	(X)	X	(X)
CCM	X	X	X
Electric Timer	X	X	X

Madison-Kipp Products Corporation
201 Waubesa Street/P.O. Box 3037
Madison, WI 53704
(608) 244-3511

signed to operate from plant air only, but can be supplied with timer control capabilities.

Trolley Wheel Lubricator

Model OP-201 –Automatically lubricates closed-bearing trolleys with precise volume of grease or oil while conveyor operates. Self-reset safety mechanism protects against damage from bent or broken trolleys. Unit has highest conveyor speed capability in the industry!

Heavy-Duty Lubricator For Flat Top Conveyors

Model OP-311—Automatic grease (or oil) application for various wheels used on flat top conveyors, car type conveyors and other heavy-duty applications.
Dispenses metered shot of grease or oil to mixed and unequally spaced trolleys while conveyor is in operation . . . without missing a single fitting.

Model OP-331—For use on inboard wheels or rollers requiring lubrication through fittings.

Free-Trolley Lubricator

Model OP-314 –Automatically lubricates (grease or oil) closed-bearing free trolleys while system is in operation. Injects metered volume of lubricant through fitting or lubrication hole—travels with free trolley for maximum on-fitting dwell time.

Free Trolley Guide Roller Lubricator

Model OP-317—Automatically lubricates guide rollers on free trolley assemblies utilizing lubricant fittings. Designed for power and free conveyor systems. Accurately dispenses a predetermined, measured shot of lubricant through the guide roller lube fittings.

All-Electric Oiler:

Model OP-40—This all-electric conveyor lubricator is designed for use where plant air is not available. Can be used on in-floor and other types of roller conveyors—as well as conveyor drive chains. Activated by a proximity switch, the OP-40 has no direct contact with the conveyor — eliminates wear problems.
It's compact and easy to install.

Counter Control Module (CCM)

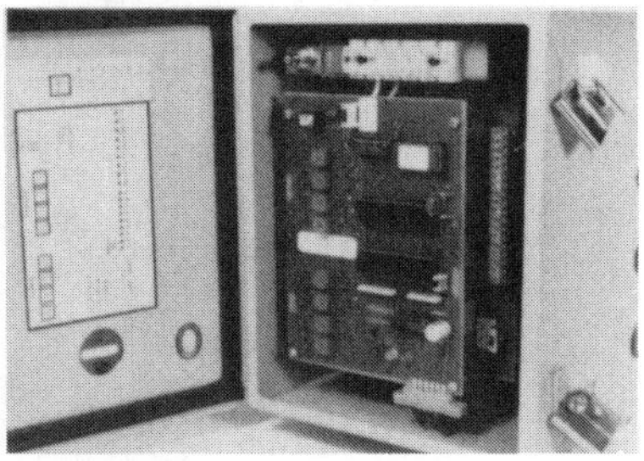

The counter control module is a microprocessor based system which controls the on-off cycle of Opco lubricators based on conveyor revolutions. Input to the counter is through a proximity switch. Electric units normally require a proximity switch and can be controlled directly by the CCM output.

Programmable Lubricator Controller (PLC)

A PLC is available to extend the capability of OPCO lubricators. It gives the unit the ability to lubricate high speed and variable speed conveyors. The PLC can be set up to provide reliable lubrication for one or more conveyors.

Power and Free Systems		
Guide Rollers	Free Trolleys	
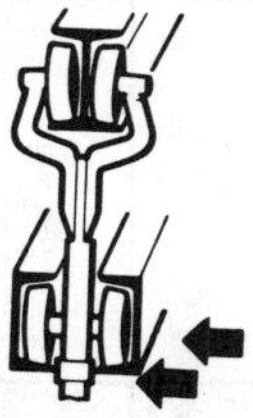	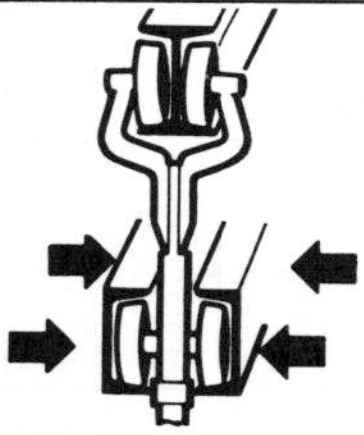	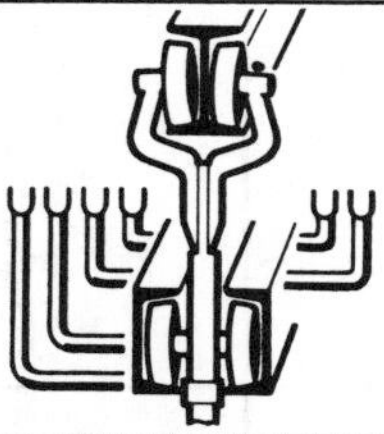
Grease (Oil) OP-317	Grease (Oil) OP-314	Oil OP-9, OP-9E
4″, 6″/Std. metrics	3″, 4″, 6″/Std. metrics	3″, 4″, 6″/Std. metrics
X	X	
(X)	(X)	X
X	X	X
X	X	X

Madison-Kipp Products Corporation
201 Waubesa Street/P.O. Box 3037
Madison, WI 53704
(608) 244-3511

Chain Pin Oiler—Model OP-4 & OP-4E

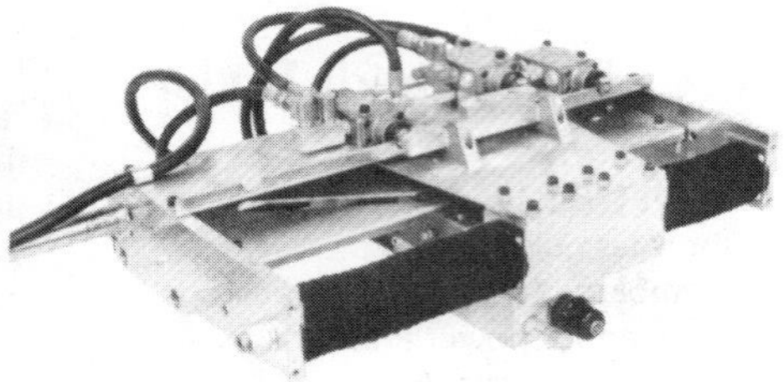
OP-311 Out-Board Wheel Lubricator

Combination Oiler—Model OP-47 & OP-47E

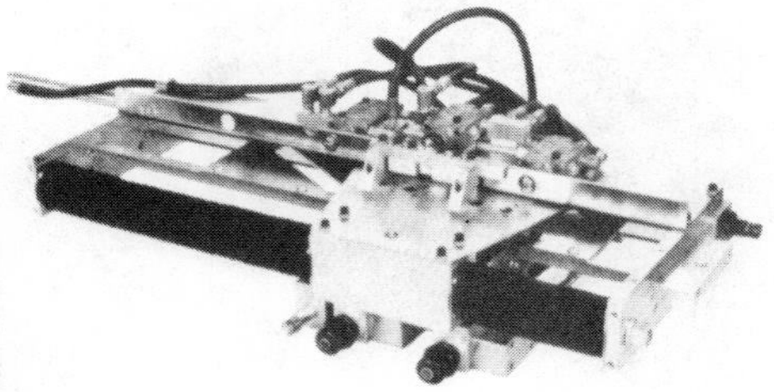
Free Trolley Wheel Lubricator—Model OP-314

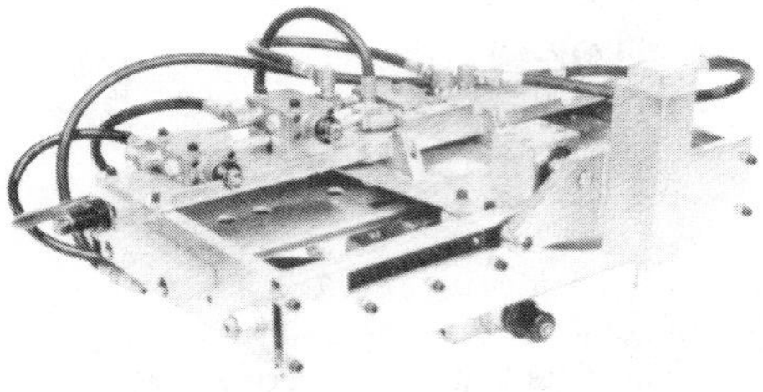
OP-331 In-Board Wheel Lubricator

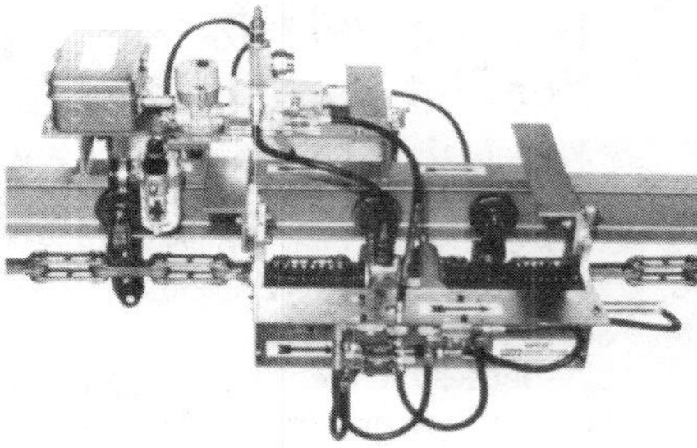
Trolley Wheel Lubricator—Model OP-201

Pneumatic Oiler—Model OP-139

Roller Chain Oiler—Model OP-34

All-Electric Oiler—Model OP-40

Overhead Monorail		In-Floor Systems	
Open Race Oil	Chain Pin Oil	With or without grease fittings	
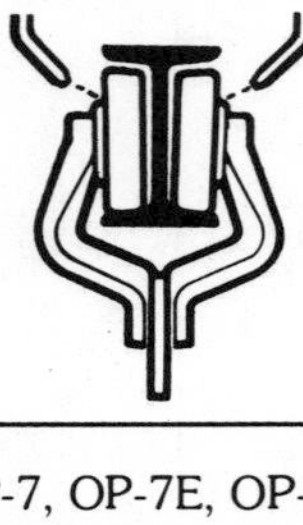	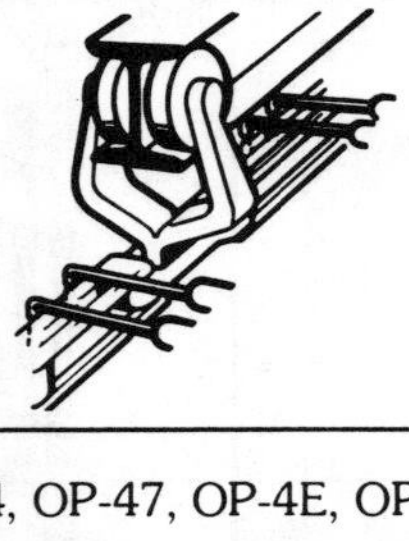	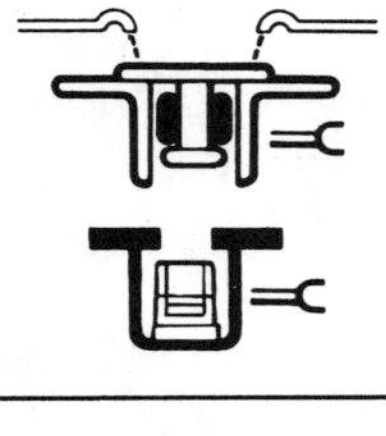	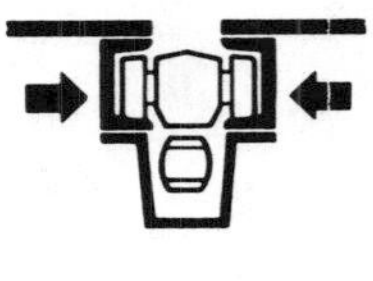
OP-7, OP-7E, OP-47	OP-4, OP-47, OP-4E, OP-47E	OP-10, OP-10E, OP-43	OP-324
2″, 3″, 4″, 6″/Std. metrics	2″, 3″, 4″, 6″/Std. metrics	3″, 4″, 6″ Wheel/Std. metrics	4″, 6″ Chain/Std. metrics
			X
X	X	X	(X)
X	X	X	X
X	X	X	X

Model OP-8
Power Brush Cleaner (Overhead)

Overhead Conveyor Powered Brush Cleaner

Model OP-8 Designed to remove all dirt, rust, scale, and chemical residue from the chain, trolley wheels, and brackets of overhead monorail and power and free conveyors.

In-Floor Conveyor Powered Brush Cleaner

Model OP-8 (In-floor) completely below floor level so that cleaning is accomplished while conveyor system is operating.

C-Hook Brush Cleaner

Model OP-13 Automatically power-scrubs C-hooks, paint hooks, load bars and other handling accessories, as well as chain, trolley and brackets.

OPCO Power Brush Cleaners offer these advantages:

- Eliminates product contamination caused by falling debris.
- Improves electro-finishing ground.
- Reduces conveyor drag.
- Fast, easy adjustment.
- Eliminates need for costly manual cleaning methods such as hand brushing, hammering; sandblasting.
- Improved ability to lubricate trolley wheels by keeping grease fittings clean.

A full range of options to meet your needs:

- Stackable brushes fit any conveyor.
- Special brushes and explosion-proof motors for hazardous areas.
- Air blowers, shrouds and vacuum systems.
- Electric timers for automatic start/stop.

OPCO Engineering, Research and Design Departments design and build custom lubrication and cleaning systems to solve unique problems. If you feel that one of our standard units will not fit your specific application, we can design a unit that will.

Model OP-8
Power Brush Cleaner (In-floor)

MADISON-KIPP PRODUCTS CORP.
201 Waubesa Street, P.O. Box 3037
Madison, Wisconsin 53704
(608/244-3511)
Telex: 265461 MADKIPP MSN

Factory Representatives in Major Cities: Atlanta, Buffalo, Cleveland, Detroit, Milwaukee, St. Louis and Toronto
Subsidiaries: OPCO SARL, Paris, France; Mega Industries, Houston, Texas
Authorized International Distributors: England, Germany, Holland, Italy, Mexico, New Zealand, South Africa, Spain and Japan.

Model OP-13 C-Hook Brush Cleaner

CLF—3-84 5M

Printed in United States of America

Application	In-Floor	Overhead Monorail — Chain, Trolley & Brackets	Overhead Monorail — Chain, Trolley Brackets & Attachments
Function	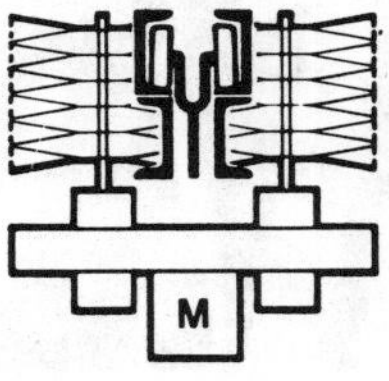	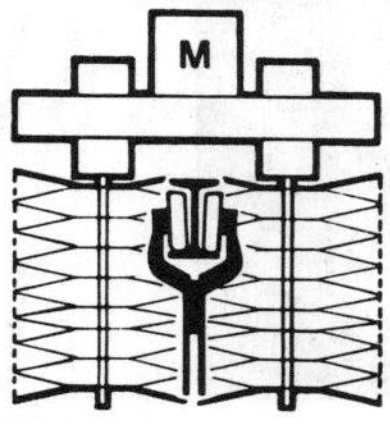	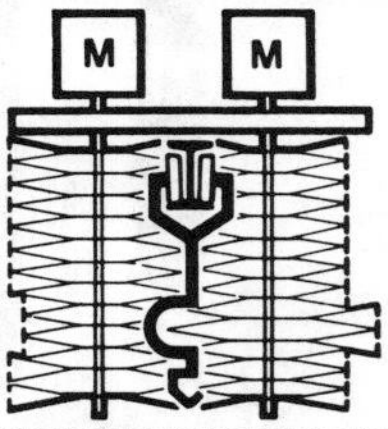
Type	OP-8	OP-8	OP-13

Renite Company
2500 East Fifth Avenue/P.O. Box 19235
Columbus, OH 43219
(614) 253-5509

RENITE MOTO-SPRAY

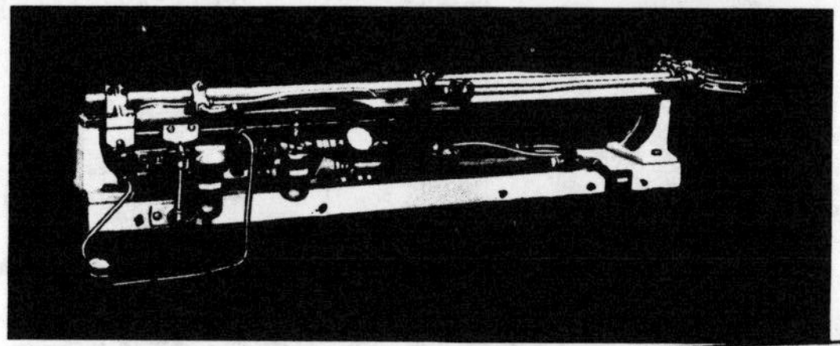

MODEL
6-A

- INTERCHANGEABLE SPRAY PATTERNS BOTH SIDES OF EACH ATOMIZER
- ANY NUMBER OF ATOMIZERS (1 TO 5) EASILY AFFIXED TO 5 STATION MANIFOLD
- ANY SPRAY VOLUME — EITHER SIDE OF ATOMIZER
- ANY ATOMIZER POSITION
- ANY STROKE LENGTH
- ANY STROKE SPEED

- MANIFOLDS ARE QUICKLY INTERCHANGEABLE FOR FREQUENT JOB CHANGES
- EACH ATOMIZER CAN BE MADE TO SPRAY IN EITHER 1 OR 2 DIRECTIONS OR CAN BE SHUT-OFF INDEPENDENT OF OTHERS
- ATOMIZERS ARE HEAT RESISTANT AND LIGHTWEIGHT — MADE OF ANODIZED ALUMINUM
- VERTICAL OR HORIZONTAL MANIFOLD POSITION
- SPRAYS THROUGH ANY PORTION OR PORTIONS OF OUTWARD AND RETURN STROKES ON 80-100 p.s.i. AIR PRESSURE
- ACTUATE MANUALLY OR AUTOMATICALLY
- MOUNT VERTICALLY, HORIZONTALLY OR UPSIDE DOWN
- GRAVITY FEED ON LUBRICANTS (NO MORE THAN 5 p.s.i.)

Here's what MOTO-SPRAY offers

Renite Company
2500 East Fifth Avenue/P.O. Box 19235
Columbus, OH 43219
(614) 253-5509

Through years of development, Renite Lubricants and release agents brought so many advantages to so many different production operations—especially forging and die casting jobs where ordinary greases and lubricants just won't work—that we *had* to develop better methods of applying them.

Swabbing was too sloppy, inefficient and slow. Standard spray-gun devices did not have the answer. Renite hand sprays, developed by us, proved to be a big step forward. But rising production speeds and labor costs called for still higher efficiency on many operations. The answer seemed to lie in working out a way to use one or more Renite spray nozzles on a fast-acting, automatic "arm" that could move in and out of position fast, spraying Renite on one or more dies and punches in any volume and pattern through any portion or portions of either the outward or return strokes, or both. *Moto-Spray is the newest advance in this specialized field.*

Already Moto-Spray with Renite has brought a new kind of controlled accuracy and uniformity to die and mold coating, in numerous fields. Die wear was reduced, die fill (metal flow) and product finish improved. Tolerances, concentricities, and contours were held much more closely. Sticking was practically eliminated. Once the best spray pattern and kind of Renite are worked out for any operation, Moto-Spray repeats them exactly, on every cycle. In addition it has made these jobs faster, safer, lower in cost.

So much so, in fact, that we've had calls for Moto-Spray units to handle jobs above and beyond the wide sphere of Renite applications.

In applying Renite, Moto-Spray "reaches" into parted dies, molds or other production machine tool areas, sprays in a pre-adjusted pattern that covers even the most irregular cavities in the most effective manner, and returns to starting position—all as fast or slow as the job requires.

You can have any fast-reach or slow-reach combination you want. Stroke is infinitely adjustable to fit any need. You can have any number of spray atomizers (from 1 to 5) in any position or combination of positions, providing a spray of just about any pattern, volume, duration, density and travel you could call for.

Moto-Spray will automatically operate through its complete cycle by pushing or hitting the palm actuating valve, or by tripping a cam valve mounted on your machine, or electrically by means of a micro switch, solenoid valve or similar actuating device.

Moto-Spray is a rugged, powerful, mechanical-pneumatic self-lubricating unit, built to cut maintenance needs to the bone. Moto-Spray has proved itself in production under all sorts of conditions and over long periods.

In short, Renite products solved a lot of manufacturing problems . . . then Moto-Spray was developed to meet new Renite application needs . . . and now Moto-Spray is ready to handle additional unanswered jobs you may have.

NOW –

What does Moto-Spray offer YOU!

Pictures on these pages give you some idea of what Moto-Spray is, how it works, what it has done and what it is doing. Different combinations of atomizers, with a selection of interchangeable nozzles or plugs, provide flexibility for most spraying needs. Does all this suggest something else it might do, for you? We do not pretend to know all the answers, but we'll be happy to work with you in "taking it from there."

If you have reason to think that Moto-Spray with or without one of the Renite lubricating, cooling or release agents, might solve some special need for you, why not let us know about it? It will cost you nothing to get the expert (but by no means know-it-all) views of our engineers, along with details on Moto-Spray itself. You won't be obligated. May we hear from you?

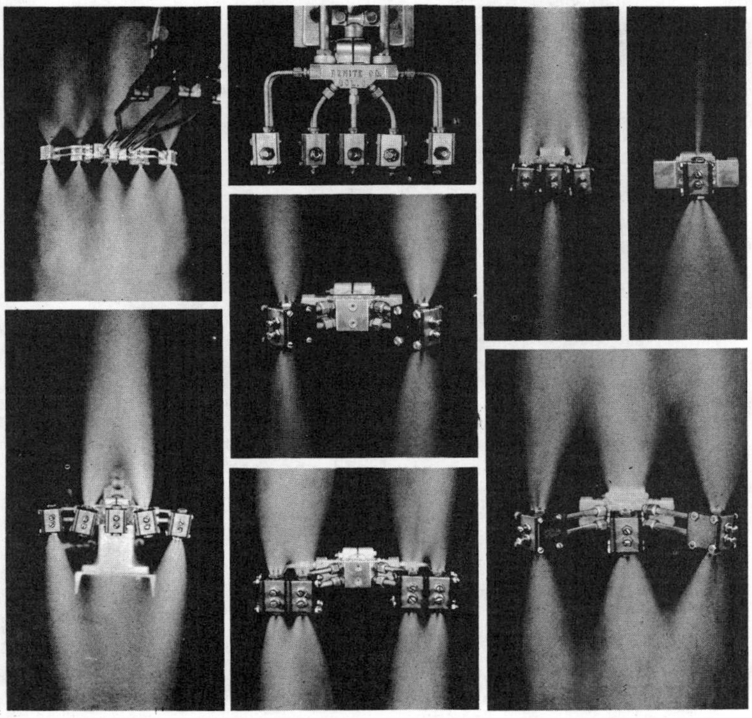

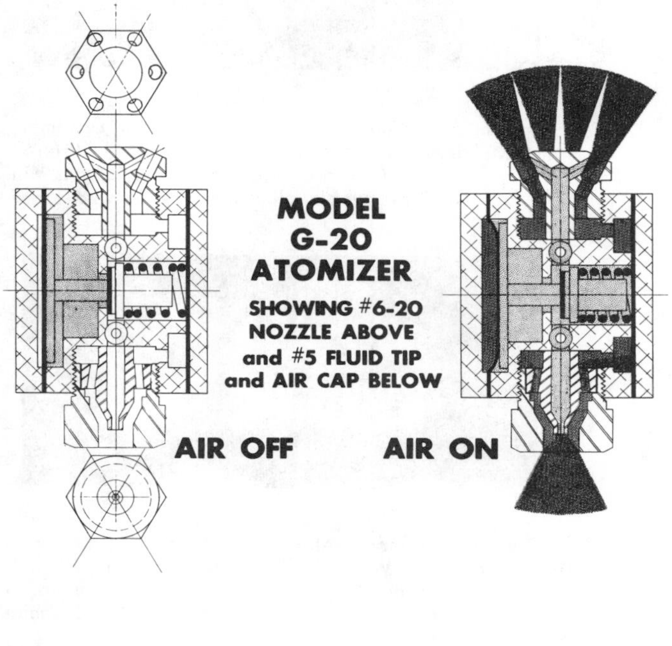

Renite Company
2500 East Fifth Avenue/P.O. Box 19235
Columbus, OH 43219
(614) 253-5509

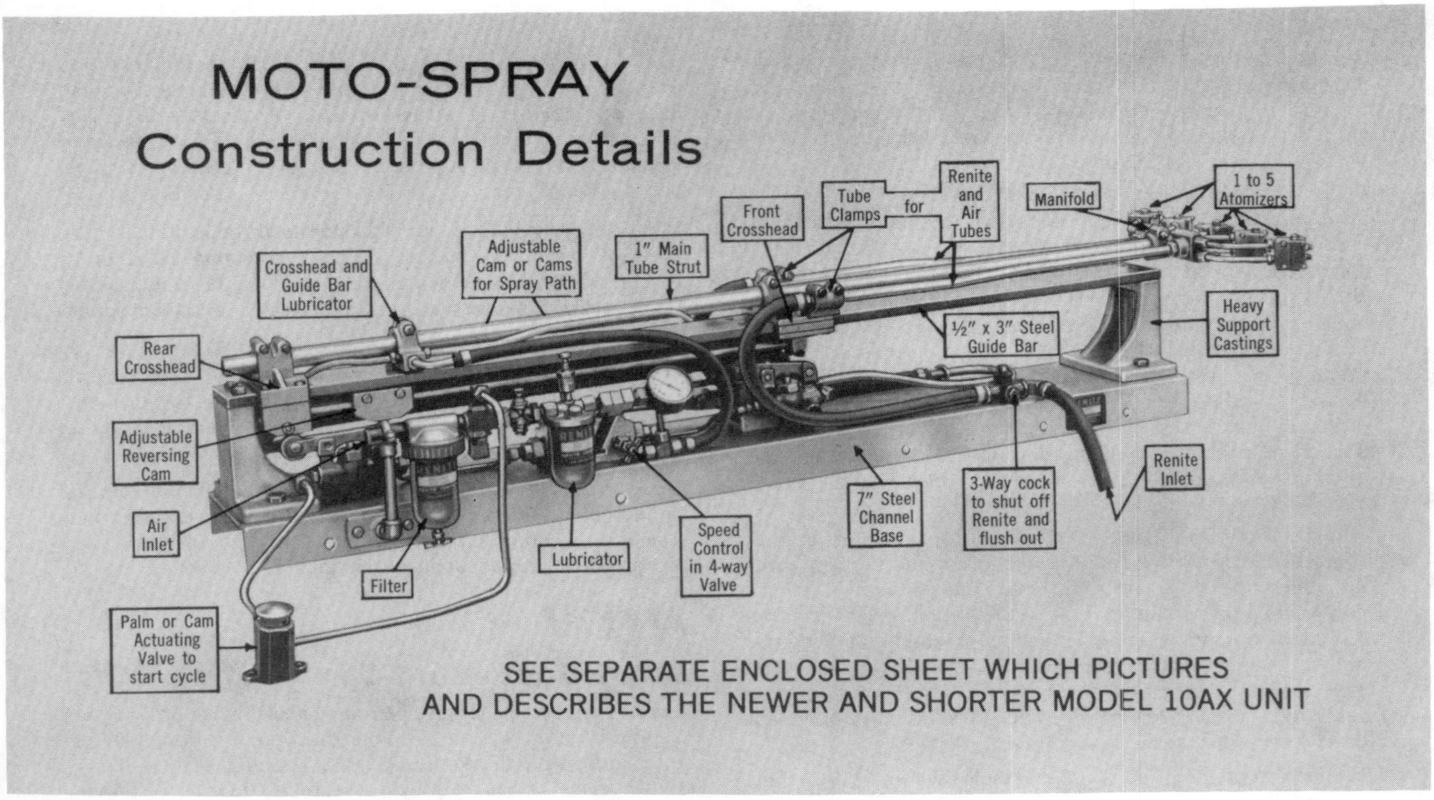

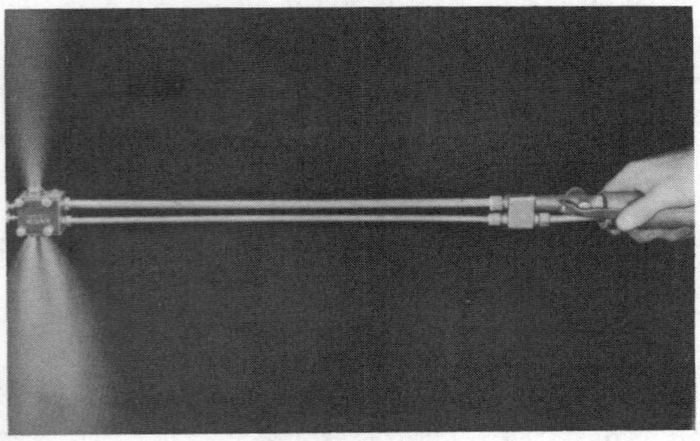

1. To better engineer results and serve our customers, we have developed and manufactured many different hand spray units. They are made in various lengths with adjustable and removable aluminum nozzles and tubes.

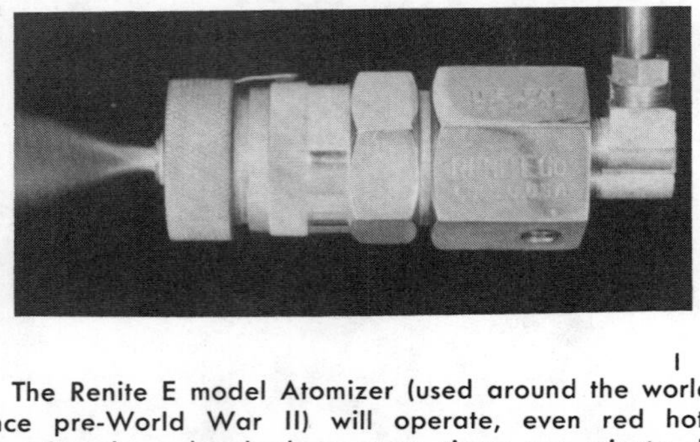

2. The Renite E model Atomizer (used around the world since pre-World War II) will operate, even red hot, intermittently a hundred or more times per minute. It provides micrometer adjustment of the amount of Renite passing through it.

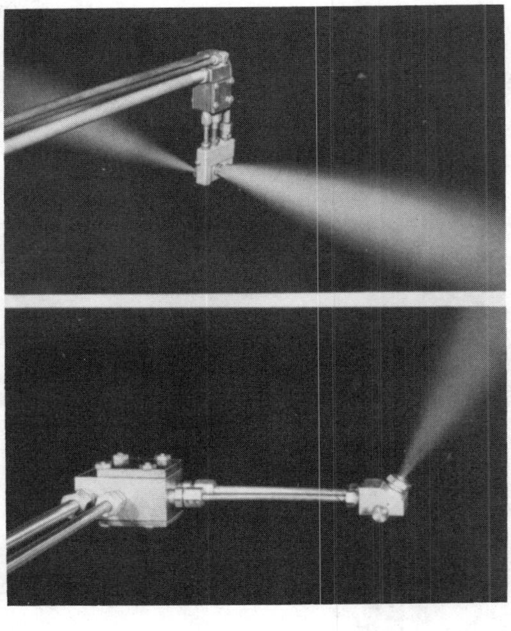

3. Where space is limited, special Renite "extension head" atomizers have been developed and manufactured for special applications. The Renite #G1-3 (top) sprays two directions while the Renite #GL-4 (bottom) sprays one direction, with interchangeable spray patterns.

RENITE ® Company *Lubrication Engineers*

2500 E. FIFTH AVE. • COLUMBUS, OHIO 43219 • U.S.A.

Phone (area 614) 253-5509

Renite Company
2500 East Fifth Avenue/P.O. Box 19235
Columbus, OH 43219
(614) 253-5509

Model 10AX MOTO-SPRAY

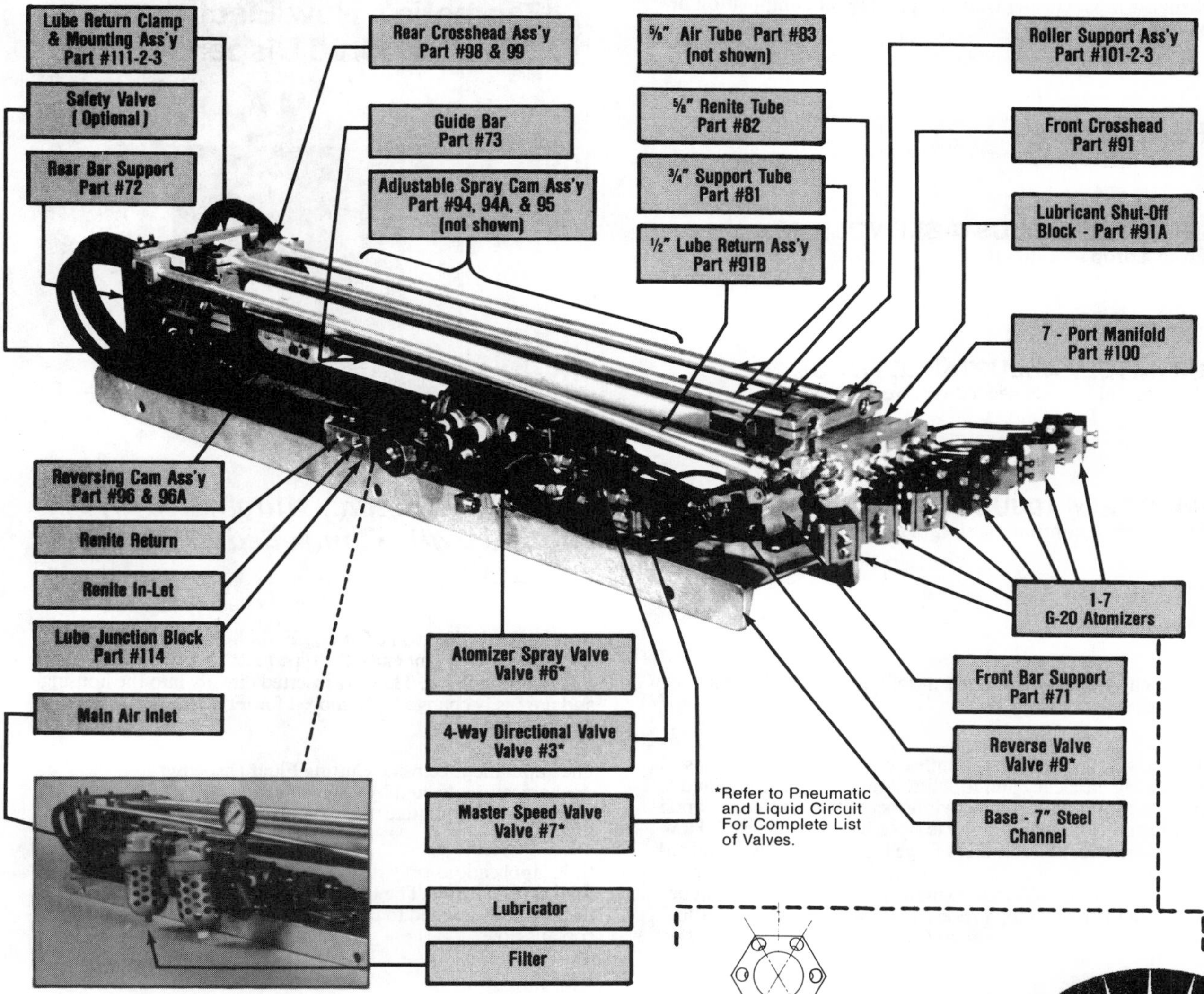

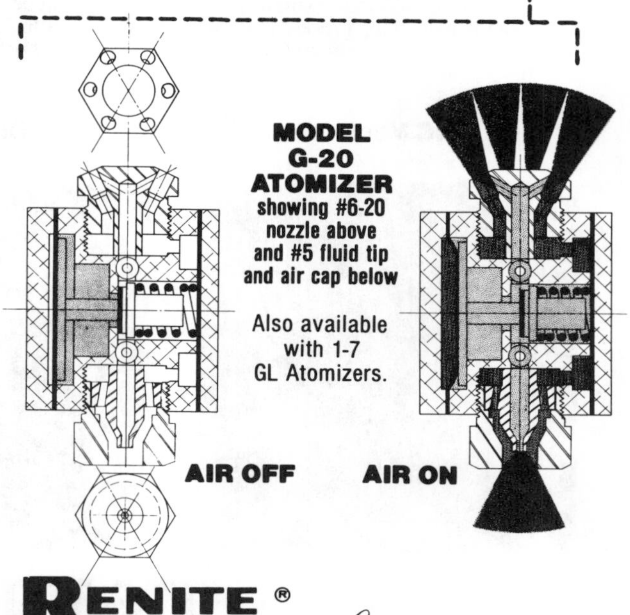

MODEL G-20 ATOMIZER
showing #6-20 nozzle above and #5 fluid tip and air cap below
Also available with 1-7 GL Atomizers.
AIR OFF
AIR ON

Renite Model 10AX Moto-Spray has been designed and engineered for vertical or horizontal mounting on Die Cast Machines, Extrusion Presses, Forging Presses, Forging Machines, etc. One to seven Model G-20 Atomizers can be affixed to the manifold which can be positioned either in a vertical or horizontal plane. (Customer to specify nozzle pattern(s), manifold position and atomizer centerline dimensions.)

The Model 10 AX Moto-Spray is a self contained unit being mechanically/pneumatically controlled. A 3-way Palm valve is supplied to initiate each cycle. Customer must install electrical actuation and interlock system if so desired.

Customer must design and fabricate the required mounting device to properly position the Moto-Spray unit on machine.

The Model 10AX Moto-Spray incorporates dual speed control, recirculating lubricant supply lines, lubricant shut-off control and quick-change spray cam assembly. If these features are not desired, specify Model 10A Moto-Spray.

RENITE® Company *Lubrication Engineers*

2500 E. FIFTH AVE. • COLUMBUS, OHIO 43219 • U.S.A.
Phone (area 614) 253-5509

Bulletin No. 282—2M

Litho in USA

Tapmatic Corporation
1851 Kettering Street
Irvine, CA 92714
(714) 979-6080

This revolutionary new low voltage device (24 volt AC) can dispense most cutting fluids, regardless of composition or viscosity.

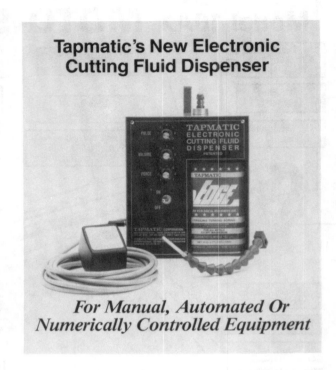

Tapmatic's New Electronic Cutting Fluid Dispenser

For Manual, Automated Or Numerically Controlled Equipment

FEATURES:

INFINITELY ADJUSTABLE VOLUME
From a drop to a stream

INFINITELY ADJUSTABLE FORCE
Applies fluid where needed without splatter

INFINITELY ADJUSTABLE PULSING
From almost continuous stream to two second intervals.

Electrical actuation simplifies installation and low voltage requirements are easily met.

This new device provides countless opportunities to increase machining efficiency and tool life. Any type of cutting fluid may be employed and the obvious benefits of automatic application and elimination of waste or spillage are achieved. However, the choice of Tapmatic's Edge Liquid provides the added advantage of being compatible with either oil or water. Now, the benefits of an efficient, extreme pressure lubricant can be added to the machining process wherever needed, alternating or cooperating with the coolant flow.

The Tapmatic Electronic Cutting Fluid Dispenser is designed for use with oval pint cans, the type most widely used for specialty cutting fluids. They are inserted directly into the housing and are easily changed or removed for refilling.

The Tapmatic Electronic Cutting Fluid Dispenser may be mechanically activated by a switch acting against a moving quill or slide on manual or automated equipment.

CNC applications only require interfacing with an N function to turn on the unit. The Electronic Cutting Fluid Dispenser can then be programmed to cooperate or alternate with normal coolant flow.

CNC Machine

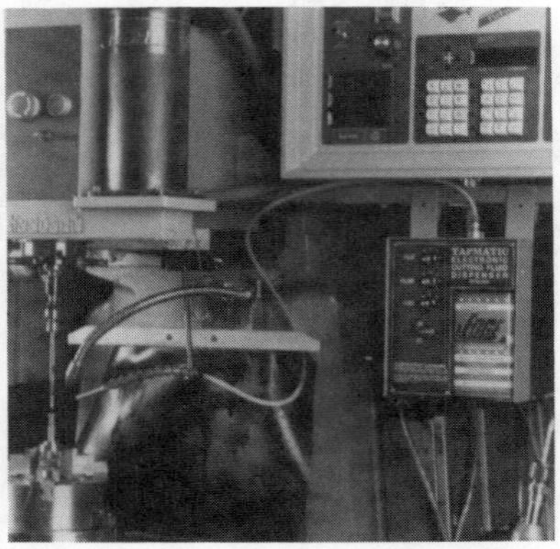

Drill Press

Milling Machine

TAPMATIC

TAPMATIC CORPORATION • 1851 Kettering Street • Irvine, California 92714-5673 • (714) 261-9302
• Cable: TAPCO Irvine • TWX: 910 5951915

Tapmatic Corporation
1851 Kettering Street
Irvine, CA 92714
(714) 979-6080

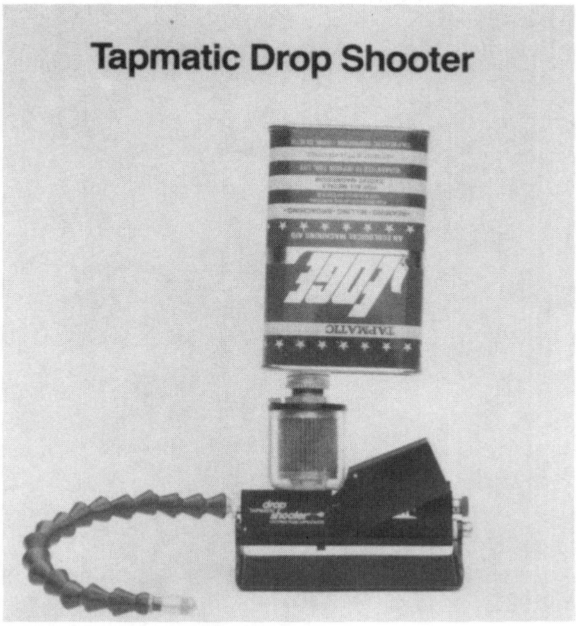

The right applicator can maximize the performance and economy of any cutting fluid/lubricant. Tapmatic Corporation offers a selection of applicators to suit your specific requirements: The *Tapmatic Fluid-Miser Applicator*, the *Tapmatic Mist Coolant Dispenser* and the newest and most revolutionary, the *Tapmatic Drop Shooter*™

Developed especially for Tapmatic Edge, which must be used sparingly for best performance, the new Drop Shooter ejects a single drop of fluid like a bullet from a gun. The volume of the drop is infinitely adjustable and can be dispensed accurately up to a distance of 24 inches.

Two styles are available: The **Hand Drop Shooter™** for manual application and the **Automatic Drop Shooter™** for CNC or automated equipment.

The Automatic Drop Shooter™ is activated by an air signal. This air signal can be triggered in a variety of ways such as a cam, hand or foot valve. It can also respond to an electric solenoid valve either directly or through the machine program.

Introducing the air to the side inlet as well as to the top inlet will atomize the drop in a minute air stream. This is essential for larger cutting tools where there is a possibility of the drop landing in the flute rather than on the cutting edge. It also serves to blow away chips.

For turning or milling where the cutting tool travels an Automatic Pulser Valve may be added to permit repeat firing at any desired interval.

75-90 P.S.I. air pressure is recommended.

A metering valve for a pint can is furnished with each Automatic Drop Shooter™ should it be desired to increase the reservoir capacity.

A Magnetic base is available in place of standard clamp type base at extra cost.

New TAPMATIC drop shooter
CUTTING FLUID APPLICATOR

Shoots Drops of Adjustable Size,
at Desired Frequency,
on Target

Other Dispensers Available from Tapmatic

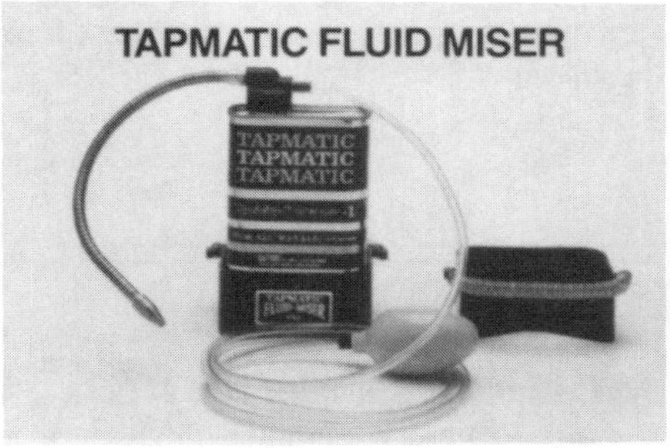

The combination of Tapmatic Dual Action Plus Cutting Fluids and Fluid-Miser applicators can actually reduce your cutting fluid cost 75% while effecting tremendous increases in production.

The Fluid-Miser applies fluid when you want it, where you want it. Waste resulting from conventional squirt-on application is eliminated.

Easily attached, the Fluid-Miser is secured to a machine column with a spring attachment. The metering bulb may be hand or foot operated.

TAPMATIC MIST COOLANT DISPENSER

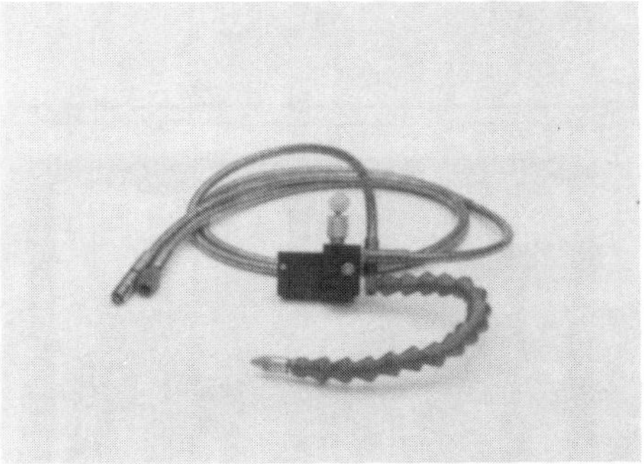

An all new mist coolant dispenser from Tapmatic. Simple, inexpensive and rugged, featuring separate air and fluid controls.

The powerful magnetic base can be mounted on flat or round surfaces. The stainless steel armored siphon hose can be dropped into any container on floor or table.

SECTION 4

METALFORMING COMPOUNDS

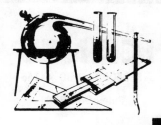

Buckeye Lubricants
20801 Salisbury Road
Bedford, OH 44146
(216) 581-3600

BUCKEYE LUBRICANTS

20801 SALISBURY RD. **BEDFORD, OHIO 44146**

Phone (216) 581-3600 Telex 98-0614

PRODUCT DATA BULLETIN

#775 PIGMENTED STAMPING & WIRE DRAWING COMPOUND

#775 is a heavy duty, paste type drawing compound, which is used on operations requiring high film strength and a superior finish.

This product adheres tenaciously to the metal allowing more than one pass in wire drawing and giving excellent die life. #775 "Rolls" in the die immediately, which is a must in wire drawing.

#775 can be diluted with either water or oil as the application requires. In addition it has excellent freeze-thaw characteristics. The product can be used if it has been frozen and subsequently thawed, without any difficulty.

#775 is 100% saponifiable and can be cleaned in alkali pressure washing machines and soak tanks.

Buckeye Lubricants will supply samples of #775 for evaluation upon request.

Buckeye Lubricants
20801 Salisbury Road
Bedford, OH 44146
(216) 581-3600

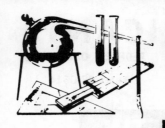

BUCKEYE LUBRICANTS

20801 SALISBURY RD. BEDFORD, OHIO 44146

Phone (216) 581-3600 Telex 98-0614

PRODUCT DATA BULLETIN

#SP40A BIODEGRADABLE DRAWING COMPOUND
WATER SOLUBLE – CONTAINS NO OIL

Buckeye Lubricants is [illegible] opment of our #SP40A Drawing Compound. Th[illegible] odegradable and contains no oil.

This product has had exce[illegible] ations, completely eliminating pick-up on the dies [illegible] e and without hardening on the dies.

#SP40A can be used "as [illegible] depending on the difficulty of the operation.

Contact Buckeye Lubric[illegible] al information.

Etna Products, Incorporated
16824 Park Circle Drive/P.O. Box 630
Chagrin Falls, OH 44022-0630
(216) 543-9845

PHONE:
(216) 543-9845
TELEX:
980131 WDMR

ETNA PRODUCTS, INC.

16824 PARK CIRCLE DRIVE
CHAGRIN FALLS, OHIO 44022 U. S. A.

MAILING ADDRESS:
P.O. BOX 630
CHAGRIN FALLS, O.
44022-0630 U.S.A.

MASTER DRAW #1179-A

Synthetic Water Soluble Stamping Lubricant

General Appearance

High viscosity, opaque fluid, pleasant mild odor.

General Characteristics

Totally synthetic compound formulated with a wide range of advanced fricitional modifiers, synthetic polymers, wetting agents and a combination of excellent ferrous and non-ferrous corrosion inhibitors along with an added broad spectrum bacteriacide.

General Use and Application

MASTER DRAW #WS1179-A has been developed for the stamping of non-ferrous and ferrous metals, ranging from 260 cartridge brass to stainless steels.

The reservoir, piping and equipment should be cleaned thoroughly before charging the systems to minimize any cross contamination. As a total synthetic MASTER DRAW #WS1179-A mixes readily with water with some mild agitation.

General Outstanding Qualities

1. Stability: The solution formed by MASTER DRAW #WS1179-A is extremely stable.

2. Water Hardness: While soft water is preferred, MASTER DRAW #WS1179-A forms an extremely stable solution in hard water.

3. Corrosion Inhibition: MASTER DRAW #WS1179-A has been specially formulated to prevent both ferrous and non-ferrous corrosion, thus preventing staining of the work piece, as well as protecting the machines.

4. Bacteriacide: MASTER DRAW #WS1179-A contains an added broad spectrum bacteriacide to minimize bacteria problems. For best control of bacteria, regular samples should be sent to our Chagrin Falls, Ohio lab for bacteria count and full analysis.

5. Recommended Concentrations: Suggested starting concentrations range from 8% (volume to volume) for non-ferrous stamping, to 18-20% for some difficult ferrous stamping applications.

Etna Products, Incorporated
16824 Park Circle Drive/P.O. Box 630
Chagrin Falls, OH 44022-0630
(216) 543-9845

PHONE:
(216) 543-9845
TELEX:
980131 WDMR

ETNA PRODUCTS, INC.
16824 PARK CIRCLE DRIVE
CHAGRIN FALLS, OHIO 44022 U. S. A.

MAILING ADDRESS:
P.O. BOX 630
CHAGRIN FALLS, O.
44022-0630 U.S.A.

TYPICAL SPECIFICATIONS

MASTER DRAW #1179-A

Acid Number	15 mg KOH/g Typical
Base Value	30 mg KOH/g Typical
Pounds Per Gallon	8.67
Specific Gravity	1.04
Flash Point	None

For further technical information or to place an order, please call 216/543-9845

FarBest Corporation
24551 Raymond Way, Suite 110
El Toro, CA 92630
(714) 855-3881

FARBEST TECHNICAL BULLETIN

PRODUCT:

FARBEST 6489 S.S. STAMPING AND DRAWING COMPOUND

FARBEST 6489 S.S. STAMPING AND DRAWING COMPOUND

DESCRIPTION:

FARBEST 6489 S.S. STAMPING AND DRAWING COMPOUND is a dark colored oil of medium-heavy viscosity which mixes readily with pale oils to perform a variety of forming, stamping, drawing and cutting operations on the full range of ferrous metals.

BENEFITS:

- Exceptionally versatile - when cut with varying amounts of pale oil, will perform most metalworking jobs.
- Economical cutting oil for heavy machining.

APPLICATION:

Cutting	Cut 10:1 with pale oil.
Stamping	Cut or use straight.
Drawing & Forming	Cut or use straight for difficult stainless steel work.

TYPICAL PROPERTIES:*

Viscosity @ 100 deg. F (SUS)	1200
Specific Gravity, @ 60 deg. F	1.016
Pounds per Gallon	8.5
Flash Point (deg. F, COC)	330

* Average values subject to minor manufacturing variances which do not affect performance.

SALES OFFICES:

24551 Raymond Way
El Toro, CA 92630
714/855-3881

6715 McKinley Avenue
Los Angeles, CA 90001
213/758-3181

1401 Greenleaf Avenue
Elk Grove Village, IL 60007
312/437-1450

Plum Street
Verona, PA 15147
412/828-5880

418 Tango Street
San Antonio, TX 78216
512/349-0321

Chemical Road
Plymouth Meeting, PA 19462
215/825-5050

FF-6489 AB020885

METALFORMING COMPOUNDS

Graphite Products Corporation
P.O. Box 29
Brookfield, OH 44403
(216) 394-1617

GRAPHITE DIE LUBRICANTS & PRECOATS

Many factors contribute to the total force generated during a closed-die forging operation — size and configuration of the part, properties of the workpiece material, die hardness and surface finish, strain rate of the deformation process, temperature of the die and workpiece, and frictional forces.

Lubrication has a major effect on forging force requirements. Use of suitable lubricants can diminish frictional forces and exert a positive influence on the other variables.

Carefully formulated to meet present day ecological demands and production requirements, the GP Metalworking Lubricants are designed to meet a variety of requirements and conditions in the metal forming process. Listed are the typical benefits that help contribute to a more cost effective metal forming process.

- Good adhesion and uniform film thickness.
- No chemical reaction to impair the surface finish of the workpiece or die cavity.
- Reduce metal contact between workpiece and die.
- Sprayable, non-pollutant and not harmful to operating personnel.
- Good thermal and physical stability during deformation.
- No build-up or residue on the die to cause defects on the surface of the workpiece.
- Economical at the point of application.
- Reduce frictional forces and improve the flow of metal.

SELECTION

PRODUCTS	PROCESS			ALLOYS							TEMP. RANGE[1]	EQUIPMENT				DIE TEMPERATURES			
DIE LUBRICANTS	HOT FORGE	HOT EXTRUDE	HOT UPSET	CARBON STEELS	STAINLESS STEELS	TITANIUM	SUPER ALLOYS	ALUMINUM ALLOYS	BRASS & COPPER ALLOYS		MECHANICAL FORGE PRESS	HYDRAULIC FORGE PRESS	HAMMER	UPSETTER	300° to 500°F.	500°F.+	750°F.+	LUBRICATING PIGMENT
GP 101	●	●	●	●	●	●	●				●			●	●			Graphite
GP 102	●	●	●	●	●	●	●	●	●		●			●	●			Graphite
GP 152	●	●	●	●	●	●	●	●	●		●	●		●		●	●	Graphite
GP 158	●	●	●	●	●	●	●	●	●		●		●	●	●			Graphite
GP 181	●	●	●	●							●		●	●		●	●	Graphite
GP 182	●	●	●	●							●		●	●		●		Graphite
GP 184	●	●	●	●				●			●	●	●	●	●	●		Graphite
GP 185	●	●	●	●					●		●	●		●	●	●		Graphite
GP 191	●	●	●	●							●			●	●			Graphite
GP 195	●	●	●	●				●	●		●		●	●	●			Graphite
GP 243		●	●	●	●		●				●	●		●	●	●		Graphite/Moly
LS 2162	●	●	●	●				●	●		●	●		●	●	●		Graphite
LS 2348	●	●	●	●							●		●	●	●			No Pigment
LS 2298	●		●	●							●				●			No Pigment
LS 2227M	●	●		●							●	●	●		●	●		Graphite
LS 2393	●		●	●							●	●	●	●		●		No Pigment
PRECOAT LUBRICANTS										TEMP. RANGE[1]								
GP 158	●	●	●	●				●		1500°F.[1]	●			●				Graphite
LS 2162	●	●	●	●					●	1700°F.[1]	●			●				Graphite

[1] Maximum temperature range for precoating workpiece. Time limitation will depend on oven atmosphere.

Graphite Products Corporation
P.O. Box 29
Brookfield, OH 44403
(216) 394-1617

SYNTHETIC DIE LUBRICANTS

A select group of non-graphited water extendable die lubricants developed for hot forging and upsetting applications.

In concentrated form the products should be diluted with water and sprayed onto the die surfaces (preheated to 300°F.) for mechanical, hydraulic press forgings including hammer and upset applications.

TYPICAL CHARACTERISTICS

Solids: Water Soluble Polymers
Carrier: Water
Diluent: Water
Weight/Gallon: 9.0 lbs. (shipping weight)

Consistencies: Creamy Fluid
Flash Points: None
Typical pH: 8.0

MIXING INSTRUCTIONS

Add water to products while slowly mixing the contents. Mild form of agitation is recommended. All products should initially be diluted with 3 parts of water to 1 part of concentrate. Dilution ratios may vary as high as 10 parts of water. Refer to PB-4 Bulletin for additional instructions.

CHART

CARRIER	TYPICAL DILUTION RATIO WITH WATER	METHOD OF APPLICATION	COMMENTS
Water	None	Spray	Precision forgings for aircraft engine parts
Water	1:9	Spray	Precision forgings for aircraft engine and automotive component parts
Water	1:5	Spray	Hot/Warm forgings of P/M preforms — Lubricant for high temp. dies above 500°F.
Water	1:9	Spray	Precision or near net shape forgings of aircraft and automotive component parts
Water	1:5	Spray	Wet out die surfaces above 750°F. — Forgings for automotive, truck and railroad component parts
Water	1:5	Spray	Wet out die surfaces above 500°F. — Forgings for automotive, truck and railroad component parts
Water	1:5	Spray	Forgings of chemical or oil field component parts such as valves, fittings, rods, cylinders
Water	1:9	Spray	Heavy duty forgings/extrusions, cylinders, valve bodies, oil field component parts
Water	1:9	Spray/Swab	Automotive, truck and railroad component parts
Water	1:10	Spray	Automotive, truck and railroad component parts — Longer forge cycles
Oil	None	Swab	Extruding precision aircraft parts
Water	1:5	Spray	High temp. die lubricant for automotive, ordnance, truck and railroad forgings
Water	1:5	Spray	Non-pigmented — Precision automotive component parts
Water	1:3	Spray	Non-pigmented — Die lubricant — heavy duty forgings
Water	1:10	Spray	Heavy duty forgings for railroad, truck, oil field and automotive component parts
Water	1:10 1:1	Spray Swab	Non-pigmented - Good release and lubricating properties on deep well forgings
Water	1:1	Precoat wk. pc.	Net shape P/M preform forgings to 1500°F. preheat furnace temperature
Water	1:1	Precoat wk. pc.	Net shape preform forgings or extrusions to 1700°F. furnace temperature

Mullen Circle Brand, Incorporated
3514 West Touhy Avenue
Chicago, IL 60645
(312) 676-1880

PRODUCT INFORMATION

FROM . . . MULLEN CIRCLE BRAND, INC. • 3514 West Touhy Avenue • Chicago, IL 60645 • 312/676-1880

MCB #3014 DRAWING OIL

#3014 DRAWING OIL is an active sulfur drawing oil suitable for the most severe ferrous metal-forming operations.

This product has been used straight for severe draws such as wheelbarrows, lawn mower decks, and snow blower housings. Diluted 25% with oil, #3014 has drawn vacuum cleaner housings and is suitable for other such moderate duty applications. Diluted 50% with oil, #3014 is a superb broaching, tapping, and heavy duty ferrous cutting oil.

In addition to high extreme pressure and anti-weld properties, #3014 also contains special additives which keeps the oil residue pliable for up to one year. This characteristic allows for easy cleaning of stacked parts. Alkaline cleaning baths perform best, however the above data was obtained using an acid wash system.

When purchased as a freight-saving base, the diluent oil should be a vigin mineral oil with a minimum flash point of 320°F. The pre-mixed blends mentioned above are available when freight costs are not a concern.

TYPICAL PROPERTIES:

Viscosity @ 100°F SUS	2000
Color	Black
Odor	Sulfurized
Specific Gravity 60/60 °F	.94
Pounds/Gallon	7.866
Flash, C.O.C. °F	400

Mullen Circle Brand, Incorporated
3514 West Touhy Avenue
Chicago, IL 60645
(312) 676-1880

PRODUCT INFORMATION

FROM... MULLEN CIRCLE BRAND, INC. • 3514 West Touhy Avenue • Chicago, IL 60645 • 312/676-1880

MUL-DRAW SUPREME

MUL-DRAW SUPREME is a non-petroleum, water soluble blend of soaps and contains no chlorine or sulfur compounds, solvents or insoluble materials. It is suitable for use on both ferrous and nonferrous metals. This product has the following advantages:

-Ease of handling - the product may be brushed, rolled, sprayed or dip coated onto the metal.
-Provides temporary indoor rust protection.
-May be applied wet or as a spray film and can be drawn wet or dry.
-Provides piercing lubrication.
-Water dilutible in proportions ranging from 1:1 to 5:1
-Easily cleaned in a conventional or low temperature cleaner.
-Will perform satisfactorily when applied over a light oil film but it is preferred to apply MUL-DRAW SUPREME over clean steel.
-Biodegradeable.
-Exceptionally high film strength.
-Mild odor.

SPECIFICATIONS

Appearance	Clear Gold Liquid
Flash Point	None
Specific Gracity	1.03
Viscosity @ 100°F SUS	200
Freeze Point	32°F, freeze/thaw stable three cycles min.
pH	8.75

Van Straaten Chemical Company
630 West Washington Boulevard
Chicago, IL 60606
(312) 454-1000

VAN STRAATEN
Chemical Company

Metal Forming Lubricant

FluidForm

PRODUCT INFORMATION

A SEMI-SYNTHETIC TUBE FORMING LUBRICANT

NITRITE AND PHENOL FREE

VAN STRAATEN FluidForm Semi-Synthetic is a water dilutable, semi-synthetic lubricant applicable for the tube forming operations of ferrous and non-ferrous metals. This product offers excellent cleanliness and in-process corrosion control. FluidForm Semi-Synthetic contains a light blue dye in the concentrate and forms a stable, translucent, blue emulsion in both hard and soft water. This product is suitable for use in both central systems and individual mill sumps.

KEY PERFORMANCE BENEFITS

Non-nitrited and Non-phenolic - to meet the most stringent industrial chemical restrictions for both operator safety and waste disposal requirements.

Outstanding Cleanliness - promotes safety, keeps mill tools working efficiently. No slippery, oily residues on either the mill tool or workpiece.

Effective Rust Control - based on our advanced technology, FluidForm Semi-Synthetic prevents in-process corrosion of steel and helps protect mill tooling from damaging bi-metallic corrosion.

Resists Rancidity and Mold Growth - ends Monday-morning odor problems. Provides extended service in recirculating central systems without requiring special maintenance procedures.

Easy to Mix - even in hard water, forming a blue, translucent mix that permits the operator to see the work in progress.

Good Operator Acceptance - mild to operators' skin, clean, nonflammable, characterized by a fresh neutral odor.

Excellent Oil Rejection Properties - tramp oil will separate out rapidly and completely from the fluid for ease of removal.

RECOMMENDED DILUTIONS

Typical operations	15:1 to 20:1

VAN STRAATEN
Chemical Company

Metal Forming Lubricant
FluidForm

PRODUCT INFORMATION

SYNTHETIC TUBE FORMING LUBRICANT

FluidForm Synthetic represents the "New Generation" of synthetic fluids. It is a clean, versatile fluid which is formulated to meet critical performance requirements, while providing easy disposability and excellent operator acceptance.

This new synthetic fluid has the ability to perform on a variety of metals while providing a clean, pleasant fluid for the operator. It will provide sufficient lubrication while being a versatile product with which to work.

FluidForm Synthetic contains an advanced synthetic lubrication package, synthetic rust preventives, and special soaps to provide an oil-free synthetic with excellent lubrication, cooling, and performance properties.

KEY PERFORMANCE BENEFITS

Unique Performance - FluidForm Synthetic incorporates a combination of non-ionic surfactant systems which offer increased lubrication over conventional synthetic products.

Excellent Separation Properties - Tramp oil will separate rapidly and completely from FluidForm Synthetic.

Exceptionally Clean - FluidForm Synthetic is a light, transparent solution with high operator acceptance. It keeps mill surfaces and tubes clean by leaving a soft, transparent film residue.

Versatile Product - Suitable for use in a wide range of water conditions - coupled with extremely low foaming features.

Long Life - Selectively chosen raw materials that have high resistance to microbiological degradation enhance the probability of long life in central systems and individual sumps.

METALS

FluidForm Synthetic is recommended for tube forming application on all ferrous metals, stainless steels and alloys.

RECOMMENDED DILUTIONS

Typical Operations	15:1 to 20:1

Van Straaten Chemical Company
630 West Washington Boulevard
Chicago, IL 60606
(312) 454-1000

VAN STRAATEN
Chemical Company

Metal Forming Lubricant 7055-C

PRODUCT INFORMATION

MEDIUM-DUTY SYNTHETIC DRAWING AND STAMPING LUBRICANT

VAN STRAATEN 7055-C is a medium-duty, synthetic drawing and stamping lubricant which does not make use of phenols or nitrites. This product is designed to be used on both ferrous and non-ferrous metals. It is a clean, safe, versatile fluid which is formulated to meet critical performance requirements while providing easy disposability and excellent operator acceptance.

KEY PERFORMANCE BENEFITS

- Residue characteristics - nonsticky, oily residue which is very soluble in water.
- Easily cleanable - easily removable with a mild alkaline cleaner at low temperatures.
- Exceptionally clean - clean, transparent solution with high operator acceptance.
- Biocide package - contains an effective biocide package.

COMPATABILITY WITH WORK MATERIAL

VAN STRAATEN 7055-C is recommmended for use on both ferrous and non-ferrous metals.

RECOMMENDED DILUTIONS

Press forming	1:1 to 10:1

Wynn Oil Company
P.O. Box 4370/2600 East Nutwood Avenue
Fullerton, CA 92634
(714) 992-2000

WYNN'S PRODUCTS FOR METALWORKING

PRODUCT	DESCRIPTION	APPLICATION	ADVANTAGES	DILUTION RATIO
CONCENTRATES SUPERKON 602	Sulfur-free concentrate for use as an improver for cutting oils. Primarily a lubricity additive with some E.P.	Can be used to improve lubricity of cutting oils. Multi-metal, will not cause staining. Can be added to used lubricating oils to make a cutting oil and can be used straight for metal forming.	Will not stain nonferrous metals. High lubricity additive for improved finishes. When used as is, product is water rinsable. Low viscosity for easy mixing. Can also be used as a lubricity additive for general lubricating oils.	Use as is for drawing or stamping. Add to other oils in increasing quantity until desired performance level is reached.
SUPERKON 603	Heavy bodied concentrate containing lubricity, anti-weld, and E.P. additives. For use as a cutting oil improver. Green in color and perfumed.	Add to oil to make a cutting oil, or add to cutting oil for difficult operations. Not recommended where staining of metal may occur. Primarily a lubricity additive with sulfur and chlorine for difficult work. Can be used straight for drawing and forming operations on ferrous metals.	Ideal for broaching, tapping, and threading operations. High lubricity gives superior finishes and E.P. additives reduce tool wear.	Use as is for metal forming operations. Mix with cutting oil until desired level of performance is reached.
SUPERKON 604	Heavy bodied concentrate containing anti-weld and E.P. additives. For use as a cutting oil improver to reduce welding in severe operations. Brown in color.	Add to oil to make a cutting oil, or add to cutting oil for difficult machining. Prevents welding from occurring on ferrous metals. Can be used straight, as a "honey" oil for drawing and forming operations on ferrous metals.	Contains chlorine for E.P. and anti-weld. Ideal for severe machining on ferrous metals. When used straight, product is hot water rinsable. Sulfur free, will not stain most metals. High E.P. and anti-weld prevents tool wear.	Use as is for metal forming operations. Mix with cutting oil until desired level of performance is reached.
SPECIALTY PRODUCTS TAPPING COMPOUND	Clear amber fluid containing ample amounts of friction reducing agents.	For use in tapping, drilling and threading operations with ferrous and nonferrous metals. Can also be used to improve finishes in machining operations, using cutting oil, by spraying tool/work surfaces.	Multi-metal safe. Will not stain nonferrous metals. Low odor and easy to apply. Provides maximum tool life by reducing seizure and galling. Excellent for blind hole tapping.	Use as is.
MACHINE CLEANER	Concentrated cleaner for removing accumulations of dirt, grit, and swarf found in coolant systems.	Add to clean water and circulate through coolant system. Flush, rinse and recharge system with new coolant. Can also be added to water for cleaning the exteriors of machines.	Helps to prolong sump life of coolants by cleaning internal coolant systems. Dissolves slime and oil deposits. Helps to prevent clogging of coolant system components.	1 qt./5-20 gls. water. 2 qt./20-40 gls. water. 3 qt./40-100 gls. water. (Circulate for at least 10 minutes.)
HEAVY DUTY CONC FOR INDUSTRY	Concentrated extreme pressure additive for industrial lubricating oils. Light viscosity additive for easy mixing with oil.	Use as an additive for gear lubricants to reduce wear. Can also be used wherever extreme pressures cause metal to metal contact of moving parts. Add to way oil, chain lubricant, grease, or any other general purpose lubricant.	Improves anti-wear and extreme pressure performance of lubricants. Increased equipment life, better efficiency, and less energy consumption. Prevents seizing and galling. Smooths pitted gears *Not recommended for automotive use.	Oil bearings/1-3% Gear boxes/3-4% Hypoid grs./3-5% Slideways/1-3% Grease/1-3%
PROTECTA-FILM	Clear liquid rust preventive. Solvent base displaces water to prevent corrosion on metal surfaces.	Use as is. Brush, spray, dip or swab protective coating onto metal surfaces. Displaces water and protects with transparent film. Effective rust protection for up to one year with indoor storage.	No lost parts due to corrosion. Displaces soluble oil and synthetic water-base compounds from machined surfaces. Parts can be made and stored for up to one year, with no corrosion. Film is soft and can be easily removed. Transparent, does not interfere with inspection.	Use as is (undiluted).

In addition to the above, Wynn's carries a full line of greases, rust penetrant, and hydraulic oil additive. Lubrication problems? Contact your local Wynn's representative.

METALFORMING COMPOUNDS

Wynn Oil Company
P.O. Box 4370/2600 East Nutwood Avenue
Fullerton, CA 92634
(714) 992-2000

WYNN'S METAL DRAW 4

wynn's
INDUSTRIAL PRODUCTS DIVISION

P. O. BOX 4370, FULLERTON, CA 92634
714/992-2000

WYNN'S METAL DRAW 4

A HEAVY DUTY, DEEP DRAWING, SEVERE STAMPING, WATER SOLUBLE, INHIBITED, CHLORINATED OIL. THIS PRODUCT IS ESPECIALLY EFFECTIVE REPLACING STRAIGHT CHLORINATED OILS USED FOR SEVERE STAMPING, DEEP DRAWING, COLD HEADING, DEEP DRILLING AND TAPPING WITH WATER AT DILUTIONS OF 1 PART PRODUCT TO 1 - 10 PARTS WATER.

DESCRIPTION

WYNN'S METAL DRAW 4 IS A BROWN, VISCOUS (HONEY-LIKE) WATER EMULSIFIABLE, CHLORINATED, DEEP DRAWING OIL WITH RUST INHIBITORS.

ADVANTAGES

- WATER SOLUBLE - USE DILUTIONS OF 1:1 - 10
- CONTAINS RUST INHIBITORS
- PREVENTS RUPTURES AND SCRATCH MARKS
- CAN BE DILUTED WITH MINERAL OIL OR SOLVENT
- READILY CLEANED BY ALKALINE DETERGENTS, SOLVENTS OR VAPOR DEGREASERS

Wynn Oil Company
P.O. Box 4370/2600 East Nutwood Avenue
Fullerton, CA 92634
(714) 992-2000

WYNN'S METAL DRAW 4

DIRECTIONS FOR USE

USE STRAIGHT FOR SEVERE DRAWING OPERATIONS. DILUTE WITH EQUAL PARTS OF EITHER WATER, MINERAL OIL OR SOLVENT FOR LESS SEVERE OPERATIONS. USE DILUTED FOR DEEP HOLE DRILLING AND TAPPING.

APPLY BY ANY CONVENTIONAL METHOD SUCH AS BRUSHING, DRIPPING, ROLLING, DIPPING OR SPRAYING.

EASILY REMOVABLE WITH ALKALINE OR SOLVENT CLEANERS VIA DIP AGITATION, SPRAY WASHING OR VAPOR DEGREASING.

TECHNICAL DATA

APPEARANCE:	CLEAR
COLOR:	4.0
SPECIFIC GRAVITY @ 60° F:	1.215
10% EMULSION IN WATER:	WHITE OPAQUE, STABLE
VISCOSITY @ 40° C (104° F):	864.7 (cST) 4033 (SUS)
VISCOSITY @ 100° C (212° F):	24.1 (cST) 116 (SUS)
FLASH (CLEVELAND OPEN CUP):	NO FLASH
POUNDS PER GALLON:	10.12

WYNN OIL COMPANY
2600 E. NUTWOOD AVE., P.O. BOX 4370, FULLERTON, CALIFORNIA 92634

SECTION 5

TREATMENT AND APPLICATION EQUIPMENT

Chem-Trend Incorporated
3205 East Grand River
Howell, MI 48843
(517) 546-4520

Metalworking Fluid

Monitoring Program

Chem-Trend's Metalworking Fluid Monitoring Program enables you to realize maximum life and performance from the Chem-Trend product used in your metalworking process. The monitoring service consists of sampling, testing, analyzing, and reporting a fluid condition. The service may be used on either a scheduled maintenance or an intermittent, problem-solving basis.

Procedure

Fluid samples are examined for:

Composition
Contamination
Biological condition
Performance capability

Testing varies, depending on the reason for the analysis. A "routine" sample is examined primarily to determine variation from standard. A "special" sample is tested for specific information. "Problem" samples command the unified efforts of our chemists, microbiologists, technical service personnel, and field engineers. Depending on the need, the sample can be analyzed with our gas chromatograph, infrared, and/or atomic absorption spectrophotometer.

Test results are analyzed and interpreted by Chem-Trend Metalworking Fluid chemists. If we determine that immediate action is required, we notify you immediately by telephone, followed by a printed report.

Samples

Fluid samples should be taken at some in-process point between the sump (or reservoir) and the first work station. In systems where this is not possible, fluid should be taken from the sump. However, the sample should be taken from well below the fluid level, as surface fluid is usually the most contaminated and not representative.

If there is a plant-wide problem in fluid performance, a sample of plant water should also be mailed to our laboratory. This water sample is analyzed and compared with the plant water standard which Chem-Trend has on file. Water supply is prone to variation in chemical content and can change fluid performance.

For your convenience, Chem-Trend will provide 8-oz. containers, mailing envelopes and Sample Analysis Request (SAR) forms, which are to be completed and mailed with each sample of metalworking fluid or water.

Chem-Trend Incorporated
3205 East Grand River
Howell, MI 48843
(517) 546-4520

Metalworking Fluid

Monitoring Program

Test Categories

Composition

Composition refers to the general makeup of the product, and samples are tested to see if their chemical and physical properties are within normal limits.

All fluid samples are tested for pH and dilution; all physical properties of the sample are tested and compared with the product standard.

The pH (a measure of acidity and/or alkalinity) is one of the most important fluid properties measured. The pH value of fluid is an excellent indicator of fluid "health." Each product has a different pH range. The variation in pH serves as a guide for both type and extent of follow-up examination.

The determination of dilution length is often difficult to assess due to contamination. Fluids assimilate some types of tramp oils which distort dilution findings. Dilution is determined by various methods. The method used is dependent on the product type and customer preference.

When a specific problem in the field is encountered, the fluid sample is compared with the retained sample of the actual shipment(s). This is possible because we keep a sample of <u>every</u> batch of metalworking fluid at the Chem-Trend Quality Control Laboratory for six months after shipment.

Contamination

Contamination is the most common cause of poor metalworking fluid life. The most prevalent causes are particles from the work process or tramp oil accumulation, usually as a result of leaks in machine lubrication and/or hydraulic systems. Factory environment, metal stock carry-in, and worker waste can also be significant factors. Contamination compromises the properties of a metalworking fluid. Unless the fluid is continuously maintained, it will deteriorate and become a risk to production and machinery.

Every fluid sample is observed for contamination and all suspect fluids are analyzed by a gas chromatograph, infrared, and/or atomic absorption spectrophotometer. Before we advise replacing or restoring a contaminated fluid, the sample is analyzed by various techniques to assure a practical and sound recommendation.

Chem-Trend Incorporated
3205 East Grand River
Howell, MI 48843
(517) 546-4520

Metalworking Fluid

Monitoring Program

Test Categories (cont'd.)

Biological Condition

We analyze the biological condition of every central system fluid sample for bacteria and fungi (mold and yeast). We report the total fungi count as "mold." Colony counts are determined and recorded; if required, proper corrective actions are recommended. However, a single sump sample is not usually analyzed for biological contamination unless it is necessary to solve a chronic problem.

Because mail-transit time of a sample is often sufficient to totally distort an actual biological situation, we sometimes conduct biological tests on site. Transit time is recorded and taken into consideration.

Performance Capability

The performance capabilities include those primary characteristics of the fluid which enable a machining process to be accomplished, and secondary characteristics which make use of the product practical or desirable. Primary characteristics include lubricating and cooling properties. Secondary characteristics include foam formation and break, corrosion resistance, and stability. Tests are conducted on the basis of the Sample Analysis Request (SAR) received.

Historical Data

Each coolant system has its own unique operating characteristics. Our analysts utilize the computer to determine the effectiveness of the control products, the generation of biological trends, and the reasons for periodic fluctuations in coolant performance for your specific system.

Similar systems can be compared and maintenance, production, and tool problems can be resolved.

Chem-Trend Incorporated
3205 East Grand River
Howell, MI 48843
(517) 546-4520

Metalworking Fluid

Monitoring Program

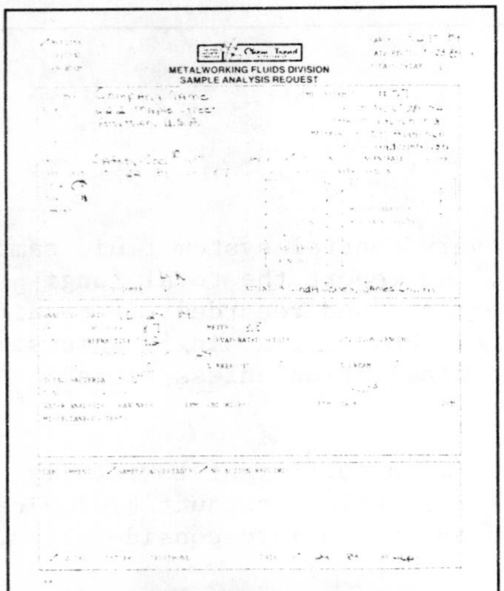

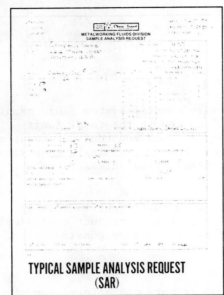

TYPICAL SAMPLE ANALYSIS REQUEST (SAR)

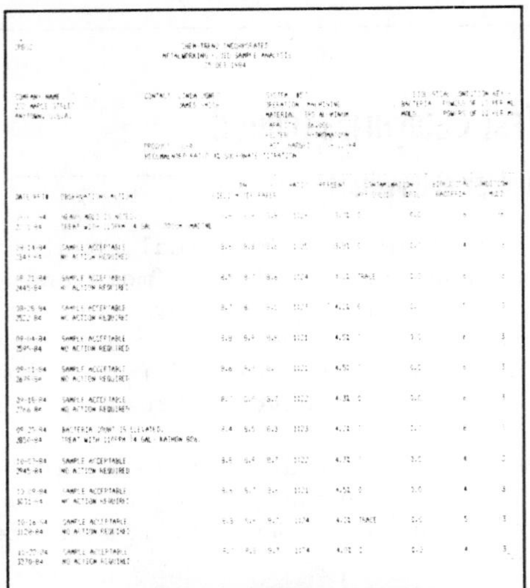

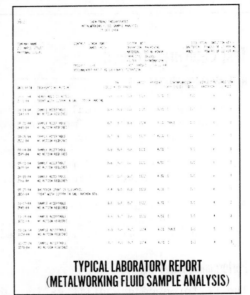

TYPICAL LABORATORY REPORT (METALWORKING FLUID SAMPLE ANALYSIS)

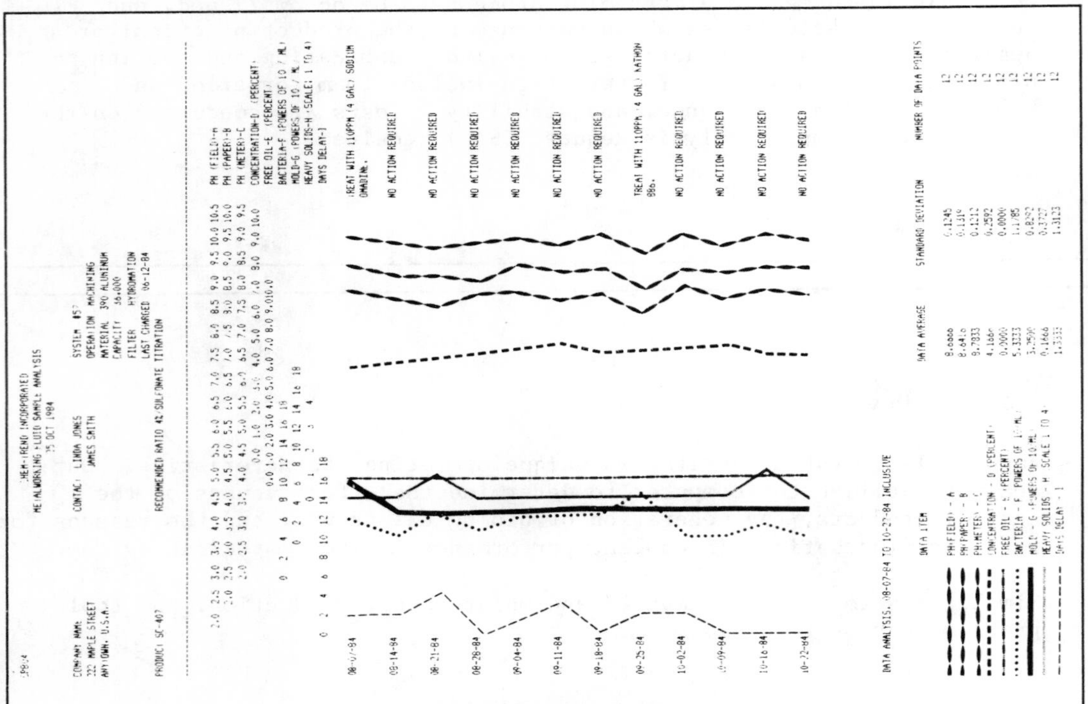

TYPICAL HISTORICAL DATA REPORT

Clinton Centrifuge Incorporated
P.O. Box 217/220 Jacksonville Road
Hatboro, PA 19040
(215) 443-7878

CLINTON CENTRIFUGE
INC.

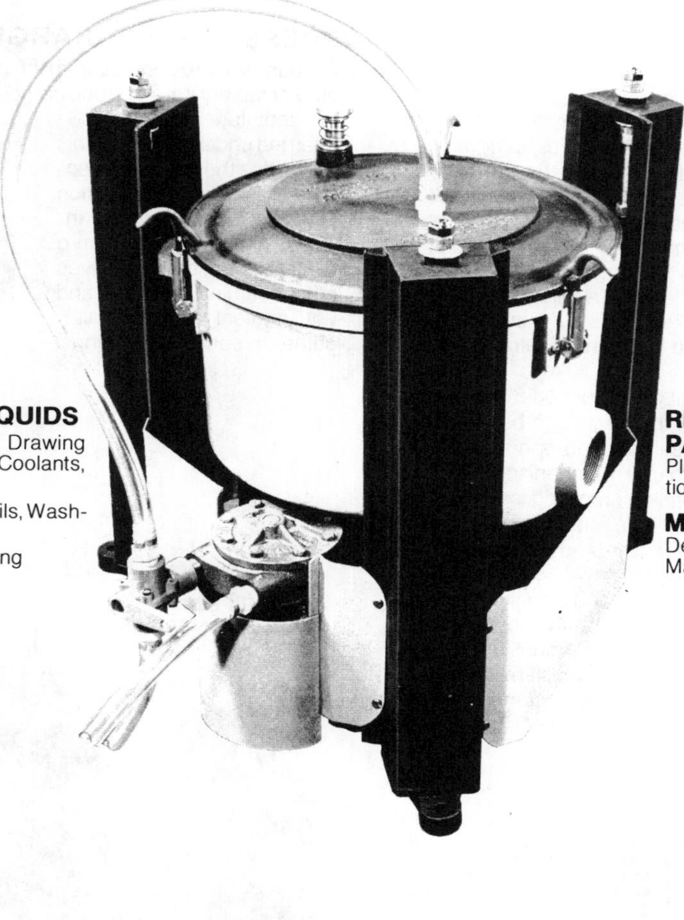

RECLAIMING COSTLY LIQUIDS
METALWORKING: Cutting Oils, Drawing Lubes, ECG, ECM, EDM, Grinding Coolants, Honing Oils, Hydraulic Fluids
HEAT TREATING: Quenching Oils, Washing Solutions
OPTICS: Cutting, Grinding, Lapping

RECOVERING VALUABLE PARTICULATES
Plating Plant Dewatering, Silver Reclamation, Precious Metal Mining

MEETING EPA REQUIREMENTS
Dewatering Sludge for Disposal, Preparing Material for Approved Disposal

SOLID & PERFORATE BASKET CENTRIFUGES

Clinton Centrifuge Incorporated
P.O. Box 217/220 Jacksonville Road
Hatboro, PA 19040
(215) 443-7878

OPERATING PRINCIPLE

The solid-wall basket centrifuge uses centrifugal force (up to 2200 gravities) to promote solid/liquid separation. The feed slurry is introduced into the rotating basket assembly and is accelerated to the basket speed. Solids/contaminants are pulled out of the liquid by centrifugal force and are collected on the basket wall. The clarified liquid flows up and out of the basket, is collected by the centrifuge housing, and then flows by gravity from the housing. Liquid clarification continues until the basket can no longer retain additional solids and must be emptied.

ADVANTAGES

LOW INITIAL COST
The price of CLINTON CENTRIFUGES start at UNDER $3000 and require no special foundations, connections or mounting.

EASY TO INSTALL AND OPERATE
Every CLINTON CENTRIFUGE is easy to install, simple to operate. Their unique anti-vibration mounting system requires

MANY SIZES AND OPTIONS FOR OPTIMUM

SLUDGE DISCHARGE

Soft or gelatinous sludges can be removed while the centrifuge continues to rotate. The removal can be fully automated for unattended operation, or can be semi-automatic.

FULLY—AUTOMATIC SLUDGE DISCHARGE utilizes a field-programmable solid-state timer to periodically empty the basket while the centrifuge operates. At the selected time interval, a self contained hydraulic pump is actuated. The hydraulic pressure closes the inlet valve to momentarily shut off the inlet flow, and simultaneously actuates the automatic discharge mechanism. Sludge is discharged through a reinforced hose to a collection drum. After about 30 seconds the pump stops, the inlet valve opens, the discharge mechanism retracts and operation continues.

SEMI—AUTOMATIC SLUDGE DISCHARGE allows periodic cleaning of the centrifuge basket by use of a hand-operated discharge mechanism. When basket emptying is desired, the operator shuts off the inlet flow and manually rotates a lever to discharge the sludge through a reinforced hose to a collection drum. After about 30 seconds the handle is released, inlet flow is resumed, and operation continues.

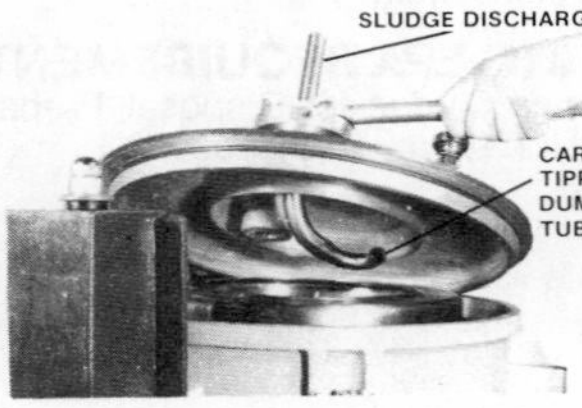

PRESSURE DISCHARGE

For pumping the centrate from the centrifuge a manually actuated or factory set skim tube carries clarified liquid to the top of the centrifuge where it is discharged under pressure. This eliminates the need for a separate pump on the collection tank and results in lower installed costs. The pressure discharge option also solves many foaming problems and will prevent foaming altogether in some applications.

DIFFUSER

For difficult to separate materials a diffuser is recommended. This device, integral to the basket, greatly improves solids capture and centrate clarity. By uniformly diffusing the feed material into the basket, rapid acceleration and separation with minimal turbulence is achieved. Uniform distribution eliminates preferential buildup of the solids and insures vibration free operation.

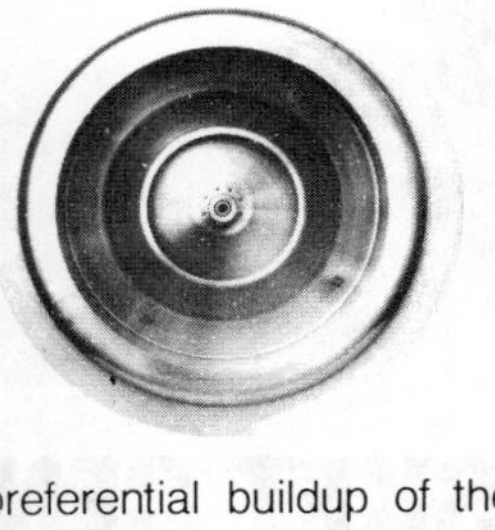

MODEL B30 WITH CONTROLS

Clinton Centrifuge Incorporated
P.O. Box 217/220 Jacksonville Road
Hatboro, PA 19040
(215) 443-7878

no anchoring. Just supply power and product connections and your separation system is ready to work for you. The centrifuge is so simple to operate that unskilled labor can be used for installation and operation.

EASY TO CLEAN
In less than two minutes a full basket can be easily removed and a clean one installed. To eliminate even this downtime, our optional sludge discharge attachment can be added to automatically empty the basket of soft or gelatinous sludge. For sludge that is too firm for automatic sludge discharge, we suggest the use of two baskets (one can be cleaned while the other is in service).

SPACE SAVING
Our smallest model, B30, occupies less than 3 square feet of floor space. Since no anchoring is necessary, units are readily portable and can be moved to various locations as required.

LONG SERVICE LIFE
Precision dynamic balancing, [illegible] with our unique, triangulated [illegible] mounting method essentially [illegible] vibration. This design improves [illegible]tion effectiveness and [illegible] all moving parts.

PERFORMANCE

- **MOTORS**
  All voltages and frequencies; explosion proof
- **VARIABLE-SPEED DRIVES**
  AC OR DC; dynamic braking
- **INTEGRAL FEED PUMP**
  Progressive cavity or gear
  7 gpm standard; others available
- **COVER SAFETY INTERLOCK**
  Prevents motor operation unless lid is properly secured
- **HIGH TEMPERATURE MODIFICATION**
  Up to 500 degrees F
- **ANTI-WAVE DEVICE**
  Recommended for B30H and B50H
- **INLET STRAINER**
  All 304 SS; recommended with integral feed pump
- **SPARE BASKETS**
  Recommended when sludge discharge option is not used
- **MOTOR CONTROLS**
  NEMA 1; NEMA 4; EXPLOSION PROOF housings for starters and controls available
- **PORTABLE ASSEMBLIES**
  Mounted on hand trucks, platform trucks, skids—with or without tanks, pumps, etc.

LOW COST
Without investing in another piece of capital equipment, installation of a perforate basket in your CLINTON CENTRIFUGE (new or existing) allows you to perform any or all of the above functions efficiently and economically.

FLEXIBLE
By interchangeably using both solid and perforate baskets in your CLINTON CENTRIFUGE, you can use it for either high-efficiency centrifuging, or for filtering, washing, drying, or wringing—especially since you can change from solid basket to perforate basket in minutes!

FILTERING/DRYING/ WASHING
In a perforate basket centrifuge, entering liquid must escape through the walls of the basket. Depending on the desired function, the liquid can be made to pass through one or more of several barriers before being discharged. FOR FILTERING the inside surface of the perforate basket can be lined with any filter paper or cloth in pre-cut strip or bag form. A wide variety of materials and micron ratings is available. FOR DRYING, liquids-solvents, oils, water, etc., are extracted by centrifugal force and discharged through the basket perforations. FOR WASHING, spray jets or nozzles thoroughly wash the basket contents. Automatic controls can provide desired washing/drying sequences.

OPTION: An optionally available sensor provides an indication of a clogged or full basket—prevents mixing of contaminated liquid with clean liquid.

Clinton Centrifuge Incorporated
P.O. Box 217/220 Jacksonville Road
Hatboro, PA 19040
(215) 443-7878

CLINTON CENTRIFUGE INC.

P.O. BOX 217
220 JACKSONVILLE ROAD, HATBORO, PA. 19040

TELEPHONE: 215/443-7878
TELEX: (ITT) 4991548 CLINCEN

Separation Specialists

SPECIFICATIONS

MODEL	B30S	B30	B30H	B50	B50H
FLOW RATE—MAX GPH	200	420	720	2000	3000
GRAVITIES (G)	2200	550	2200	550	1400
BASKET					
SLUDGE CAPACITY—$IN.^3$	157	413	413	1160	1160
DIAMETER—IN.	12	12	12	20	20
HEIGHT—IN.	2.5	6	6	9	9
MOTOR H.P.	1	1.5	1.5	5	5
STANDARD MATERIALS					
BODY AND LID	ALUMINUM				
BASKET	304 SS				
OPTIONAL MATERIALS					
BODY AND LID	316 SS				
BODY AND LID	CAST IRON				
BASKET	316 SS				
BODY AND LID COATINGS	A wide selection of epoxy or fluorocarbon coatings is available to meet specific requirements.				
INSTALLATION DATA					
WIDTH X LENGTH X HEIGHT—IN.	24x24x24	24x24x24	24x24x24	36x50x24	36x50x24
INLET SIZE—NPTF	1/2	1/2	1/2	2	2
OUTLET SIZE—NPTF	2	2	2	3	3
NET WEIGHT—LBS.	60	80	80	300	300
SHIPPING WEIGHT—LBS.	75	100	100	400	400
OPTIONS	Basket, motor and attachment options described in the brochure are available for all models. Please call for optimum configuration in your application.				

SAMPLE TESTING

Know in advance a CLINTON CENTRIFUGE will do your job. Send us a 10-gallon sample of fluid to our testing laboratory in Hatboro, PA. For $150, creditable toward your purchase, we will process the sample and send you a report, plus the clarified liquid and sludge.

RENTAL SERVICE

Try a CLINTON CENTRIFUGE under your own actual operating conditions for a very nominal rental fee. Minimum rental period—3 months, renewable month-by-month. 75% of the first three months rental will be credited to your purchase. Call for details. Rental units are not available for toxic or radioactive applications.

Printed in U.S.A.

Clinton Centrifuge Incorporated
P.O. Box 217/220 Jacksonville Road
Hatboro, PA 19040
(215) 443-7878

CLINTON CENTRIFUGE

LIQUID/LIQUID SEPARATOR

CONTAMINATED LIQUID

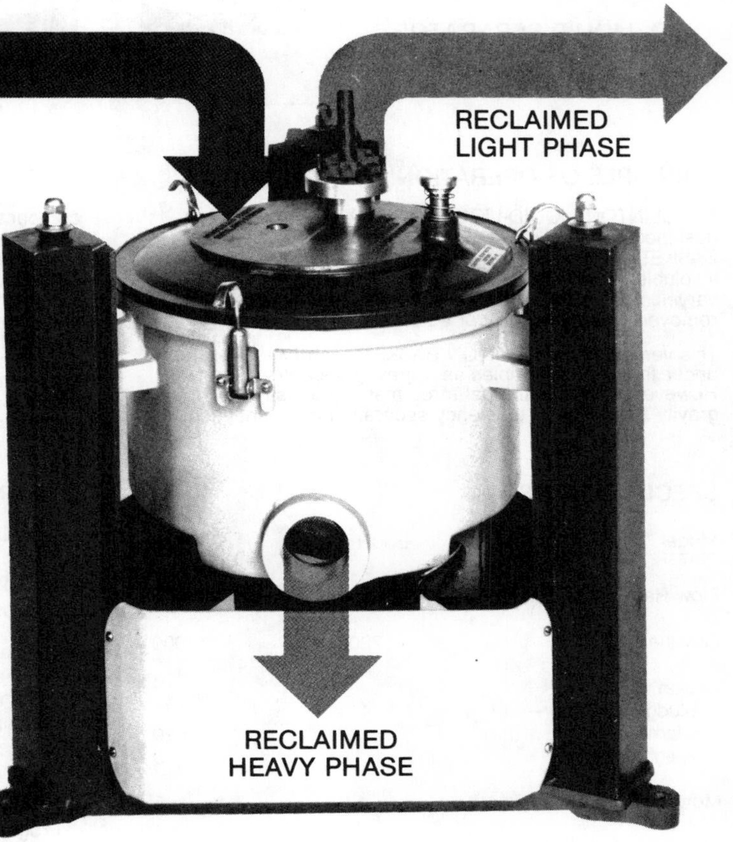

- Coolant/Tramp Oil
- Lubricant/Tramp Water
- Quench Oil/Tramp Water
- Aqueous Cleaning Solution/Tramp Oil
- Cutting Oil/Tramp Water
- Hydraulic Oil/Tramp Water

Remcves solids to 10 microns

Removes up to 99% of contaminants

Easily adjusted for a wide range of specific gravities.

INSTALLATION RECOMMENDATIONS

For use on individual machine sumps, a dedicated or portable unit can recirculate the sump through the centrifuge periodically. Complete portable systems can be supplied.

For central systems, the separator should operate in a bypass mode. Systems in excess of 20,000 gallons can often be serviced with a single B50H-3.

PERFORMANCE

The efficiency of separation attained is determined by the specific gravity difference between the two liquids and the centrifuge feed rate. A minimum specific gravity difference of 1.5% is required to achieve a successful separation. Large specific gravity differentials and low flow rates make it possible to attain 99% removal of contaminants. Typically, solids down to 10 microns are removed, although submicron separations can be achieved under ideal conditions.

Clinton Centrifuge Incorporated
P.O. Box 217/220 Jacksonville Road
Hatboro, PA 19040
(215) 443-7878

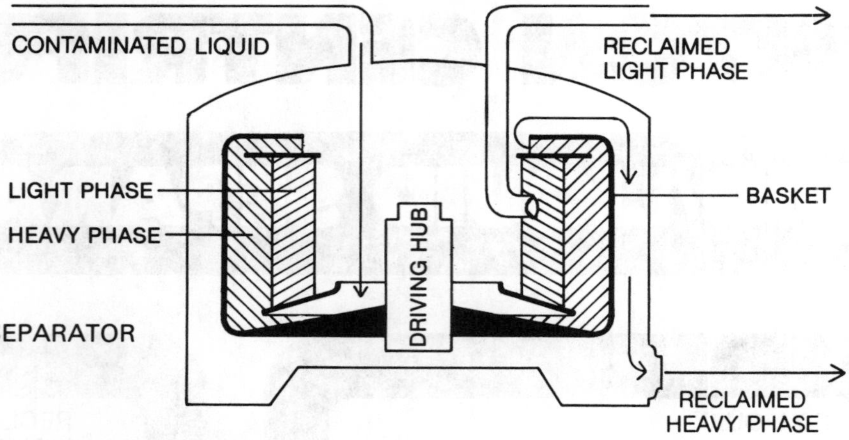

LIQUID/LIQUID SEPARATOR

PRINCIPLE OF OPERATION

The CLINTON LIQUID-LIQUID SEPARATOR is a specially designed variation of the proven high-gravity CLINTON BASKET unit which allows continuous separation of insoluble liquids. The unit will not clog due to high or varying solids, and accumulated solids may be easily removed either manually or automatically.

This version of the CLINTON BASKET centrifuge operates under the same principles as a gravity decant/settling tank. However, using centrifugal force many times stronger than gravity permits high efficiency separation in a continuous compact installation. The feed is introduced through the accelerator/diffuser at the bottom of the rotating basket. Here the material is accelerated rapidly to the basket speed, where centrifugal force causes the light liquid phase to "float" on the heavy liquid phase. As the separated components flow toward the top of the basket further clarification of the individual components takes place. At the top of the basket, the heavy phase overflows the lip into the centrifuge body for gravity discharge while the light phase is skimmed off the free surface to discharge under pressure through the machine cover.

SPECIFICATIONS

Model	B30H-3	B50H-3
Flow Rate—Max GPH	240	600
Gravities (G)	2200	900
Basket		
Sludge Capacity—In.^{3}	413	1160
Diameter—In.	12	20
Height—In.	6	9
Motor H.P.	1.5	5
Standard Materials		
Body and Lid	Aluminum	Aluminum
Basket	304SS	304SS

OPTIONS

- OPTIONAL MATERIALS
  316 s.s. and cast iron available.
- SLUDGE DISCHARGE
  Soft or gelatinous sludges can be removed while the centifuge continues to rotate. The removal can be fully automated for unattended operation, or can be semi-automatic.
- MOTORS
  All voltages and frequencies; explosion proof
- HIGH TEMPERATURE MODIFICATION
  Up to 500 degrees F
- INTEGRAL FEED PUMP
  Progressive cavity or gear
  7 gpm standard; others available
- PORTABLE ASSEMBLIES
  Mounted on hand trucks, platform trucks, skids—with or without tanks, pumps, etc.

Filtra-Systems Company
23900 Haggerty, Bldg. B
Farmington, MI 48024
(313) 477-7322

PERMA-FLOW

A Rotary Indexing,
Wedge Wire Filter

Part Of The Filtration Systems Technology of **FILTRA-SYSTEMS**

PERMA-FLOW offers complete versatility in liquid-solids separation

- **Strainer type separation using either continuous or timed backwash operation**
- **Finer filtration utilizing the filter cake by backwashing on vacuum differential**
- **Fine filtration to less than one micron by precoating the wedge wire with bulk media**

The features of PERMA-FLOW offer many distinct benefits —

Unique, pressurized incremental backwash gives continuous system operation and positive cake removal.

Cylinder I.D. and O.D. wedge wire provides high density filter area.

Modular design provides for parallel operation on large systems, and for total spare capability.

A complete product line with units from 3.5 square feet up to 225 square feet means that your needs can be matched exactly.

Filtra-Systems Company
23900 Haggerty, Bldg. B
Farmington, MI 48024
(313) 477-7322

PERMA-FLOW filters are unmatched in flexibility, enable a wide range of applications and installations.

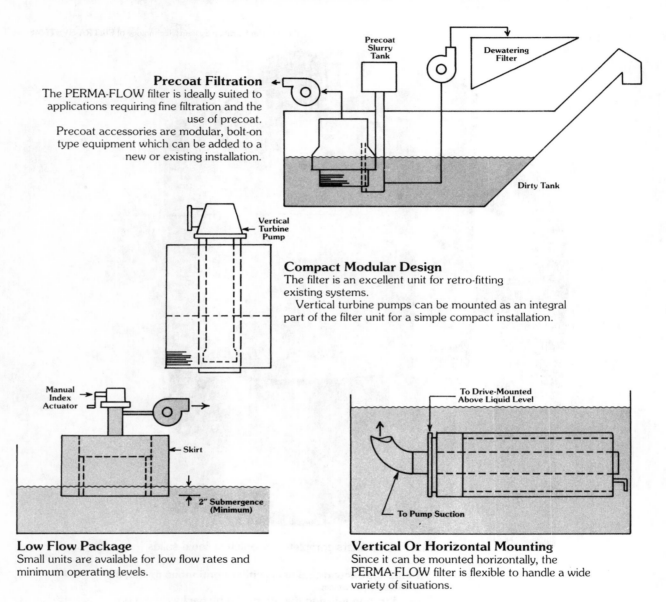

Precoat Filtration

The PERMA-FLOW filter is ideally suited to applications requiring fine filtration and the use of precoat.

Precoat accessories are modular, bolt-on type equipment which can be added to a new or existing installation.

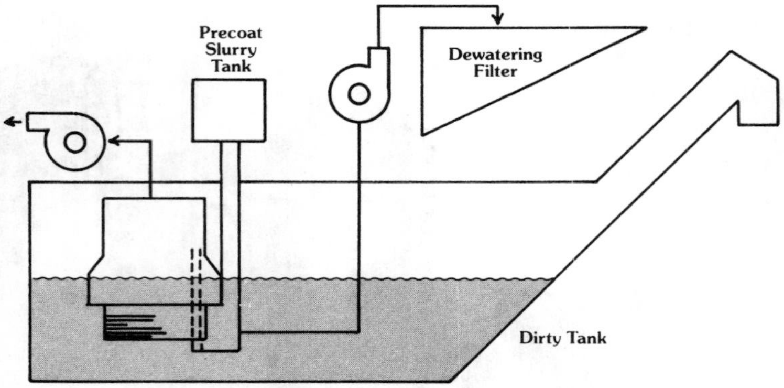

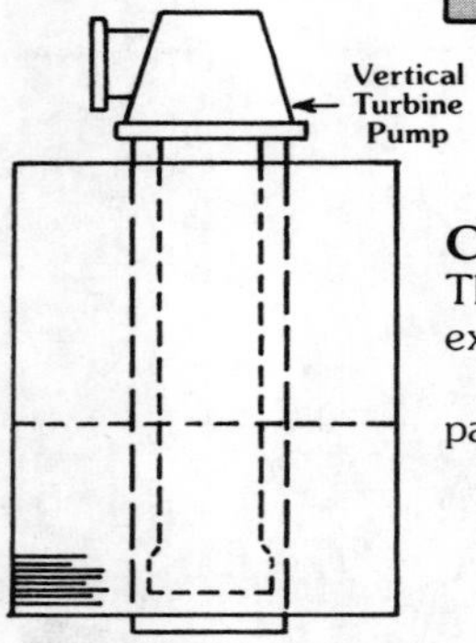

Compact Modular Design

The filter is an excellent unit for retro-fitting existing systems.

Vertical turbine pumps can be mounted as an integral part of the filter unit for a simple compact installation.

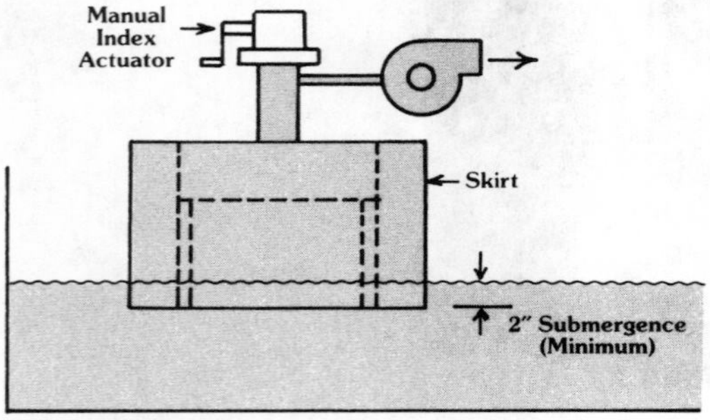

Low Flow Package

Small units are available for low flow rates and minimum operating levels.

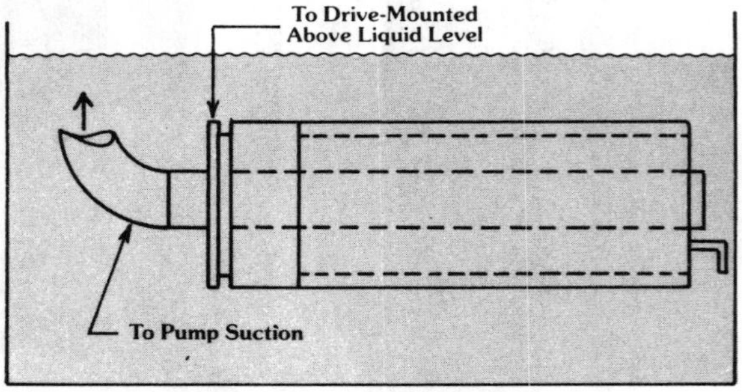

Vertical Or Horizontal Mounting

Since it can be mounted horizontally, the PERMA-FLOW filter is flexible to handle a wide variety of situations.

The PERMA-FLOW filter is another example of our continuing commitment to meet your filtration needs of today. We have complete laboratory facilities to evaluate your samples and a full size, fully operational unit set up for your observation and evaluation. Call or write.

FILTRA-SYSTEMS™
FILTRATION SYSTEMS TECHNOLOGY

23900 HAGGERTY RD.
FARMINGTON, MICHIGAN 48024
(313) 477-7322

Lincoln St. Louis/McNeil Corporation
4010 Goodfellow Boulevard
St. Louis, MO 63120
(314) 383-5900

there's a Centro-Matic system for you

. . . because we specifically design it to fit your application!

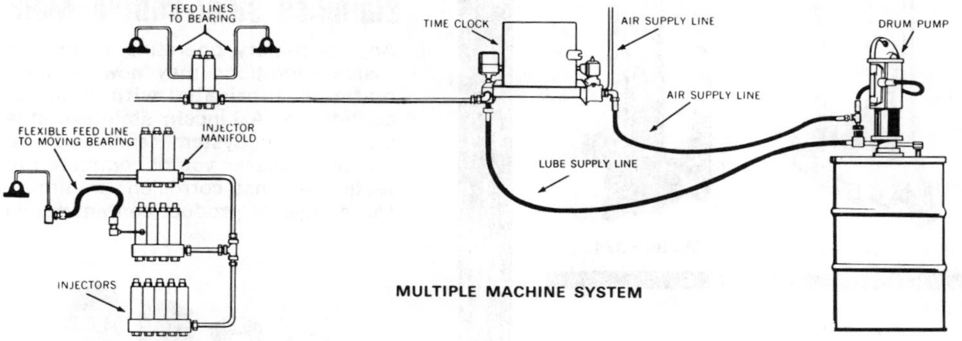

MULTIPLE MACHINE SYSTEM

Each Centro-Matic system is different. However, the basic idea of a centralized lubrication system is the same in every case. A pump delivers lubricant through delivery lines to an injector (metering device). Each injector measures out the correct amount of lubricant required for each bearing.

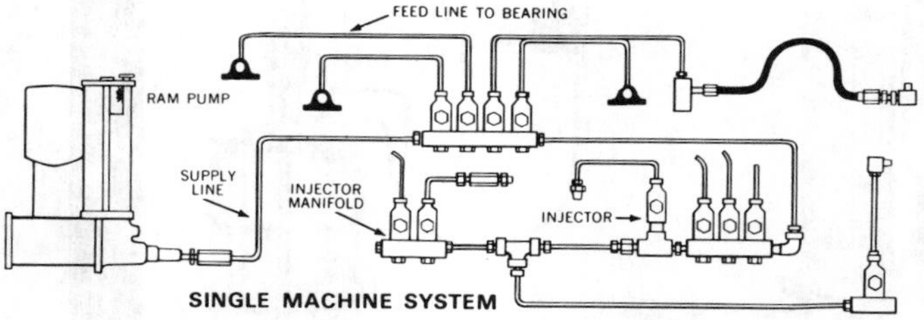

SINGLE MACHINE SYSTEM

BASIC OPERATING PRINCIPLES OF CENTRO-MATIC INJECTORS

Each Lincoln Centro-Matic Injector may be adjusted to discharge just the right amount of lubricant to meet actual bearing requirements. Injectors may be mounted singly at each bearing, or grouped in manifold at one location. In either case, injectors are supplied with lubricant under pump pressure, through a single supply line.

Two types of injectors are available—with top adjustment and side adjustment. They may be used in the same circuit and their selection is made on the basis of bearing lubricant requirements.

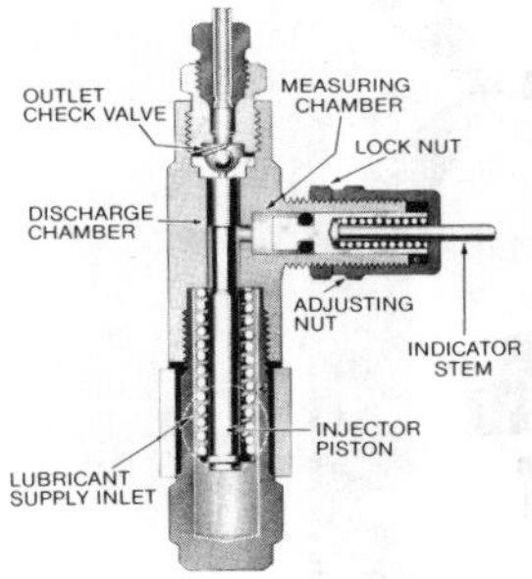

SL-32, SL-33, SL-42, and SL-43

Stage 1

Incoming lubricant—under pressure from the supply line—moves the injector piston forward. The piston forces a precharge of lubricant from the discharge chamber through the outlet check valve to the feed line.

Stage 2

When the system is vented (pressure relieved), the piston returns to rest position, transferring lubricant in the measuring chamber to the discharge chamber.

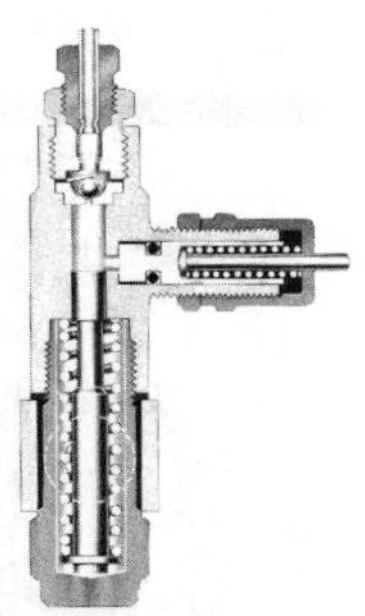

SL-1, SL-11, SL-41, and SL-44

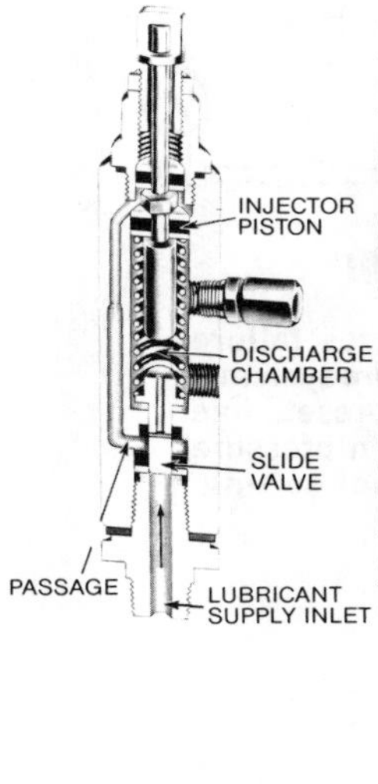

Stage 1

The injector piston is in its normal, or rest, position. The discharge chamber is filled with lubricant from the previous cycle. Under the pressure of incoming lubricant, the slide valve is about to open the passage leading to the piston.

Stage 2

When the slide valve uncovers the passage, lubricant is admitted to the top of the piston, forcing the piston down. The piston forces lubricant from the discharge chamber through the outlet port to the bearing.

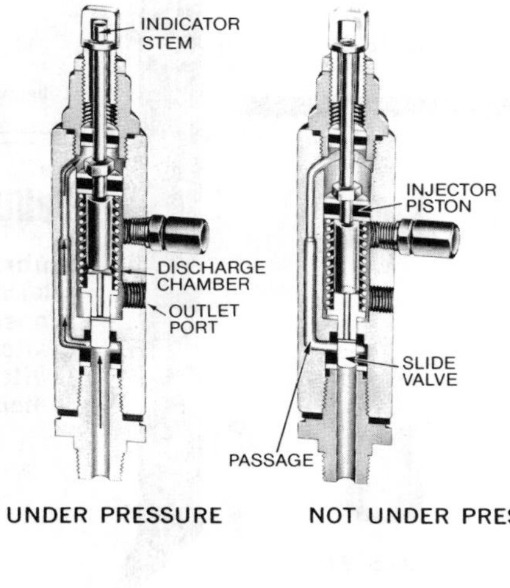

UNDER PRESSURE   NOT UNDER PRESSURE

Stage 3

As the piston completes its stroke, it pushes the slide valve past the passage, cutting off further admission of lubricant to the passage. Piston and slide valve remain in this position until lubricant pressure in the supply line is vented (relieved) at the pump.

Stage 4

After pressure is relieved, the compressed spring moves the slide valve to closed position. This opens the port from the measuring chamber and permits the lubricant to be transferred from the top of the piston to the discharge chamber.

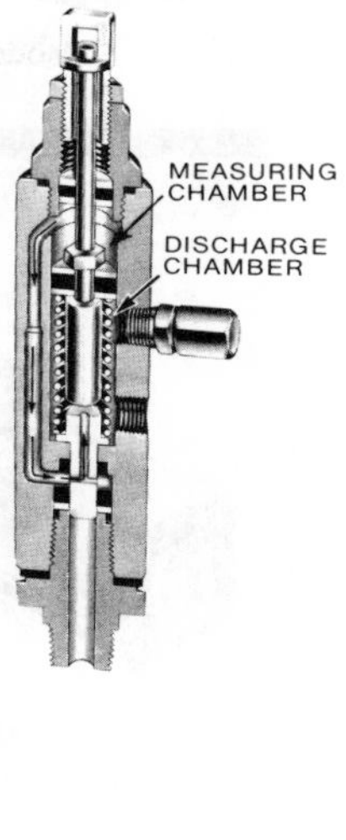

Lincoln St. Louis/McNeil Corporation
4010 Goodfellow Boulevard
St. Louis, MO 63120
(314) 383-5900

TYPICAL PUMPING UNITS

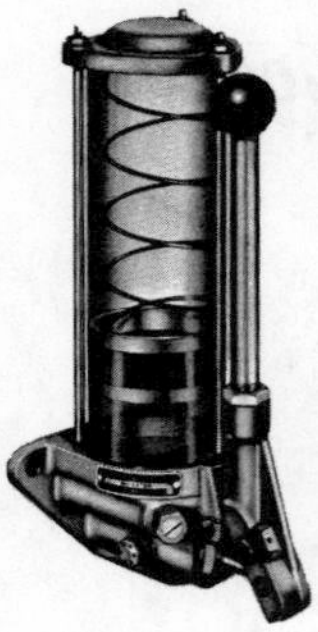

Model 1810

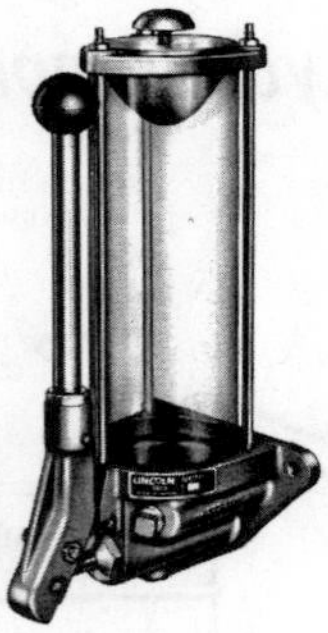

Model 1812

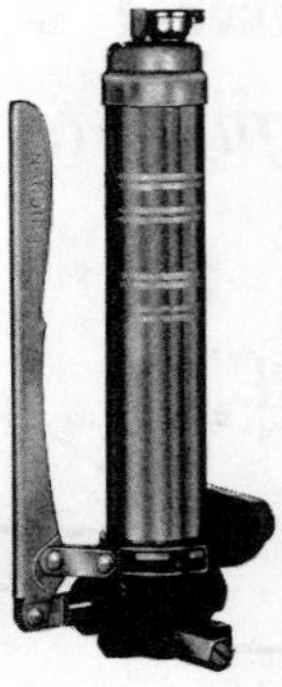

Model 83742

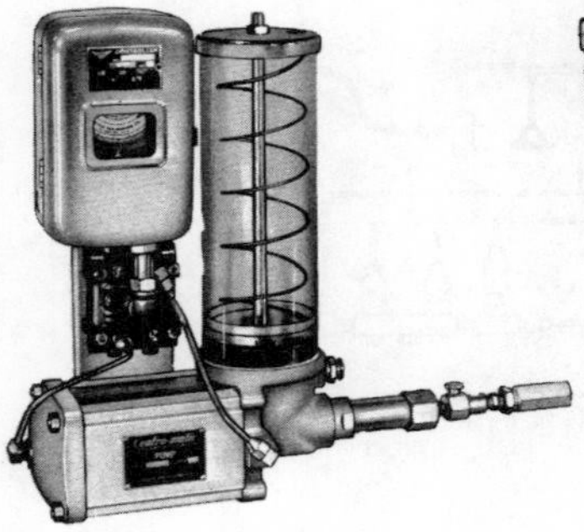

Model 82655

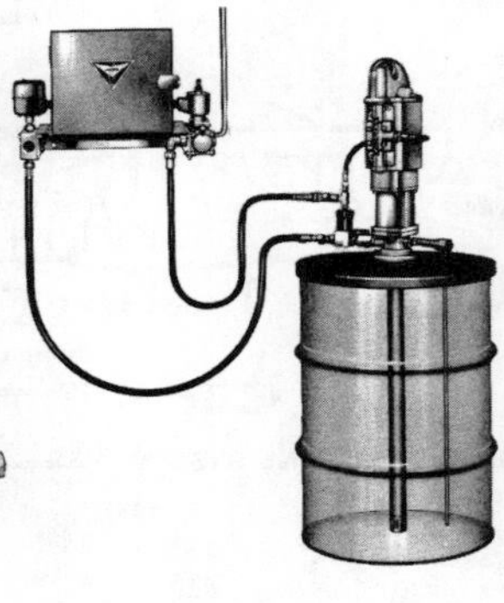

Model 1827

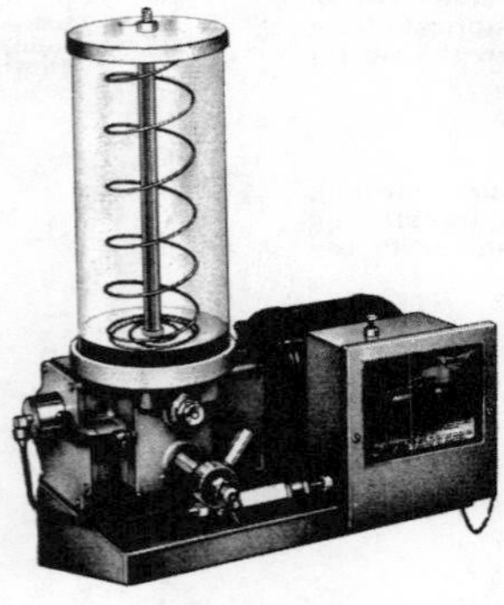

Model 1849

Model 1837

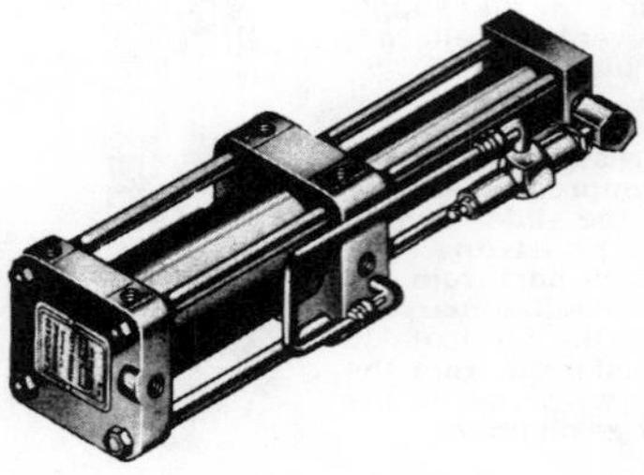

Model 1820

Model 83846

stainless steel Centro-Matic

Any machinery operating under corrosive conditions may now be automatically lubricated with complete confidence. A Lincoln stainless steel Centro-Matic System (exclusive with Lincoln) assures you of complete protection against corrosion; minimizes the danger of product contamination.

Bank of Lincoln injectors comparing heavy corrosion of carbon steel with unaffected stainless steel injectors after four months' use on can-filling machine in food processing plant (Unretouched photo.)

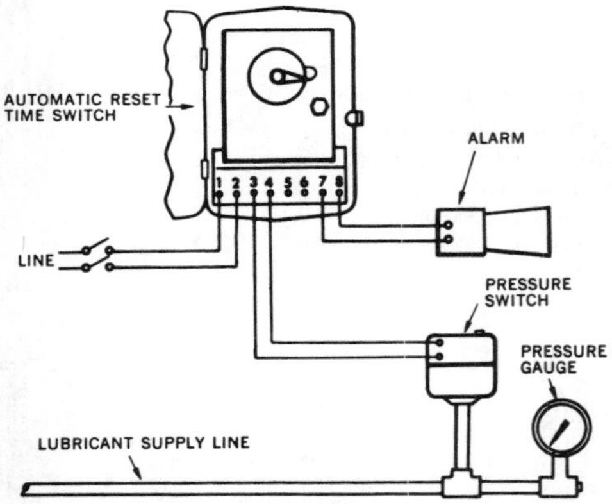

failure alarm system

Lubricant pressure actuates failure alarm for low or high pressure systems. Consists of an automatic reset time switch and a normally open pressure switch. 5-minute minimum setting. 4-hour maximum setting.

Lincoln St. Louis/McNeil Corporation
4010 Goodfellow Boulevard
St. Louis, MO 63120
(314) 383-5900

MANUAL —Designed for smaller individual machines. This type of pump provides a low cost, highly efficient method of pumping lubricant to injectors. To cycle a complete bank of injectors takes only a few seconds. Visual indicator signals the completion of the cycle. In Manually-Operated Systems, the lubricant pump is of the hand-operated type and intervals of operation are normally performed by the operator.

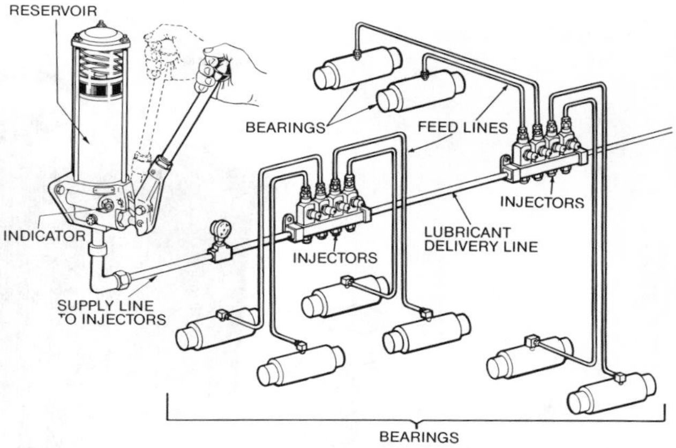

AIR-OPERATED —Pumps are actuated automatically by compressed air at various pre-determined intervals. Air-operated pump delivers lubricant under pressure to injectors. When all injectors have cycled, pump shuts off automatically and vents itself. Systems available for pumping lubricant to single machines or over long distances to large machine groups. Available with automatic, manual, mechanical, or electrical controls.

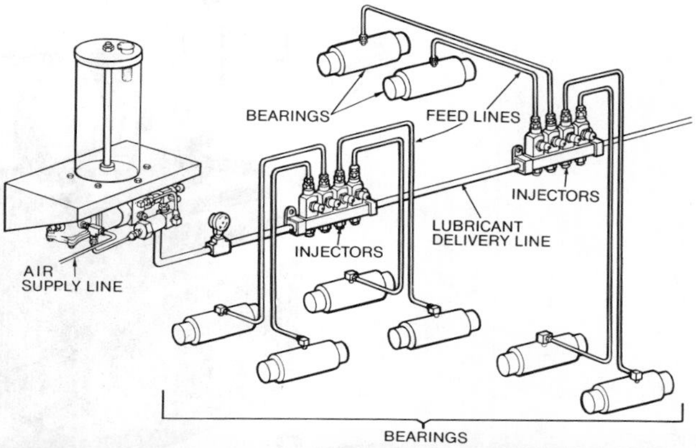

ELECTRIC —Highly efficient lubricant pumps make centralized lubrication available to many plants, mills and mines where compressed air is not available or where electrical operation is preferred. Totally enclosed motor supplies the power requirements of the pumping mechanism. Time control is adjustable to provide pre-determined frequency of lubrication.

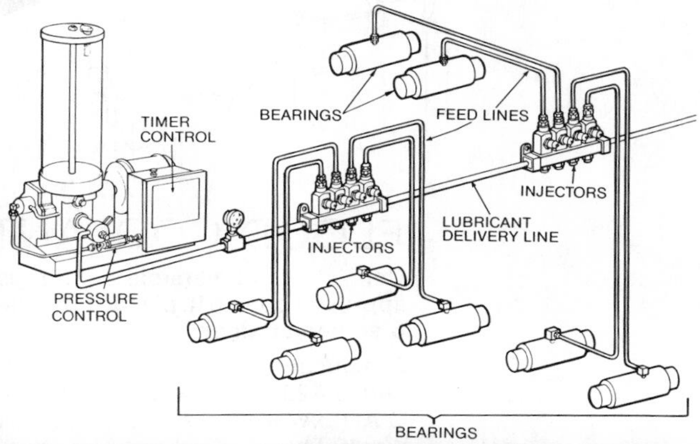

HYDRAULIC —A complete hydraulically-powered pumping unit for centralized lubrication of individual machines; or for installation on machinery, such as coal mining and earth moving equipment, which utilizes a hydraulic pressure system for operation of various movements of the machine, and requires periodic high pressure application of fluid lubricants. Available with automatic, manual, mechanical, or electrical controls.

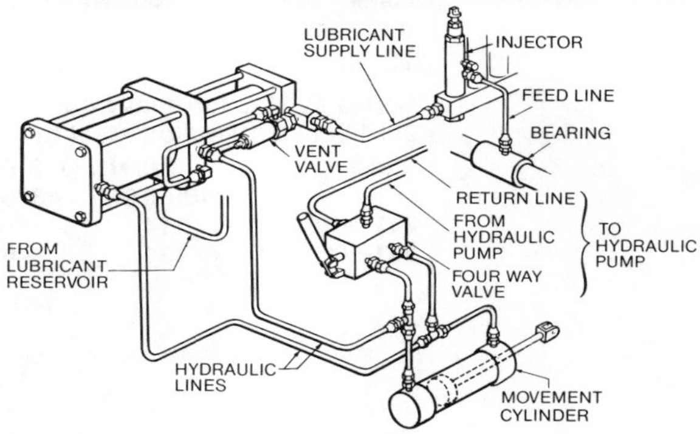

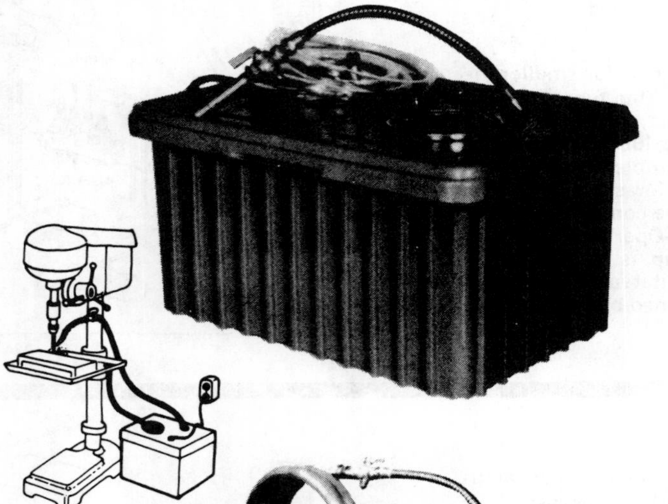

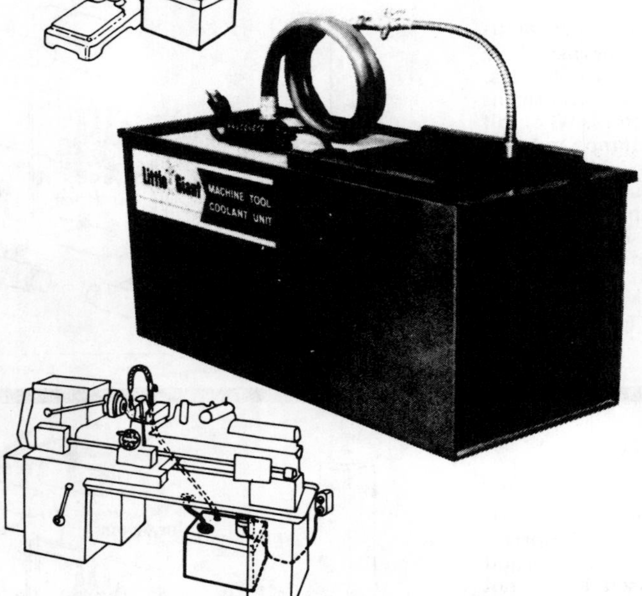

FLOOD COOLANT UNITS

Consist of submersible centrifugal pump, 6 ft. U.L. approved cord, 6 ft. plastic hose, flexible nozzle, valve & screen strainer.

Specifications:
Series #231-1 for water soluble & light cutting oils. O.A. (LxWxHt.) 17x11x8", 5 gallon polyethylene tank, Wt. 19 Lb.

Cat.#	Gal./Hr. @ 3 Ft.	Price
231-6908	84	**$150.**
231-6909	166	**$200.**

Series #231-2 for heavy cutting oils. O.A. (LxWxHt.) 25x11x11", 10 gallon steel tank, Wt. 46 Lb.

Cat.#	Gal/Hr. @ 3 Ft.		Price
	#100 SSU	#300 SSU	
231-6910	114	92	**$250.**
231-6911	131	106	**$300.**

Rowald Refrigeration & Heating Systems Incorporated
1121 First Avenue
Rockford, IL 61104-1295
(815) 962-7733

LIQUID COOLANT CHILLERS

Refrigeration

Rowald Refrigeration & Heating Systems Incorporated
1121 First Avenue
Rockford, IL 61104-1295
(815) 962-7733

Rowald LIQUID COOLANT CHILLERS

FOR MORE THAN A QUARTER CENTURY, Rowald liquid coolant chillers have demonstrated their superiority in machine tool operations throughout the United States. Actual on-the-job experience has proved repeatedly that these chillers reduce operating costs by more than 20%! Adaptable to all types of machine tools and plastic molding equipment, Rowald chillers increase production, together with abrasive and tool life, and help to maintain rigid tolerances required in precision machining.

BENEFITS

- Available in both water- and air-cooled models
- Eliminate the need for extra-large coolant tanks
- Retard bacteria growth
- Cooler machined parts provide better handling conditions for machine operator
- Control vapor fumes to a minimum
- Eliminate burning of parts, tools and abrasives
- Save time on gaging when coolant is maintained at a uniform temperature
- Insure that bonding on abrasive wheels remains firm—resulting in longer periods between down-time for redressing of wheels
- Eliminate expansion and contraction of machine and tooling
- Increase productivity

AXI (Air-Cooled) Chillers are equipped with an air-cooled condenser. The heat is dissipated into the air. Especially economical when water shortage is a big factor.

WXI (Water-Cooled) Chillers constructed with a water-cooled condenser dissipate heat with use of water. Available for tap-water or cooling-tower applications.

NOTE: Approximately 3 GPM of water per horsepower of refrigeration are required when using tower water. Tap-water consumption is approximately 1 GPM per horsepower.

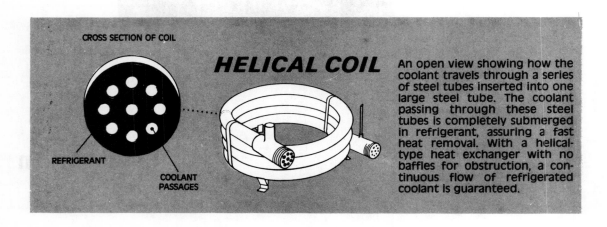

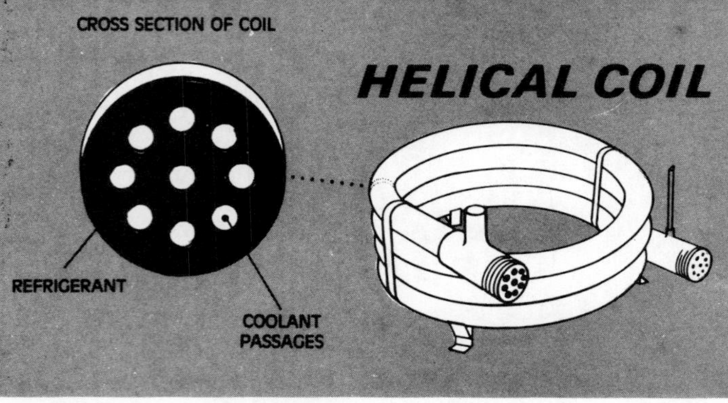

HELICAL COIL

An open view showing how the coolant travels through a series of steel tubes inserted into one large steel tube. The coolant passing through these steel tubes is completely submerged in refrigerant, assuring a fast heat removal. With a helical-type heat exchanger with no baffles for obstruction, a continuous flow of refrigerated coolant is guaranteed.

AIR COOLED

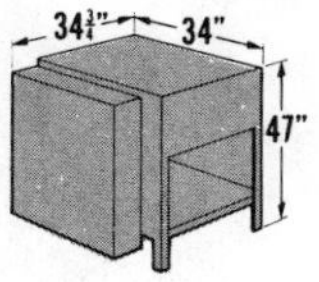

Model AXI-300
3 H.P.

Max. BTUH 39,500
Coolant Flow 4 to 35 GPM
Coolant Circulating Pump Optional
Pipe Sizes for Coolant "In" and Coolant "Out" Standard 2" NPT

Power	**Unit Full Load**
208-230/3/60	17 AMP
460/3/60	8.5 AMP

Shipping Weight 800 lbs. approx.

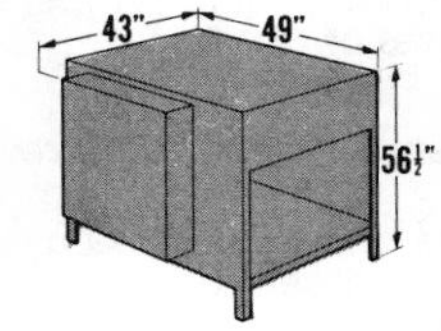

Model AXI-500
5 H.P.

Max. BTUH 64,000
Coolant Flow 9 to 50 GPM
Coolant Circulating Pump Optional
Pipe Sizes for Coolant "In" and Coolant "Out" Standard 2" NPT

Power	**Unit Full Load**
208-230/3/60	23 AMP
460/3/60	11.5 AMP

Shipping Weight 1000 lbs. approx.

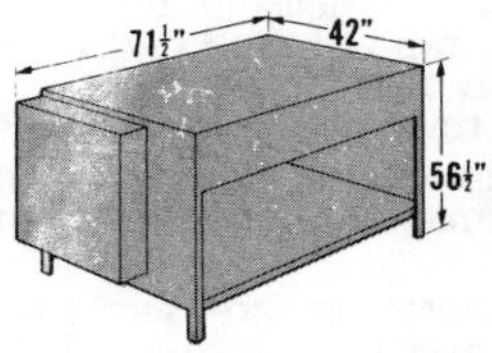

Model AXI-750
7.5 H.P.

Max. BTUH 94,000
Coolant Flow 12 to 75 GPM
Coolant Circulating Pump Optional
Pipe Sizes for Coolant "In" and Coolant "Out" Standard 2" NPT

Power	**Unit Full Load**
208-230/3/60	38 AMP
460/3/60	19 AMP

Shipping Weight 1200 lbs. approx.

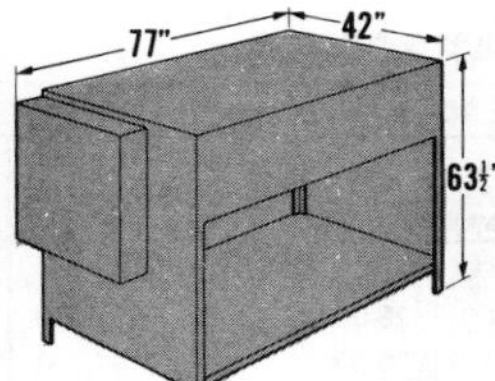

Model AXI-1000
10 H.P.

Max. BTUH 137,000
Coolant Flow 13 to 100 GPM
Coolant Circulating Pump Optional
Pipe Sizes for Coolant "In" and Coolant "Out" Standard 2" NPT

Power	**Unit Full Load**
208-230/3/60	62 AMP
460/3/60	31 AMP

Shipping Weight 1800 lbs. approx.

WATER COOLED

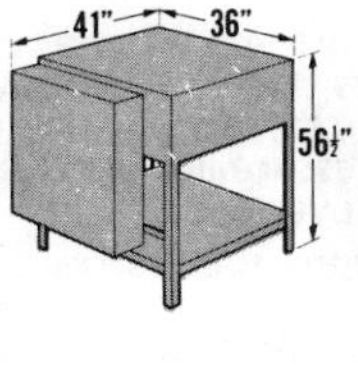

Model WXI-300
3 H.P.

Max. BTUH 46,000
Coolant Flow 4 to 35 GPM
Coolant Circulating Pump Optional
Pipe Sizes for Coolant "In" and Coolant "Out" Standard 2" NPT
Water Connections 1" NPT

Power	**Unit Full Load**
208-230/3/60	14 AMP
460/3/60	7 AMP

Shipping Weight 800 lbs. approx.

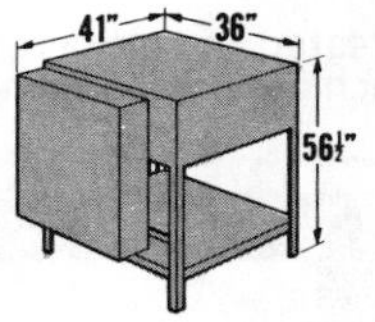

Model WXI-500
5 H.P.

Max. BTUH 73,000
Coolant Flow 9 to 50 GPM
Coolant Circulating Pump Optional
Pipe Sizes for Coolant "In" and Coolant "Out" Standard 2" NPT
Water Connections 1" NPT

Power	**Unit Full Load**
208-230/3/60	23 AMP
460/3/60	11.5 AMP

Shipping Weight 1000 lbs. approx.

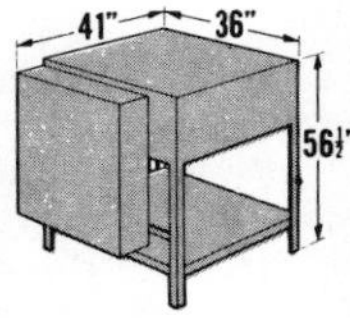

Model WXI-750
7.5 H.P.

Max. BTUH 108,000
Coolant Flow 12 to 75 GPM
Coolant Circulating Pump Optional
Pipe Sizes for Coolant "In" and Coolant "Out" Standard 2" NPT
Water Connections 1" NPT

Power	**Unit Full Load**
208-230/3/60	33 AMP
460/3/60	16.5 AMP

Shipping Weight 1200 lbs. approx.

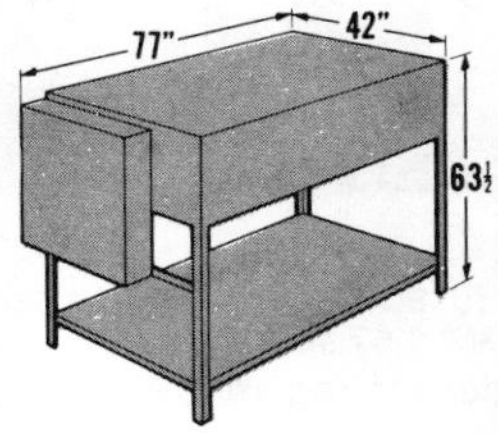

Model WXI-1000
10 H.P.

Max. BTUH 158,000
Coolant Flow 13 to 100 GPM
Coolant Circulating Pump Optional
Pipe Sizes for Coolant "In" and Coolant "Out" Standard 2" NPT
Water Connections 1" NPT

Power	**Unit Full Load**
208-230/3/60	60 AMP
460/3/60	30 AMP

Shipping Weight 1500 lbs. approx.

Rowald Refrigeration & Heating Systems Incorporated
1121 First Avenue
Rockford, IL 61104-1295
(815) 962-7733

CONDITIONS AT RATED CAPACITIES

Water-Cooled Units*

The capacities are based on refrigerant liquid temperatures of 100° F and suction gas temperatures of 65° F.

Air-Cooled Units*

The capacities are based on 90° F ambient temperatures and 90° F suction gas temperatures.

*Increase/decrease the capacities of 6% for each 10° lower/higher ambient or condensing temperatures.

RANGE is the difference between the outlet coolant temperatures and the inlet coolant temperatures.

APPROACH is the difference between the outlet coolant temperatures and refrigerant temperatures. (This will vary with flow rate and coolant used.)

Water-Cooled Units—approximate water consumption:

Tap water—One (1) GPM/HP
Tower water—Three (3) GPM/HP

Maximum continuous coolant IN temperature is 100° F. However, special designs are available for higher coolant temperatures. All listed capacities are for 60 Hz operation.

COOLING CAPACITIES

Coolant Leaving Temp.	BTUH @ 20° Approach							
	AXI—0300 XXX-XXX	AXI—0500 XXX-XXX	AXI—0750 XXX-XXX	AXI—1000 XXX-XXX	WXI—0300 XXX-XXX	WXI—0500 XXX-XXX	WXI—0750 XXX-XXX	WXI—1000 XXX-XXX
40° F.	24,000	35,000	54,000	85,000	22,000	35,000	50,000	80,000
50° F.	28,000	44,000	65,000	98,000	28,000	44,000	65,000	99,000
60° F.	33,000	52,000	78,000	116,000	35,000	54,000	81,000	121,000
70° F.	37,000	60,000	89,000	130,000	42,500	66,500	99,000	146,000
75-100° F.	39,500	64,000	94,000	137,000	46,000	73,000	108,000	158,000

METRIC CAPACITY CONVERSION TABLE

- BTUH X .252 = Kilogram Calories
- °F. -32 X 5/9 = °C.
- Inches X 25.4 = Millimeters
- Pounds X .454 = Kilograms
- BTU @ 60 HZ X 0.8333 = BTU @ 50 HZ.
- BTU @ 60 HZ X 0.21 = KCAL @ 50 HZ.

UNIT CONSTRUCTION

The AXI and WXI cabinets are constructed on an angle-iron frame. The upper and lower shelves are of 3/16" or 1/4" steel plates. All frame members are electrically welded together. The cabinet enclosurer panels are 20-gauge cold-rolled steel.

All sections of the cabinet are acid treated and primered with zinc chromate. The standard paint is machinery gray enamel.

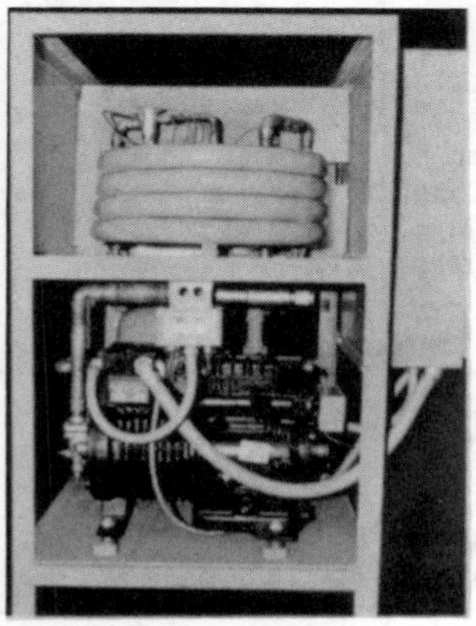

ELECTRICAL

A NEMA 12 enclosurer, including fused disconnect, is used. The electrical is built primarily to JIC-EMP-1-67 specifications.

EXCEPTIONS: Some small HP condenser fan motors are of the open type.

Pressure switches are not available as oil tight.

All wiring is in seal-tight conduit except for open-type motors, which are armored cable.

Compressor starters are definite-purpose contactors with overload relays.

Electrics can be built to customer's specifications.

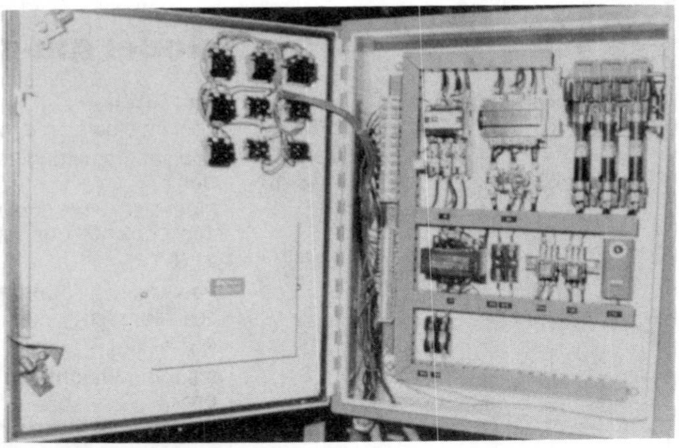

REFRIGERATION SYSTEM

WXI—The condensers are of shell and tube or shell and coil type. The units are piped for tap water. (If tower water is to be used, this should be indicated on the purchase order. We will pipe the condenser water for a tower application, if so indicated.)

AXI—The condensers are copper tube with aluminum fins. They are protected from damage to their face by an expanded metal grill.

COMPRESSOR—The compressors are semi-hermetic type, protected with high/low-pressure controls and inherent overload protectors. The larger sizes, 5 HP and up, may have a lube oil-pressure failure safety switch and motor temperature solid-state protection, depending on the manufacturer. All compressors have crankcase heaters to prevent refrigeration dilution of the lube oil.

EVAPORATOR—The evaporator is a helical coil tube in tube using no baffles. The refrigerant surrounds the coolant tubes and the counter-flow design assures good heat transfer. Evaporators are available in copper and steel.

COPPER is used for chilling water or water and ethylene-glycol solutions. Also, any other liquid which is compatible with copper may be used.

STEEL is used for most coolants, including water-soluble oil and all oil products.

CONTROLS—An adjustable temperature control senses the coolant temperature as it enters the chiller. A flow switch stops the compressor if the flow is interrupted; this prevents the coil from freezing up. The pressure-limiting refrigerant feed valve eliminates compressor overloads.

Special-application designs, such as low temperatures, larger HP, belt-driven compressors and special evaporators are available. Please contact the plant for prices and delivery.

OPTIONAL EQUIPMENT

- Coolant circulating Pumps and controls.
- Refrigerant pressure gauges.
- Coolant pressure gauges.
- Temperature indicators.
- Paint—other than machinery gray enamel.

Rowald Refrigeration

1121 Charles Street • Rockford, Illinois 61104 • 815/962-7733

INDEX OF PARTICIPATING COMPANIES

U

V

W